T0178052

Lecture Notes in Computer Science 13761

More information about this series at https://link.springer.com/bookseries/558

Philippe Fournier-Viger · Ahmed Hassan ·
Ladjel Bellatreche (Eds.)

Model and Data Engineering

11th International Conference, MEDI 2022
Cairo, Egypt, November 21–24, 2022
Proceedings

Editors
Philippe Fournier-Viger
Shenzhen University
Shenzhen, Guangdong, China

Ahmed Hassan
Nile University
Giza, Egypt

Ladjel Bellatreche
ISAE-ENSMA
Poitiers, France

ISSN 0302-9743 ISSN 1611-3349 (electronic)
Lecture Notes in Computer Science
ISBN 978-3-031-21594-0 ISBN 978-3-031-21595-7 (eBook)
https://doi.org/10.1007/978-3-031-21595-7

This Springer imprint is published by the registered company Springer Nature Switzerland AG
The registered company address is: Gewerbestrasse 11, 6330 Cham, Switzerland

Preface

The International Conference on Model and Data Engineering (MEDI) is a yearly conference that provides a platform for researchers and practitioners to present research advances on modeling and data management, including topics such as database theory, database systems technology, data models, advanced database applications, and data processing. MEDI is a well-established conference, founded by researchers from Euro-Mediterranean countries, which has been a starting point for numerous international scientific collaborations and projects, as well as research visits and exchanges by students and faculty members from various institutions. MEDI has been held in various countries over the years, including France, Morocco, Spain, Greece, Cyprus, Italy, France, Estonia, and Portugal.

This year is the 11th edition of MEDI, held during November 21–24, 2022, in Cairo, Egypt. A total of 65 submissions were received. Each submission was rigorously evaluated and received three to five single blind reviews from an international Program Committee consisting of researchers from 20 different countries. Based on the result of the evaluation, it was decided to accept 18 papers, which represents an acceptance rate of 27.6%, for full presentation at the conference and 12 papers for short presentation. The 18 full papers are published in this proceedings book, while short papers are published in a separate volume. The accepted papers are from authors in 11 countries and include topics such as database systems, data stream analysis, knowledge graphs, machine learning, model-driven engineering, image processing, diagnosis, natural language processing, optimization, and advanced applications such as the Internet of Things and healthcare.

At MEDI 2022, two well-renowned researchers were keynote speakers. Vincent S. Tseng from the National Yang Ming Chiao Tung University gave a talk entitled "Broad and Deep Learning of Heterogeneous Health Data for Medical AI: Opportunities and Challenges". The second keynote talk was given by Athman Bouguettaya from the University of Sydney and was titled "A Service-based Approach to Drone Service Delivery in Skyway Networks".

MEDI 2022 was held in hybrid mode (in person and online) due to the special circumstances related to the COVID-19 pandemic. The organizers would like to thank all authors who submitted research papers for evaluation at MEDI 2022, as well as all members of the Program Committee and external reviewers, who carefully evaluated all contributions. Moreover, we extend our special thanks to the Local Organizing Committee members who were a key reason for the success of this year's edition. We also appreciated using the EasyChair conference management system for handling all tasks related to handling submission and the reviewing process.

October 2022

Philippe Fournier-Viger
Ahmed Hassan
Ladjel Bellatreche

Organization

General Chairs

Ahmed Hassan	Nile University, Egypt
Ladjel Bellatreche	ISAE-ENSMA, France

Program Committee Chairs

Ladjel Bellatreche	ISAE-ENSMA, France
Philippe Fournier-Viger	Shenzhen University, China

Workshop Chair

Ahmed Awad	Tartu University, Estonia

Proceedings Chair

Walid Al-Atabany	Nile University, Egypt

Financial Chair

Hala Zayed	Nile University, Egypt

Program Committee

Antonio Corral	University of Almeria, Spain
Mamoun Filali-Amine	IRIT, France
Flavio Ferrarotti	Software Competence Centre Hagenberg, Austria
Sofian Maabout	University of Bordeaux, France
Yannis Manolopoulos	Open University of Cyprus, Cyprus
Milos Savic	University of Novi Sad, Serbia
Alberto Cano	Virginia Commonwealth University, USA
Essam Houssein	Minia University, Egypt
Moulay Akhloufi	Université de Moncton, Canada
Neeraj Singh	University of Toulouse, France
Dominique Mery	Université de Lorraine, Loria, France
Duy-Tai Dinh	Japan Advanced Institute of Science and Technology, Japan
Giuseppe Polese	University of Salerno, Italy

M. Saqib Nawaz	Peking University, China
Jérôme Rocheteau	Icam Nantes, France
Mourad Nouioua	Hunan University, China
Ivan Luković	University of Belgrade, Serbia
Jaroslav Frnda	University of Zilina, Slovakia
Radwa El Shawi	Tartu University, Estonia
Enrico Gallinucci	University of Bologna, Italy
Anirban Mondal	University of Tokyo, Japan
Pinar Karagoz	Middle East Technical University (METU), Turkey
El Hassan Abdelwahed	Cadi Ayyad University, Morocco
Irena Holubova	Charles University in Prague, Czech Republic
Georgios Evangelidis	University of Macedonia, Greece
Panos Vassiliadis	University of Ioannina, Greece
Mohamed Mosbah	LaBRI, University of Bordeaux, France
Patricia Derler	Palo Alto Research Center, USA
Idir Ait Sadoune	LRI, CentraleSupélec, France
Goce Trajcevski	Iowa State University, USA
Jerry Chun-Wei Lin	Western Norway University of Applied Sciences, Norway
Yassine Ouhammou	LIAS, ISAE-ENSMA, France
Srikumar Krishnamoorthy	Indian Institute of Management Ahmedabad, India
Mirjana Ivanovic	University of Novi Sad, Serbia
Yves Ledru	Université Grenoble Alpes, France
Raju Halder	Indian Institute of Technology Patna, India
Orlando Belo	University of Minho, Portugal
Stefania Dumbrava	ENSIIE Paris-Evry, France
Chokri Mraidha	CEA LIST, France
Amirat Hanane	Universiy of Laghoaut, Algeria
Javier Tuya	Universidad de Oviedo, Spain
Luis Iribarne	University of Almería, Spain
Elvinia Riccobene	University of Milan, Italy
Regine Laleau	Paris-Est Créteil University, France
Jaroslav Pokorný	Charles University in Prague, Czech Republic
Oscar Romero	Universitat Politècnica de Catalunya, Spain

Organization Committee

Mohamed El Helw	Nile University, Egypt
Islam Tharwat	Nile University, Egypt
Sahar Selim	Nile University, Egypt
Passant El Kafrawy	Nile University, Egypt

Sahar Fawzy	Nile University, Egypt
Nashwa Abdelbaki	Nile University, Egypt
Wala Medhat	Nile University, Egypt
Heba Aslan	Nile University, Egypt
Mohamed El Hadidi	Nile University, Egypt
Mostafa El Attar	Nile University, Egypt

Abstracts of Invited Talks

A Service-Based Approach to Drone Service Delivery in Skyway Networks

Athman Bouguettaya

University of Sydney, Australia
athman.bouguettaya@sydney.edu.au

Abstract. We propose a novel *service framework* to effectively provision drone-based delivery services in a skyway network. This service framework provides a high-level service-oriented architecture and an abstraction to model the drone service from both *functional* and *non-functional* perspectives. We focus on *spatio-temporal* aspects as key parameters to query the drone services under a range of requirements, including drone capabilities, flight duration, and payloads. We propose to *reformulate* the problem of drone package delivery as finding an optimal composition of drone delivery services from a designated take-off station (e.g., a warehouse rooftop) to a landing station (e.g., a recipient's landing pad). We select and compose those drone services that provide the best quality of delivery service in terms of payload, time, and cost under a range of *intrinsic* and *extrinsic* environmental (i.e., context-aware) factors, such as battery life, range, wind conditions, drone formation, etc. This talk will overview the key challenges and propose solutions in the context of single drones and swarms of drones for service delivery.

Bio: Athman Bouguettaya is Professor and previous Head of School of Computer Science, at the University of Sydney, Australia. He was also previously Professor and Head of School of Computer Science and IT at RMIT University, Melbourne, Australia. He received his PhD in Computer Science from the University of Colorado at Boulder (USA) in 1992. He was previously Science Leader in Service Computing at the CSIRO ICT Centre (now DATA61), Canberra. Australia. Before that, he was a tenured faculty member and Program director in the Computer Science department at Virginia Polytechnic Institute and State University (commonly known as Virginia Tech) (USA). He is a founding member and past President of the Service Science Society, a non-profit organization that aims at forming a community of service scientists for the advancement of service science. He is or has been on the editorial boards of several journals including, the IEEE Transactions on Services Computing, IEEE Transactions on Knowledge and Data Engineering, ACM Transactions on Internet Technology, the International Journal on Next Generation Computing, VLDB Journal, Distributed and Parallel Databases Journal, and the International Journal of Cooperative Information Systems. He is also the Editor-in-Chief of the Springer-Verlag book series on Services Science. He served as a guest editor of a number of special issues including the special issue of the ACM Transactions on Internet Technology on Semantic Web services, a special issue the IEEE Transactions on Services Computing on Service Query Models, and a special issue of IEEE Internet Computing on Database Technology on the Web. He was the General Chair of the IEEE ICWS for 2021 and 2022. He was also General Chair of ICSOC for 2020. He served as a Program Chair of the 2017 WISE Conference, the 2012 International Conference on Web and Information System Engineering, the 2009 and 2010 Australasian Database Conference, 2008 International Conference on Service Oriented Computing (ICSOC) and the IEEE RIDE Workshop on Web Services for E-Commerce and E-Government (RIDE-WS-ECEG'04). He also served on the IEEE Fellow Nomination Committee. He has published more than 300 books, book chapters, and articles in journals and conferences in the area of databases and service computing (e.g., the IEEE Transactions on Knowledge and Data Engineering, the ACM Transactions on the Web, WWW Journal, VLDB Journal, SIGMOD, ICDE, VLDB, and EDBT). He was the recipient of several federally competitive grants in Australia (e.g., ARC), the US (e.g., NSF, NIH), Qatar (NPRP). EU (FP7), and China (NSFC). He also won major industry grants from companies like HP and Sun Microsystems (now Oracle). He is a Fellow of the IEEE, Member of the Academia Europaea (Honoris Causa) (MAE) (HON), WISE Fellow, AAIA Fellow, and Distinguished Scientist of the ACM.

Broad and Deep Learning of Big Heterogeneous Health Data for Medical AI: Opportunities and Challenges

Vincent S. Tseng

National Yang Ming Chiao Tung University, Taiwan
vtseng@cs.nctu.edu.tw

Abstract. In healthcare domains, large-scale heterogeneous types of data like medical images, vital signs, electronic health records (EHR), genome, etc., have been collected constantly, forming the valuable big health data. Broad and deep learning of these big heterogeneous biomedical data can enable innovative applications for Medical AI with rich research lines/challenges arisen. In this talk, I will introduce recent developments and ongoing projects on the topic of Medical AI, especially in intelligent diagnostic decision support and disease risk prediction by using various advanced data mining/deep learning techniques including image analysis(for medical images), multivariate time-series analysis(for vital signs like ECG/EEG), patterns mining (for EHR), text mining (for medical notes), sensory analysis (for sensory data like air quality) as well as fusion methods for integrated modelling. Some innovative applications on Medical AI with breakthrough results based on the developed techniques, as well as the underlying challenging issues and open opportunities, will be addressed too at the end.

Bio: Vincent S. Tseng is currently a Chair Professor at Department of Computer Science in National Yang Ming Chiao Tung University (NYCU). He served as the founding director for Institute of Data Science and Engineering in NYCU during 2017–2020, chair for IEEE CIS Tainan Chapter during 2013–2015, the president of Taiwanese Association for Artificial Intelligence during 2011–2012 and the director for Institute of Medical Informatics of National Cheng Kung University during 2008 and 2011. Dr. Tseng received his Ph.D. degree with major in computer science from National Chiao Tung University, Taiwan, in 1997. After that, he joined Computer Science Division of EECS Department in University of California at Berkeley as a postdoctoral research fellow during 1998–1999. He has published more than 400 research papers, which have been cited by more than 13,000 times with H-Index 60 by Google Scholar. He has been on the editorial board of a number of top journals including *IEEE Transactions on Knowledge and Data Engineering (TKDE)*, *IEEE Journal of Biomedical and Health Informatics (JBHI)*, *IEEE Computational Intelligence Magazine (CIM)*, *ACM Transactions on Knowledge Discovery from Data (TKDD)*, etc. He has also served as chairs/program committee members for a number of premier international conferences related to data mining/machine learning, and currently he is the Steering Committee Chair for *PAKDD*. Dr. Tseng has received a number of prestigious awards, including IICM Medal of Honor (2021), Outstanding

Research Award (2019 & 2015) by Ministry of Science and Technology Taiwan, 2018 Outstanding I.T. Elite Award, 2018 FutureTech Breakthrough Award, and 2014 K. T. Li Breakthrough Award. He is also a Fellow of IEEE and Distinguished Member of ACM.

Contents

Image Processing and Diagnosis

Chaos-Based Image Encryption Using DNA Manipulation and a Modified Arnold Transform ... 3
Marwan A. Fetteha, Wafaa S. Sayed, Lobna A. Said, and Ahmed G. Radwan

Rice Plant Disease Detection and Diagnosis Using Deep Convolutional Neural Networks and Multispectral Imaging ... 16
Yara Ali Alnaggar, Ahmad Sebaq, Karim Amer, ElSayed Naeem, and Mohamed Elhelw

A Novel Diagnostic Model for Early Detection of Alzheimer's Disease Based on Clinical and Neuroimaging Features ... 26
Eyad Gad, Aya Gamal, Mustafa Elattar, and Sahar Selim

Machine Learning and Optimization

Benchmarking Concept Drift Detectors for Online Machine Learning ... 43
Mahmoud Mahgoub, Hassan Moharram, Passent Elkafrawy, and Ahmed Awad

Computational Microarray Gene Selection Model Using Metaheuristic Optimization Algorithm for Imbalanced Microarrays Based on Bagging and Boosting Techniques ... 58
Rana Hossam Elden, Vidan Fathi Ghoneim, Marwa M. A. Hadhoud, and Walid Al-Atabany

Fuzzing-Based Grammar Inference ... 72
Hannes Sochor, Flavio Ferrarotti, and Daniela Kaufmann

Natural Language Processing

In the Identification of Arabic Dialects: A Loss Function Ensemble Learning Based-Approach ... 89
Salma Jamal, Salma Khaled, Aly M. Kassem, Ayaalla Eltabey, Alaa Osama, Samah Mohamed, and Mustafa A. Elattar

Emotion Recognition System for Arabic Speech: Case Study Egyptian Accent ... 102
Mai El Seknedy and Sahar Ali Fawzi

Modelling

Towards the Strengthening of Capella Modeling Semantics by Integrating Event-B: A Rigorous Model-Based Approach for Safety-Critical Systems ... 119
Khaoula Bouba, Abderrahim Ait Wakrime, Yassine Ouhammou, and Redouane Benaini

A Reverse Design Framework for Modifiable-off-the-Shelf Embedded Systems: Application to Open-Source Autopilots ... 133
Soulimane Kamni, Yassine Ouhammou, Emmanuel Grolleau, Antoine Bertout, and Gautier Hattenberger

Efficient Checking of Timed Ordered Anti-patterns over Graph-Encoded Event Logs ... 147
Nesma M. Zaki, Iman M. A. Helal, Ehab E. Hassanein, and Ahmed Awad

Trans-Compiler-Based Database Code Conversion Model for Native Platforms and Languages ... 162
Rameez Barakat, Moataz-Bellah A. Radwan, Walaa M. Medhat, and Ahmed H. Yousef

MDMSD4IoT a Model Driven Microservice Development for IoT Systems ... 176
Meriem Belguidoum, Aya Gourari, and Ines Sehili

Database Systems

Parallel Skyline Query Processing of Massive Incomplete Activity-Trajectories Data ... 193
Amina Belhassena and Wang Hongzhi

Compact Data Structures for Efficient Processing of Distance-Based Join Queries ... 207
Guillermo de Bernardo, Miguel R. Penabad, Antonio Corral, and Nieves R. Brisaboa

Towards a Complete Direct Mapping from Relational Databases to Property Graphs ... 222
Abdelkrim Boudaoud, Houari Mahfoud, and Azeddine Chikh

A Matching Approach to Confer Semantics over Tabular Data Based on Knowledge Graphs 236
Wiem Baazouzi, Marouen Kachroudi, and Sami Faiz

τJUpdate: A Temporal Update Language for JSON Data 250
Zouhaier Brahmia, Fabio Grandi, Safa Brahmia, and Rafik Bouaziz

Author Index 265

Image Processing and Diagnosis

Chaos-Based Image Encryption Using DNA Manipulation and a Modified Arnold Transform

Marwan A. Fetteha[1(✉)], Wafaa S. Sayed[2], Lobna A. Said[1], and Ahmed G. Radwan[2,3]

[1] Nanoelectronics Integrated Systems Center (NISC), Nile University, Giza 12588, Egypt
M.Ahmed2129@nu.edu.eg

[2] Engineering Mathematics Department, Faculty of Engineering, Cairo University, Giza 12613, Egypt

[3] School of Engineering and Applied Sciences, Nile University, Giza 12588, Egypt

Abstract. Digital images, which we store and communicate everyday, may contain confidential information that must not be exposed to others. Numerous researches are interested in encryption, which protects the images from ending up in the hands of unauthorized third parties. This paper proposes an image encryption scheme using chaotic systems, DNA manipulation, and a modified Arnold transform. Both DNA manipulation and hyperchaotic Lorenz system are utilized in the substitution of the images' pixel values. An additional role of hyperchaotic Lorenz system is that it generates the random numbers required within the DNA manipulation steps. DNA cycling is implemented based on simple DNA coding rules and DNA addition and subtraction rules with modulus operation. The modified Arnold transform alters the pixels' positions, where it guarantees effective pixel permutation that never outputs the same input pixels arrangement again. The proposed design is simple and amenable for hardware realization. Several well established performance evaluation tests including statistical properties of the encrypted image, key space, and differential attack analysis were conducted for several images. The proposed scheme passed the tests and demonstrated good results compared to several recent chaos-based image encryption schemes.

Keywords: Arnold transform · Chaos · DNA · Image encryption

1 Introduction

Communication methods have undergone significant changes in the recent few decades due to the quick development of computer and network technology. The need for secure communication of media and exchanged information has gradually developed [1]. Specifically, image encryption has been the topic of numerous researches to protect the user's privacy [2]. The strong correlation and redundancy between neighbouring pixels of an image require devising new encryption schemes rather than the typical ones [3].

P. Fournier-Viger et al. (Eds.): MEDI 2022, LNAI 13761, pp. 3–15, 2023.
https://doi.org/10.1007/978-3-031-21595-7_1

Chaotic systems are good candidates for image encryption systems because of their pseudorandomness, initial value sensitivity, parameter sensitivity, and unpredictability, among other qualities, which increase the security level [4–6]. Both Deoxyribonucleic acid (DNA) encoding and Arnold permutation have appeared in recent works as well. In [7], an image encryption algorithm based on bit-level Arnold transform and hyperchaotic maps was proposed. The algorithm divides the grayscale image into 8 binary images. Then, a chaotic sequence is used to shift the images. Afterwards, Arnold transform is applied. Finally, image diffusion is applied using the hyper chaotic map. The system requires image division, which increases the system's complexity and may halt it from being optimized to applicable hardware design. In [6], Luo et al. used double chaotic systems, where two-dimensional Baker chaotic map is used to set the state variables and system parameters of the logistic chaotic map. In [8], Ismail et al. developed a generalised double humped logistic map, which is used in gray scale image encryption. In [9], a chaotic system and true random number generator were utilized for image encryption. The presence of both the chaotic system and true random number generator increases the system's complexity making it less suitable for hardware implementation. In [1], a plaintext-related encryption scheme that utilises two chaotic systems and DNA manipulation was presented. The system depends on the values of some pixels for the encryption process, which threatens image restoration if they are changed.

This paper proposes an image encryption algorithm that uses hyperchaotic Lorenz system, an optimized DNA manipulation system and a new method for applying Arnold transform, which is more suitable for encryption applications. The rest of the paper is organized as follows: Sect. 2 provides a brief explanation of the utilized methods. Section 3 demonstrates the proposed encryption and decryption algorithms. Section 4 validates their good performance. Finally, Sect. 5 concludes the work.

2 Preliminaries

Generally, encryption systems require a source of randomness that can be regenerated in the decryption process. This section explains the main sources of randomness that are employed in the proposed scheme.

2.1 Hyperchaotic Lorenz System

Hyperchaotic Lorenz system [10] provides the randomness needed for encryption. The system is solved using Euler's method:

$$x_{i+1} = x_i + h(a(y_i - x_i) + w_i), \tag{1a}$$
$$y_{i+1} = y_i + h(cx_i - y_i - x_i z_i), \tag{1b}$$
$$z_{i+1} = z_i + h(x_i y_i - bz_i), \tag{1c}$$
$$w_{i+1} = w_i + h(y_i z_i + rw_i), \tag{1d}$$

where $h = 0.01$, $a = 10$, $b = 8/3$, $c = 28$, and $r = -1$. Figure 1 shows the output results with initial conditions $x_0 = 0.23$, $y_0 = 0$, $z_0 = 0.7$, and $w_0 = 0.11$.

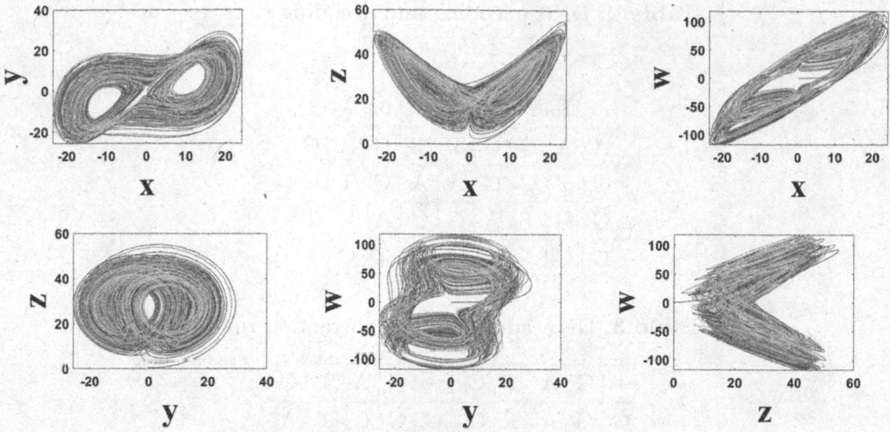

Fig. 1. Output of hyperchaotic Lorenz system.

Table 1. DNA binary codes

DNA base	Binary code
G	00
A	01
T	10
C	11

2.2 DNA Coding

DNA coding [11] is used to change the bit values according to some set of rules. This is done to enhance the security of the algorithm. DNA consists of 4 bases, which are Adenine (A), Thymine (T), Cytosine (C), and Guanine (G). The relation between these bases is that 'A' is complementary to 'T' and 'G' is complementary to 'C'. Table 1 shows the used binary code for each DNA base.

Based on these relations, we can apply rules to manipulate the data as long as the relation between these bases does not change. Table 2 shows the list of all possible rules that are used in the encryption algorithm, where a random number is used to select the rule and then the two input bits are replaced with the corresponding DNA base. For example, if the chosen rule is 6 and the input is 'T', then the output will be 'C', which is equal to '11'.

Table 3 shows the results of DNA addition and subtraction, which can be done using simple operations on the DNA bases if the binary representation of Table 1 is used. The DNA sequence has a cyclic behavior, where each base is repeated every 4 cycles (i.e., T, C, G, A, T, C, ...). This enables performing 'DNA cycling' by dividing the number of cycles by 4 and then referring to Table 4.

Table 2. DNA encoding and decoding rules

	Rules							
	1	2	3	4	5	6	7	8
A	A	C	C	T	T	G	A	G
T	T	G	G	A	A	C	T	C
G	G	T	A	G	C	A	C	T
C	C	A	T	C	G	T	G	A

Table 3. DNA addition and subtraction rules

+	G	A	T	C	−	G	A	T	C
G	G	A	T	C	G	G	C	T	A
A	A	T	C	G	A	A	G	C	T
T	T	C	G	A	T	T	A	G	C
C	C	G	A	T	C	C	T	A	G

Table 4. DNA cycling

Number of cycles	T	C	G	A
4n+0	T	C	G	A
4n+1	C	G	A	T
4n+2	G	A	T	C
4n+3	A	T	C	G

2.3 Arnold Transform

Arnold transform [12] is used to permute the pixels positions of the image. Arnold transform and the inverse operation are defined as follows:

$$\begin{bmatrix} x' \\ y' \end{bmatrix} = mod\left(\begin{bmatrix} 1 & 1 \\ 1 & 2 \end{bmatrix}\begin{bmatrix} x \\ y \end{bmatrix}, M^2\right), \tag{2a}$$

$$\begin{bmatrix} x \\ y \end{bmatrix} = mod\left(\begin{bmatrix} 2 & -1 \\ -1 & 1 \end{bmatrix}\begin{bmatrix} x' \\ y' \end{bmatrix}, M^2\right). \tag{2b}$$

where (x, y) represents the original pixel position and $\left(x', y'\right)$ represents the new pixel position after applying the transform on an $M \times M$ image.

Arnold transform is a periodic transform [12], which means that at a specific iteration or cycle, the permuted image becomes the same as the original image. The period of the transform depends on M as shown in Table 5 [12].

If the number of cycles of Arnold transform is random, the image will not be permuted if this random number happens to be 0 or P, where P is the Arnold transform period of image dimensions $M \times M$ shown in Table 5. To overcome

Table 5. Arnold transform (2) period with different M

M	32	64	128	256	512
Period	24	48	96	192	384

this periodicity, we propose a modified Arnold transform, where the image will be permuted for any number of cycles chosen.

3 Proposed Algorithm

The proposed algorithm for encryption and decryption is shown in Fig. 2. The proposed modified Arnold transform is explained, after that the encryption and decryption process.

3.1 Modified Arnold Transform

To guarantee image permutation for any number of cycles, the number of cycles (Cyc) of the Arnold transform must not equal to 0 or P. Hence, we apply the following equation:

$$G = mod\,(Cyc, P - 2) + 1. \tag{3}$$

This will make the effective number of cycles G be in the range of $1 \rightarrow (P - 1)$, which avoids these two cases and eliminates the chances of periodicity.

3.2 Encryption Process

Step 1: The 4 input sub keys (K_1, K_2, K_3, and K_4) are converted from hexadecimal to decimal representation to set the initial state of each variable of the hyperchaotic Lorenz system (1), x_0, y_0, z_0, and w_0. To make the initial conditions bounded by the chaotic system's basin of attraction, they are computed as:

$$x_0 = \left(\frac{K_1}{A/40}\right) - 20, \tag{4a}$$

$$y_0 = \left(\frac{K_2}{A/40}\right) - 20, \tag{4b}$$

$$z_0 = \left(\frac{K_3}{A/50}\right), \tag{4c}$$

$$w_0 = \left(\frac{K_4}{A/200}\right) - 100, \tag{4d}$$

where $A = 2^{52}$. Then, the 4 chaotic sequences x, y, z and w are generated with length equals $M^2 + 1000$.

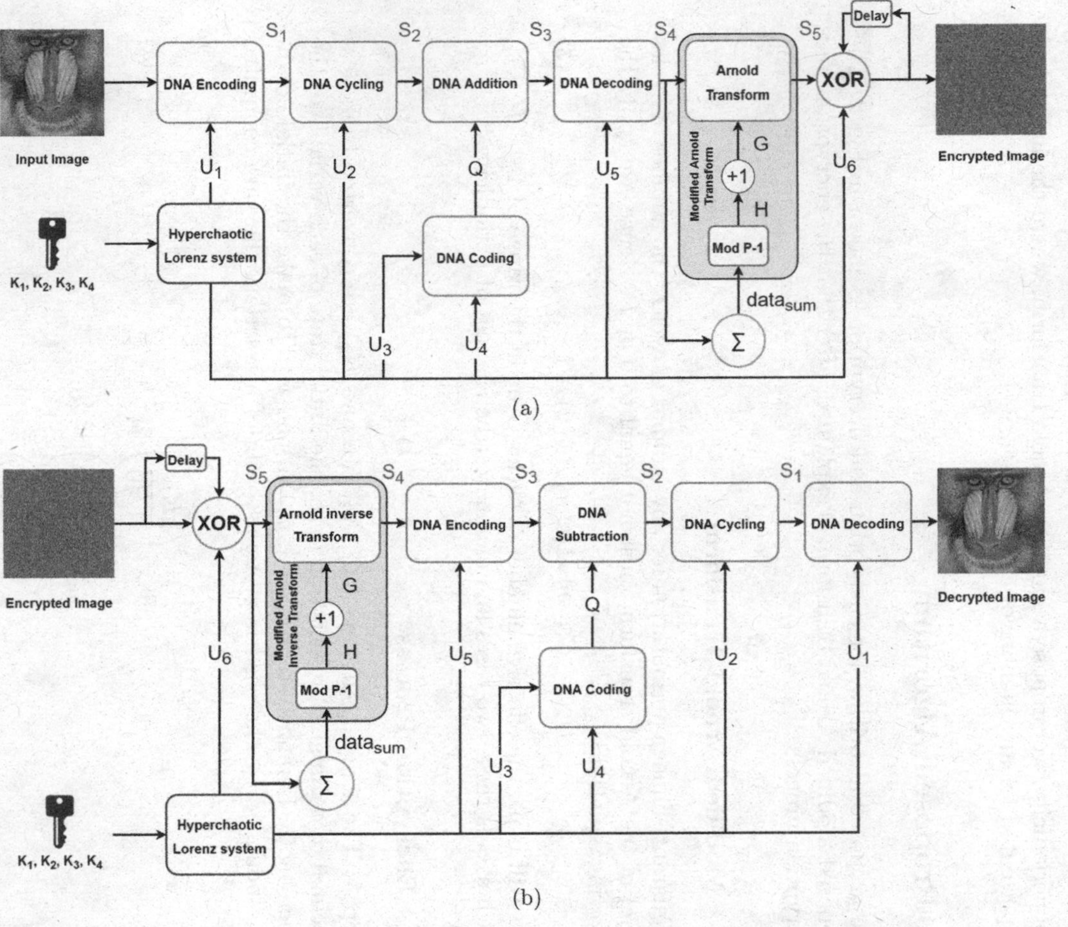

Fig. 2. (a) Encryption and (b) Decryption block diagrams.

Step 2: The first 1000 iterations are removed from the four chaotic sequences to generate X_h, Y_h, Z_h and W_h. Then, the vectors U_1, U_2, U_3, U_4, U_5, and U_6 are generated by the following equations:

$$U_1 = mod(\lceil X_h \times 10^{13} \rceil, 8) + 1, \tag{5a}$$
$$U_2 = mod(\lceil (U - \lceil U \rceil) \times 10^{13} \rceil, M^2), \tag{5b}$$
$$U_3 = mod(\lceil W_h \times 10^{13} \rceil, 8) + 1, \tag{5c}$$
$$U_4 = mod(\lceil (X_h + Y_h) \times 10^{13} \rceil, 256) + 1, \tag{5d}$$
$$U_5 = mod(\lceil Y_h \times 10^{13} \rceil, 8) + 1, \tag{5e}$$
$$U_6 = mod(\lceil (W_h + Z_h) \times 10^{13} \rceil, 256), \tag{5f}$$

where $\lceil\ \rceil$ is the ceiling operator, and $U = [X_h, Y_h, Z_h, W_h]$.
Step 3: U_1 is used to select the DNA rule to encode the input image according to Table 2.
Step 4: U_2 is used to perform DNA cycling on S_1. The result of $mod(U_2, 4)$ chooses how many times the data is shifted according to Table 4.
Step 5: U_3 is used to DNA encode U_4 to generate Q. Then, according to Table 3, the following equations are applied on S_2:

$$q = Q(1) - Q(M^2), \tag{6a}$$
$$S_3(1) = S_2(1) + Q(1) + q, \tag{6b}$$
$$S_3(i) = S_2(i-1) + S_2(i) + Q(i). \tag{6c}$$

Step 6: U_5 is used to select the rule for DNA decoding for S_3 according to Table 2.
Step 7: Every byte of S_4 is accumulated to calculate '$data_{sum}$'. The proposed modified Arnold transform (3) is applied on S_4 to generate S_5, where $cyc = data_{sum}$.
Step 8: S_5 is then XORed with the U_6 to generate the encrypted image.

3.3 Decryption Process

Steps 1 and 2: The same as the encryption process.
Step 3: The input encrypted image is XORed with U_6.
Step 4: The same as step 7 in the encryption process. The only difference is using Arnold inverse transform (2b), instead of Arnold transform. This step is possible even though we are taking the '$data_{sum}$' before the Arnold inverse transform, which is not symmetric with the encryption process. This is because Arnold Transform does not change the pixels values, it only changes their positions.
Step 5: U_5 is used to select the DNA coding rule for S_4.
Step 6: U_3 is used to DNA encode U_4 to generate Q. Then, according to Table 3, the following equations are applied:

$$q = Q(1) - Q(M^2), \tag{7a}$$
$$S_2(1) = S_3(1) - Q(1) - q, \tag{7b}$$
$$S_2(i) = S_3(i) - Q(i) - S_3(i-1). \tag{7c}$$

Step 7: U_2 is used to cyclic shift S_2, which is done by checking the result of $mod(U_2, 4)$ to choose how many times the data is shifted.
Step 8: U_1 is used to select the DNA decoding rule for S_1 to restore the original image.

4 Performance Evaluation

The proposed system is tested using the gray-scale 'Lena' (256 × 256), 'Baboon' (512 × 512), and 'Pepper' (512 × 512) images.

4.1 Encryption Quality Metrics

Figure 3 shows the histogram of the original and encrypted images, which indicate flat and uniform distribution. Mean Square Error (MSE) [13] and Peak Signal-to-Noise Ratio (PSNR) [14] are used to test encryption quality and are given by:

$$MSE = \frac{1}{M^2}\sum_{i=1}^{M}\sum_{j=1}^{M}\left[O_{i,j} - E_{i,j}\right]^2, \tag{8a}$$

$$PSNR = 10\log_{10}\frac{(2^n - 1)^2}{MSE}, \tag{8b}$$

where $O_{i,j}$ and $E_{i,j}$ are the original and encrypted image at position (i, j) respectively and n is the number of bits per pixel. MSE and PSNR $\in [0, \infty]$, where high MSE and low PSNR values indicate huge difference between the original and encrypted images. Table 6 shows that the proposed system gives similar MSE and PSNR values compared to other researches.

4.2 Correlation Analysis

The correlation coefficient is given by:

$$\rho = \frac{Cov(x, y)}{\sqrt{D(x)}\sqrt{D(y)}}, \tag{9a}$$

$$Cov(x, y) = \frac{1}{M^2}\sum_{i=1}^{M^2}(x_i - \frac{1}{M^2}\sum_{i=j}^{M^2} x_j)(y_i - \frac{1}{M^2}\sum_{i=j}^{M^2} y_j), \tag{9b}$$

$$D(x) = \frac{1}{M^2}\sum_{i=1}^{M^2}(x_i - \frac{1}{M^2}\sum_{i=j}^{M^2} x_j)^2, \tag{9c}$$

$$D(y) = \frac{1}{M^2}\sum_{i=1}^{M^2}(y_i - \frac{1}{M^2}\sum_{i=j}^{M^2} y_j)^2, \tag{9d}$$

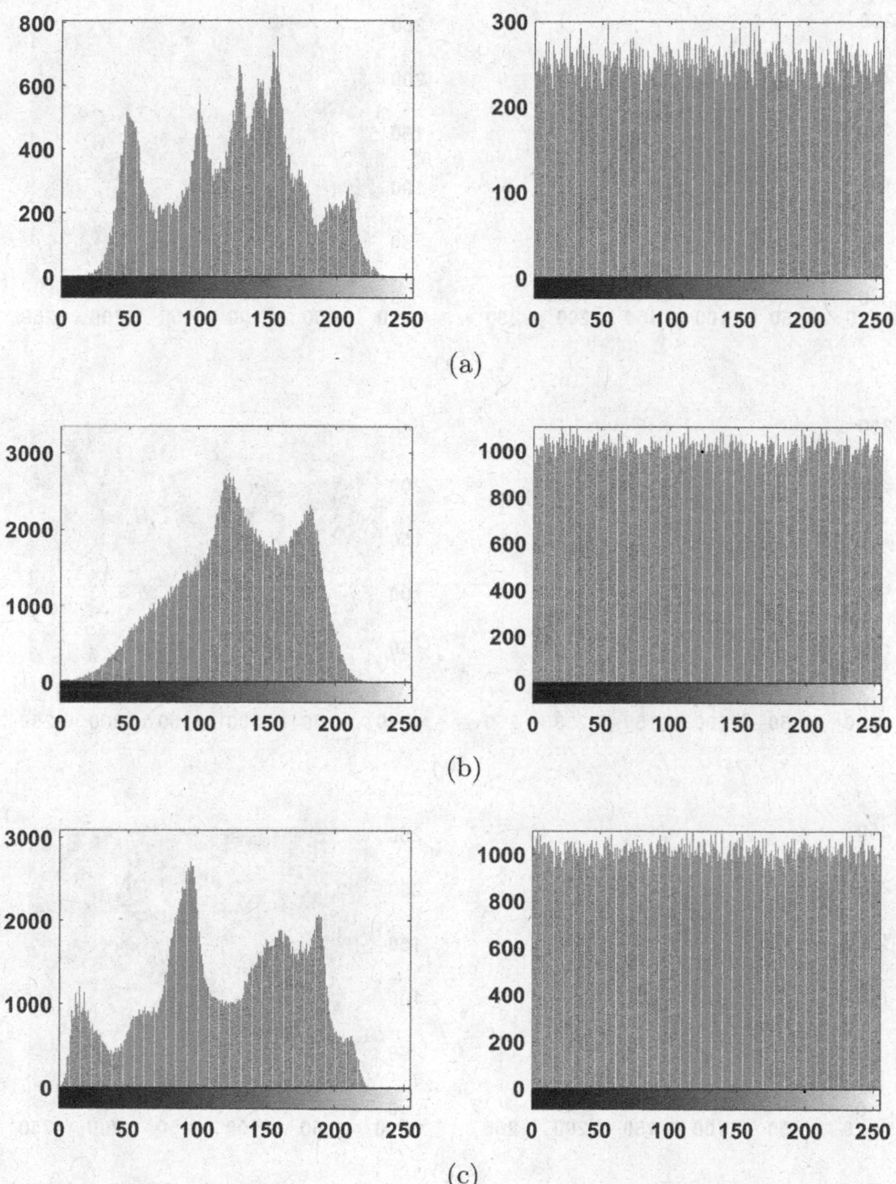

Fig. 3. Histogram of the original image (left), and encrypted image (right) for Lena, Baboon, and Pepper in (a), (b), and (c), respectively.

where $Cov(x, y)$ is the covariance between pixels x and y, and D is the standard deviation. The values of the correlation coefficients for the encrypted images must be close to 0, which means that even the neighbouring pixels are uncorrelated. The results in Table 6 show that the correlation coefficients are close to 0 and

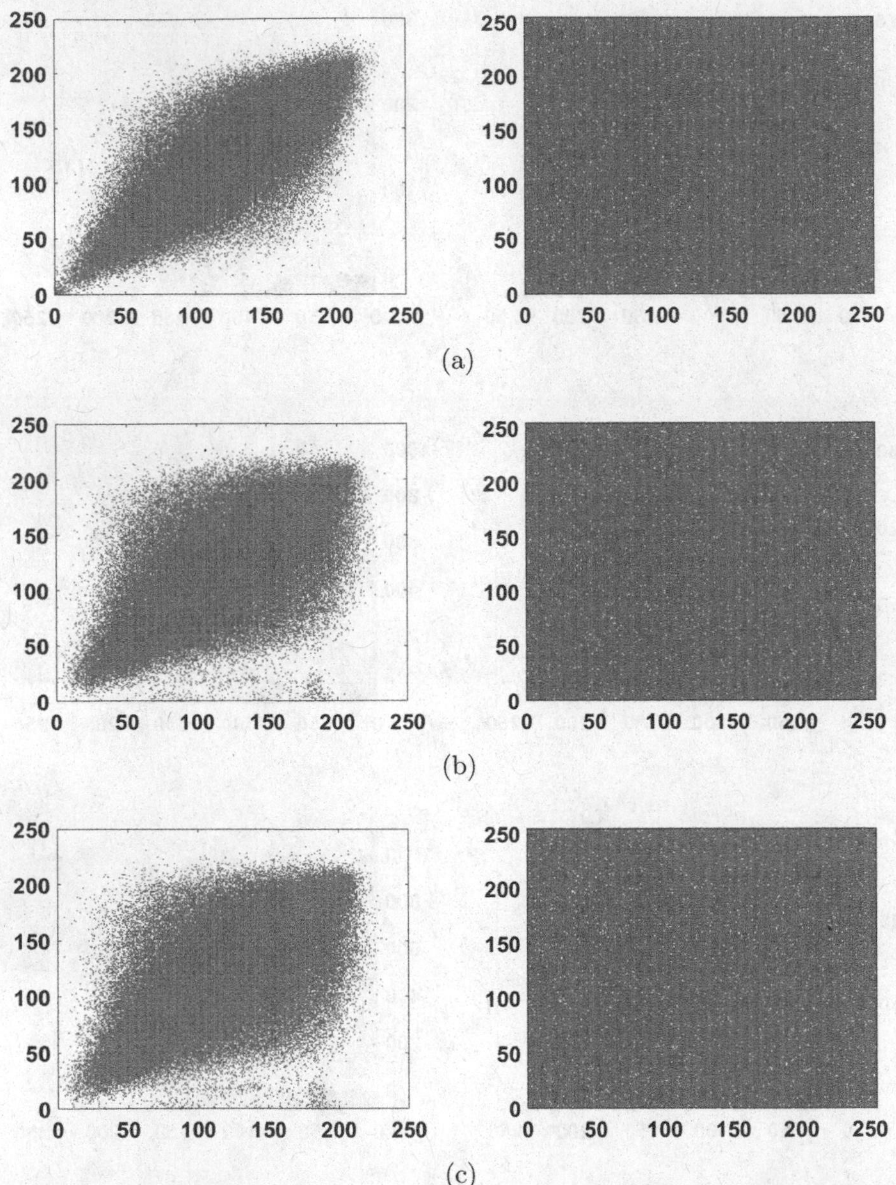

Fig. 4. (a) Horizontal, (b) vertical, and (c) diagonal correlation of Baboon image (left) and encrypted Baboon image (right).

comparable to other works. Figure 4 further indicates that the original image pixel values are grouped in a region, which shows that they are correlated. On the contrary, the encrypted image pixel values are spread all over.

4.3 Information Entropy

Information entropy is the average amount of information conveyed by each pixel [14] and is given by:

$$Entropy = -\sum_{i=0}^{255} P(i)log_2P(i), \tag{10}$$

where $P(i)$ is the probability of occurrence of i. For an 8 bit image, the ideal value is 8, which means that the information is distributed uniformly over all pixel values. The results in Table 6 shows that the entropy of the encrypted images successfully approach 8.

4.4 Key Space and Sensitivity Analysis

The proposed system has a total number of 4 sub keys, each represented by 52 bits, where K_1 = $(FF123FF0567EF)_{16}$, K_2 = $(F655FF000FFFF)_{16}$, K_3 = $(FFAB0957FFFFF)_{16}$ and K_4 = $(46FF0108F214F)_{16}$ are the values for the sub keys used. This results in a key space equals $2^{208} \approx 10^{63}$, which is large enough to resist brute force attacks [1,15]. In addition, the key must have high sensitivity such that any slight change in the decryption key (single bit) prevents recovering the original image. Figure 5 shows the original image of 'Baboon' and the wrong decrypted image when changing the least significant bit of the first sub key.

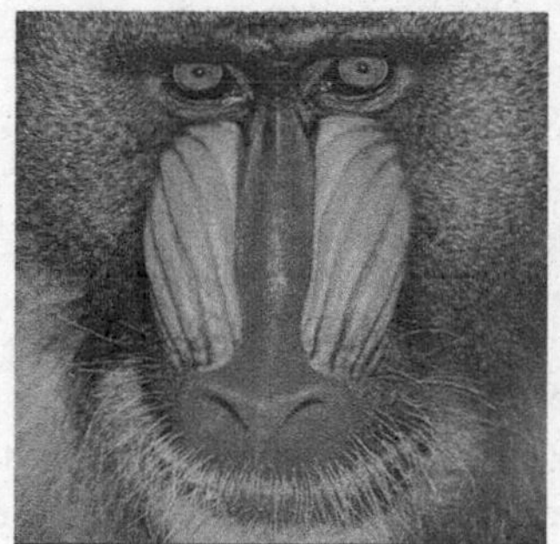

Fig. 5. Original Baboon image (left) and wrong decrypted image (right).

4.5 Robustness Against Differential Attacks

This test is done by changing the least significant bit of a random pixel in the original image and comparing the newly encrypted image to the original encrypted image using Number of Pixels Change Rate (NPCR) and Unified Average Changing Intensity (UACI) [16], which are given by:

$$NPCR = \frac{1}{M^2} \sum_{i=1}^{M} \sum_{j=1}^{M} DE(i,j) \times 100\%, \tag{11a}$$

$$UACI = \frac{1}{M^2} \sum_{i=1}^{M} \sum_{j=1}^{M} \frac{\mid E_1(i,j) - E_2(i,j) \mid}{255} \times 100\%, \tag{11b}$$

$$DE(i,j) = \begin{cases} 0, & \text{if } E_1(i,j) = E_2(i,j), \\ 1, & \text{if } E_1(i,j) \neq E_2(i,j), \end{cases} \tag{11c}$$

where the difference between corresponding pixels in the encrypted versions of the original image $E_1(i,j)$ and the modified image $E_2(i,j)$ is $DE(i,j)$. The NPCR and UACI values are calculated as the average values of 50 iterations and given in Table 6. They are close to the ideal values 99.61% and 33.46%, respectively, [17] and comparable to recent works.

Table 6. Performance analysis

Ref.	Encrypted image	Encryption quality metrics		Correlation ($\times 10^2$)			Entropy	Robustness against differential attacks	
		MSE	PSNR	H	V	D		NPCR (%)	UACI (%)
This paper	Lena	**7828**	**9.1943**	0.42	0.12	**0.01**	7.9973	**99.6042**	33.4204
	Baboon	**7289**	**9.5041**	**−0.18**	−0.07	**−0.01**	**7.9993**	**99.6091**	33.4791
	Pepper	8390	8.8931	**0.17**	−0.08	**0.39**	7.9991	99.6086	**33.4612**
[1]	Lena	7793	9.21	−0.18	0.11	−0.09	7.9975	99.6147	33.4723
	Baboon	7285	9.52	0.19	−0.41	−0.99	7.9992	99.6063	**33.4565**
	Pepper	**8436**	**8.86**	−0.63	**−0.06**	−0.46	**7.9993**	**99.6112**	33.4776
[9]	Lena	–	9.2645	**−0.03**	**−0.07**	**−0.01**	7.9977	99.60	**33.45**
[7]	Lena	–	–	−0.06	−0.39	0.16	**7.9978**	–	–
	Baboon	–	–	−0.23	**−0.00**	−0.15	7.9982	99.6056	33.4282

5 Conclusion

This paper presented an encryption algorithm, utilizes hyperchaotic system, DNA manipulation, and a modified Arnold transform. The modified Arnold transform enhances the encryption process by eliminating the cases at which pixel permutation is cancelled. The performance evaluation for the proposed system shows that it is reliable for image encryption compared to recent similar schemes. The design is simple and amenable for real life application hardware realization. For future work, it can be applied on colored images for each channel separately rather than grayscale images only.

Acknowledgment. This paper is based upon work supported by Science, Technology, and Innovation Funding Authority (STIFA) under grant number (#38161).

References

1. Li, M., Wang, M., Fan, H., An, K., Liu, G.: A novel plaintext-related chaotic image encryption scheme with no additional plaintext information. Chaos, Solitons Fractals **158**, 111989 (2022)
2. Xian, Y., Wang, X.: Fractal sorting matrix and its application on chaotic image encryption. Inf. Sci. **547**, 1154–1169 (2021)
3. Li, T., Shi, J., Li, X., Wu, J., Pan, F.: Image encryption based on pixel-level diffusion with dynamic filtering and DNA-level permutation with 3D Latin cubes. Entropy **21**(3), 319 (2019)
4. Alawida, M., Samsudin, A., Teh, J.S., Alkhawaldeh, R.S.: A new hybrid digital chaotic system with applications in image encryption. Sig. Process. **160**, 45–58 (2019)
5. Belazi, A., Abd El-Latif, A.A., Belghith, S.: A novel image encryption scheme based on substitution-permutation network and chaos. Sig. Process. **128**, 155–170 (2016)
6. Luo, Y., Yu, J., Lai, W., Liu, L.: A novel chaotic image encryption algorithm based on improved baker map and logistic map. Multimed. Tools Appl. **78**(15), 22023–22043 (2019). https://doi.org/10.1007/s11042-019-7453-3
7. Ni, Z., Kang, X., Wang, L.: A novel image encryption algorithm based on bit-level improved Arnold transform and hyper chaotic map. In: 2016 IEEE International Conference on Signal and Image Processing (ICSIP), pp. 156–160. IEEE (2016)
8. Ismail, S.M., Said, L.A., Radwan, A.G., Madian, A.H., Abu-Elyazeed, M.F.: Generalized double-humped logistic map-based medical image encryption. J. Adv. Res. **10**, 85–98 (2018)
9. Zhou, S., Wang, X., Zhang, Y., Ge, B., Wang, M., Gao, S.: A novel image encryption cryptosystem based on true random numbers and chaotic systems. Multimed. Syst. **28**(1), 95–112 (2022). https://doi.org/10.1007/s00530-021-00803-8
10. Wang, X., Wang, M.: A hyperchaos generated from Lorenz system. Phys. A **387**(14), 3751–3758 (2008)
11. Wu, J., Liao, X., Yang, B.: Image encryption using 2D Hénon-Sine map and DNA approach. Sig. Process. **153**, 11–23 (2018)
12. Wu, L., Zhang, J., Deng, W., He, D.: Arnold transformation algorithm and anti-Arnold transformation algorithm. In: 2009 First International Conference on Information Science and Engineering, pp. 1164–1167. IEEE (2009)
13. Mehra, I., Nishchal, N.K.: Optical asymmetric image encryption using gyrator wavelet transform. Opt. Commun. **354**, 344–352 (2015)
14. Kaur, M., Kumar, V.: A comprehensive review on image encryption techniques. Arch. Comput. Methods Eng. **27**(1), 15–43 (2020). https://doi.org/10.1007/s11831-018-9298-8
15. Ghebleh, M., Kanso, A., Noura, H.: An image encryption scheme based on irregularly decimated chaotic maps. Sig. Process. Image Commun. **29**(5), 618–627 (2014)
16. Wu, Y., Noonan, J.P., Agaian, S., et al.: NPCR and UACI randomness tests for image encryption. Cyber J. Multidisc. J. Sci. Technol. J. Sel. Areas Telecommuni. (JSAT) **1**(2), 31–38 (2011)
17. Alghafis, A., Munir, N., Khan, M., Hussain, I.: An encryption scheme based on discrete quantum map and continuous chaotic system. Int. J. Theor. Phys. **59**(4), 1227–1240 (2020). https://doi.org/10.1007/s10773-020-04402-7

Rice Plant Disease Detection and Diagnosis Using Deep Convolutional Neural Networks and Multispectral Imaging

Yara Ali Alnaggar[1], Ahmad Sebaq[1], Karim Amer[1(✉)], ElSayed Naeem[2], and Mohamed Elhelw[1]

[1] Center for Informatics Science, Nile University, Giza, Egypt
{y.ali,a.sebaq,k.amer,melhelw}@nu.edu.eg
[2] Rice Research Institute, Kafr ElSheikh, Egypt

Abstract. Rice is considered a strategic crop in Egypt as it is regularly consumed in the Egyptian people's diet. Even though Egypt is the highest rice producer in Africa with a share of 6 million tons per year [5], it still imports rice to satisfy its local needs due to production loss, especially due to rice disease. Rice blast disease is responsible for 30% loss in rice production worldwide [9]. Therefore, it is crucial to target limiting yield damage by detecting rice crops diseases in its early stages. This paper introduces a public multispectral and RGB images dataset and a deep learning pipeline for rice plant disease detection using multi-modal data. The collected multispectral images consist of Red, Green and Near-Infrared channels and we show that using multispectral along with RGB channels as input archives a higher F1 accuracy compared to using RGB input only.

Keywords: Deep learning · Computer vision · Multispectral imagery

1 Introduction

In Egypt, rice is important in Egyptian agriculture sector, as Egypt is the largest rice producer in Africa. The total area used for rice cultivation in Egypt is about 600 thousand ha or approximately 22% of all cultivated area in Egypt during the summer. As a result, it is critical to address the causes of rice production loss to minimize the gap between supply and consumption. Rice plant diseases contribute mostly to this loss, especially rice blast disease. According to [9], rice blast disease causes 30% worldwide of the total loss of rice production. Thus, rice crops diseases detection, mainly rice blast disease, in the early stages can play a great role in restraining rice production loss.

Early detection of rice crops diseases is a challenging task. One of the main challenges of early detection of such disease is that it can be misclassified as the

Supported by Data Science Africa.

P. Fournier-Viger et al. (Eds.): MEDI 2022, LNAI 13761, pp. 16–25, 2023.
https://doi.org/10.1007/978-3-031-21595-7_2

brown spot disease by less experienced agriculture extension officers (as both are fungal diseases and have similar appearances in their early stage) which can lead to wrong treatment. Given the current scarcity of experienced extension officers in the country, there is a pressing need and opportunity for utilising recent technological advances in imaging modalities and computer vision/artificial intelligence to help in early diagnosis of the rice blast disease. Recently, multispectral photography has been deployed in agricultural tasks such as precision agriculture [3], food safety evaluation [11]. Multispectral cameras could capture images in Red, Red-Edge, Green and Near-Infrared bands wavebands, which captures what the naked eye can't see. Integrating the multispectral technology with deep learning approaches would improve crops diseases identification capability. However, it would be required to collect multispectral images in large numbers.

In this paper, we propose a public multispectral and RGB images dataset and a deep learning pipeline for rice plant disease detection. First, the dataset we present contains 3815 pairs of multispectral and RGB images for rice crop blast, brown spot and healthy leaves. Second, we developed a deep learning pipeline trained on our dataset which calculates the Normalised Difference Vegetation Index (NDVI) channel from the multispectral image channels and concatenates it along its RGB image channels. We show that using NDVI+RGB as input archives a higher F1 score by 1% compared to using RGB input only.

2 Literature Review

Deep learning has emerged to tackle problems in different tasks and fields. Nowadays, it is being adopted to solve the challenge of crop disease identification. For example, Mohanty et al. [8] trained a deep learning model to classify plant crop type and its disease based on images. Furthermore, [1] proposed a deep learning-based approach for banana leaf diseases classification.

Furthermore, multispectral sensors have proven its capability as a new modality to detect crop fields issues and diseases. Some approaches use multispectral images for disease detection and quantification. Cui et al. [4] developed an image processing-based method for quantitatively detecting soybean rust severity using multi-spectral images. Also, [12] utilize digital and multispectral images captured using quadrotor unmanned aerial vehicles (UAV) to collect high-spatial resolution imagery data to detect the ShB disease in rice.

After the reliable and outstanding results deep learning models could achieve on rgb images, some approaches were developed to use deep learning on multispectral images, especially of crops and plants. [10] proposed a deep learning-based approach for weed detection in lettuce crops trained on multispectral images. In addition, Ampatzidis et al. [2] collects multispectral images of citrus fields using UVA for crop phenotyping and deploys a deep learning detection model to identify trees.

3 Methodology

3.1 Hardware Components

We used a MAPIR Survey3N camera, shown in Fig. 1 to collect our dataset. This camera model captures ground-level multispectral images of red, green and NIR channels. It was chosen in favour of its convenient cost and easy integration with smartphones. In addition, we used the Samsung Galaxy M51 mobile phone camera to capture RGB images, paired with the MAPIR camera.

Fig. 1. MAPIR Survey3N camera.

We Designed a holder gadget to combine the mobile phone, MAPIR camera and a power bank in a single tool, as seen in Fig. 2, to facilitate the data acquisition operation for the officers. It was designed using SolidWorks software and manufactured by a 3D printer.

3.2 Data Collection Mobile Application

An android frontend application was also developed to enable the officers who collect the dataset to control the multispectral and the smartphone cameras for capturing dual RGNIR/RGB images simultaneously while providing features such as image labelling, imaging session management, and Geo-tagging. The mobile application is developed with Flutter and uses Firebase real-time database to store and synchronise the captured data including photos and metadata. Furthermore, Hive local storage database is used within the application to maintain a local backup of the data.

Fig. 2. Holder gadget.

3.3 Analytics Engine Module

Our engine is based on ResNet18 [6] architecture which consists of 18 layers and it utilize the power of residual network, see Fig. 3, residual network help us avoid the vanishing gradient problem.

We can see how layers are configured in the ResNet-18 architecture. The architecture starts with a convolution layer with 7 × 7 kernel size and stride of 2. Next we begin with the skip connection. The input from here is added to the output that is achieved by 3 × 3 max pool layer and two convolution layers with kernel size 3 × 3, 64 kernels each. This is the first residual block.

The output of this residual block is added to the output of two convolution layers with kernel size 3 × 3 and 128 such filters. This constituted the second residual block. Then the third residual block involves the output of the second block through skip connection and the output of two convolution layers with filter size 3 × 3 and 256 such filters. The fourth and final residual block involves output of third block through skip connections and output of two convolution layers with same filter size of 3 × 3 and 512 such filters.

Finally, average pooling is applied on the output of the final residual block and received feature map is given to the fully connected layers followed by softmax function to receive the final output.

The vanishing gradient is a problem which happens when training artificial neural networks that involved gradient based learning and backpropagation. We use gradients to update the weights in a network. But sometimes what happens is that the gradient becomes very small, effectively preventing the weights to be updated. This leads to network to stop training. To solve such problem, residual neural networks are used.

Layer Name	Output Size	ResNet-18
conv1	112 × 112 × 64	7 × 7, 64, stride 2
conv2_x	56 × 56 × 64	3 × 3 max pool, stride 2
		$\begin{bmatrix} 3 \times 3, 64 \\ 3 \times 3, 64 \end{bmatrix} \times 2$
conv3_x	28 × 28 × 128	$\begin{bmatrix} 3 \times 3, 128 \\ 3 \times 3, 128 \end{bmatrix} \times 2$
conv4_x	14 × 14 × 256	$\begin{bmatrix} 3 \times 3, 256 \\ 3 \times 3, 256 \end{bmatrix} \times 2$
conv5_x	7 × 7 × 512	$\begin{bmatrix} 3 \times 3, 512 \\ 3 \times 3, 512 \end{bmatrix} \times 2$
average pool	1 × 1 × 512	7 × 7 average pool
fully connected	1000	512 × 1000 fully connections
softmax	1000	

Fig. 3. ResNet18 original architecture

Residual neural networks are the type of neural network that applies identity mapping. What this means is that the input to some layer is passed directly or as a shortcut to some other layer. If x is the input, in our case its an image or a feature map, and $F(x)$ is the output from the layer, then the output of the residual block can be given as $F(x) + x$ as shown in Fig. 4.

We changed the input shape to be 256 × 256 instead of 224 × 244, also we replaced the last layer in the original architecture with a fully connected layer where the output size was modified to three to accommodate our task labels.

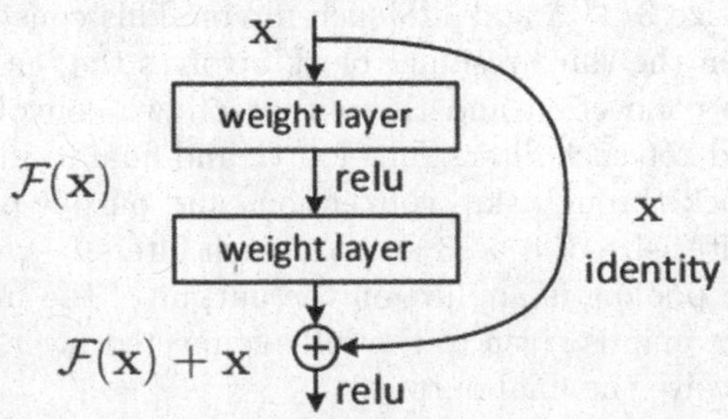

Fig. 4. Residual block

4 Experimental Evaluation

4.1 Dataset

We have collected 3815 samples of rice crops of three labels: blast disease, brown spot disease and healthy leaves distributed, shown in Fig. 5, as the following: 2135, 1095 and 585, respectively. Each sample is composed of a pair of (RGB) and (R-G-NIR) images as seen in Fig. 6, which were captured simultaneously. Figure 7 shows samples of the three classes in our dataset.

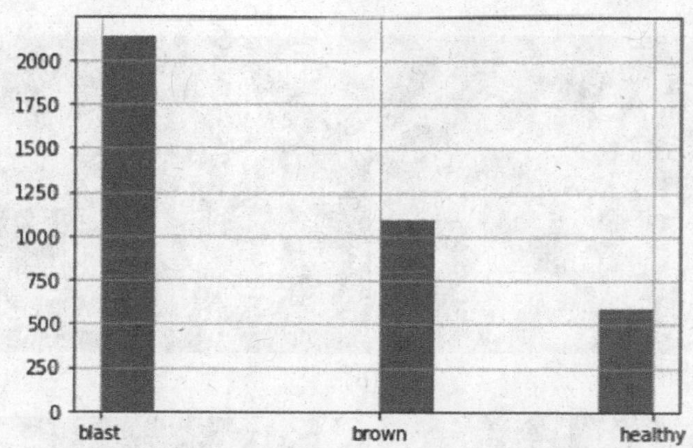

Fig. 5. Collected dataset distribution.

Fig. 6. On the left is the RGB image and on the right is its R-G-NIR pair.

(a)

(b)

(c)

Fig. 7. (a) Blast class sample. (b) Brown spot class sample. (c) Healthy class sample.

4.2 Training Configuration

In this section, we explain our pipeline for training data preparation and preprocessing. Also, we mention our deep learning models training configuration for loss functions and hyperparameters.

Data Preparation

RGB Images Registration. Since the image sample of our collected dataset consists of a pair of RGB and R-G-NIR images, the two images are expected to have a similar field of view. However, the phone and MAPIR camera have different field of view parameters that the mapir camera has a 41° FOV compared to the phone camera with 123° FOV. As a result, we register the rgb image to the r-g-nir image using the OpenCV library. The registration task starts by applying an ORB detector over the two images to extract 10K features. Next, we use a brute force with Hamming distance matcher between the two images extracted features. Based on the calculated distances for the matches, we sort them descendingly and drop the last 10%. Finally, the homography matrix is calculated using the matched points in the two images to be applied over the RGB images. Figure 8 shows an RGB image before and after registration.

Fig. 8. On the left is an RGB image before calibration and on the right is after registration.

MAPIR Camera Calibration. The MAPIR camera sensor captures the reflected light which lies in the Wavelengths in the Visible and Near Infrared spectrum from about 400–1100 n and saves the percentage of reflectance. After this step, calibration of each pixel is applied to ensure that it is correct. This calibration is performed before every round of images captured using the MAPIR Camera Reflectance Calibration Ground Target board, which consists of 4 targets with known reflectance values, as shown in Fig. 9.

Models Training Configuration. We trained our models for 50 epochs with a batch size of 16 using Adam optimizer and Cosine Annealing with restart scheduler [7] with cycle length 10 epochs and learning rate of 0.05. For the loss function, we used a weighted cross entropy to mitigate the imbalance of the training dataset. Images were resized to dimension 256 × 256.

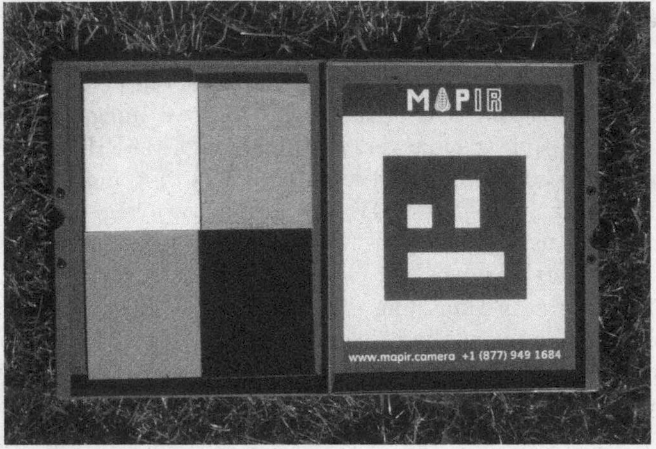

Fig. 9. MAPIR camera reflectance calibration ground target board.

Results. For training the deep learning model using RGB and R-G-NIR pairs, we generate a NDVI channel, using Eq. 1, and concatenate it to the RGB image. Our study shows that incorporating the NDVI channel improves the model capability to classify the rice crops diseases. Our model could achieve a F1 score with 5-kFold of 84.9% when using RGB+NDVI as input compared to using only RGB image which could obtain a F1 score of 83.9%. Detailed results are presented in Table 1.

$$NDVI = \frac{NIR - Red}{NIR + Red} \tag{1}$$

Table 1. F1 score over our collected dataset achieved by using RGB as input versus RGB+NDVI.

Class	RGB	RGB+NDVI
Blast	89.64%	90.02%
Spot	82.64%	83.26%
Healthy	79.08%	81.54%

5 Conclusion

We presented our public dataset and deep learning pipeline for rice plant disease detection. We showed that employing multispectral imagery with RGB improves the model capability of disease identification by 1% compared to using solely RGB imagery. We believe using a larger number of images for training would enhance current results also considering a larger number of images when using a deeper model this will result in better results. In addition, more investigation on how to fuse multispectral imagery with RGB for training could be applied, for example we can calculate NDVI from the blue channel instead of the red this may also boost the model performance.

Acknowledgements. This work has been done with the Data Science Africa support.

References

1. Amara, J., Bouaziz, B., Algergawy, A.: A deep learning-based approach for banana leaf diseases classification. Datenbanksysteme für Business, Technologie und Web (BTW 2017)-Workshopband (2017)
2. Ampatzidis, Y., Partel, V.: UAV-based high throughput phenotyping in citrus utilizing multispectral imaging and artificial intelligence. Remote Sens. **11**(4), 410 (2019)
3. Candiago, S., Remondino, F., De Giglio, M., Dubbini, M., Gattelli, M.: Evaluating multispectral images and vegetation indices for precision farming applications from UAV images. Remote Sens. **7**(4), 4026–4047 (2015)
4. Cui, D., Zhang, Q., Li, M., Hartman, G.L., Zhao, Y.: Image processing methods for quantitatively detecting soybean rust from multispectral images. Biosys. Eng. **107**(3), 186–193 (2010)
5. Elbasiouny, H., Elbehiry, F.: Rice production in Egypt: the challenges of climate change and water deficiency. In: Ewis Omran, E.-S., Negm, A.M. (eds.) Climate Change Impacts on Agriculture and Food Security in Egypt. SW, pp. 295–319. Springer, Cham (2020). https://doi.org/10.1007/978-3-030-41629-4_14
6. He, K., Zhang, X., Ren, S., Sun, J.: Deep residual learning for image recognition. In: Proceedings of the IEEE Conference on Computer Vision and Pattern Recognition, pp. 770–778 (2016)
7. Loshchilov, I., Hutter, F.: SGDR: stochastic gradient descent with warm restarts. arXiv preprint arXiv:1608.03983 (2016)
8. Mohanty, S.P., Hughes, D.P., Salathé, M.: Using deep learning for image-based plant disease detection. Front. Plant Sci. **7**, 1419 (2016)
9. Nalley, L., Tsiboe, F., Durand-Morat, A., Shew, A., Thoma, G.: Economic and environmental impact of rice blast pathogen (magnaporthe oryzae) alleviation in the united states. PLoS ONE **11**(12), e0167295 (2016)
10. Osorio, K., Puerto, A., Pedraza, C., Jamaica, D., Rodríguez, L.: A deep learning approach for weed detection in lettuce crops using multispectral images. AgriEngineering **2**(3), 471–488 (2020)
11. Qin, J., Chao, K., Kim, M.S., Lu, R., Burks, T.F.: Hyperspectral and multispectral imaging for evaluating food safety and quality. J. Food Eng. **118**(2), 157–171 (2013). https://doi.org/10.1016/j.jfoodeng.2013.04.001, https://www.sciencedirect.com/science/article/pii/S0260877413001659
12. Zhang, D., Zhou, X., Zhang, J., Lan, Y., Xu, C., Liang, D.: Detection of rice sheath blight using an unmanned aerial system with high-resolution color and multispectral imaging. PLoS ONE **13**(5), e0187470 (2018)

A Novel Diagnostic Model for Early Detection of Alzheimer's Disease Based on Clinical and Neuroimaging Features

Eyad Gad[1(✉)], Aya Gamal[2], Mustafa Elattar[2,3], and Sahar Selim[2,3(✉)]

[1] School of Engineering and Applied Sciences, Nile University, Giza, Egypt
e.gad@nu.edu.eg

[2] Medical Imaging and Image Processing Research Group, Center for Informatics Science, Nile University, Giza, Egypt
sselim@nu.edu.eg

[3] School of Information Technology and Computer Science, Nile University, Giza, Egypt

Abstract. Alzheimer's Disease (AD) is a dangerous disease that is known for its characteristics of eroding memory and destroying the brain. The classification of Alzheimer's disease is an important topic that has recently been addressed by many studies using Machine Learning (ML) and Deep Learning (DL) methods. Most research papers tackling early diagnosis of AD use these methods as a feature extractor for neuroimaging data. In our research paper, the proposed algorithm is to optimize the performance of the prediction of early diagnosis from the multimodal dataset by a multi-step framework that uses a Deep Neural Network (DNN) as an optimization technique to extract features and train these features by Random Forest (RF) classifier. The results of the proposed algorithm showed that using only demographic and clinical data results in a balanced accuracy of 88% and an area under the curve (AUC) of 94.6. Ultimately, combining clinical and neuroimaging features, prediction results improved further to a balanced accuracy of 92% and an AUC of 97%. This study successfully outperformed other studies for both clinical and the combination of clinical and neuroimaging data, proving that multimodal data is efficient in the early diagnosis of AD.

Keywords: Alzheimer's disease (AD) · Machine learning (ML) · Deep learning (DL) · Random Forest (RF) · Mild cognitive impairment (MCI)

1 Introduction

Alzheimer's Disease is an extremely dangerous disease that is known for its properties of memory erosion and brain destruction [1]. The disease itself is the degradation of grey matter and memory function in the brain causing a person with Alzheimer's disease to exhibit "abnormal behavior" compared to their normal self. Grey matter degradation will lead to mild cognitive impairment (MCI) on several levels, most notably the inability to concentrate and short-term memory loss if the disease is in its early stages to complete personality distortion and halting of response to external stimuli. Alzheimer's disease

P. Fournier-Viger et al. (Eds.): MEDI 2022, LNAI 13761, pp. 26–39, 2023.
https://doi.org/10.1007/978-3-031-21595-7_3

is existed throughout human history, mainly known for its effect on the elderly in terms of cognitive impairment, dementia, and other symptoms such as violent outbreaks and severe short- and long-term memory loss [2]. Many of the disease's causes, such as contact sports head injuries, and ageing, lead to the deterioration of gray matter in the brain. AD occurs in old age due to a lack of neurogenesis or an increase in brain cells. After a certain age, the brain stops producing new cells, as a result of which all brain cells remain active for most of their lives, which can lead to deterioration due to the virtue of time. As the disease normally manifests in 20 million people per year making it dangerous as it cannot be contracted nor easily predicted at an early age [2].

Anticipating this disease before it causes any harm is a necessity in daily life, as without predicting the possibility of developing the disease, countermeasures to reduce symptoms will be greatly delayed. The classification of AD has been a very active topic in the past decade, based on various methodologies mainly using ML and DL approaches [3]. This is due to the importance of the research as with newer experiments and research results the possibility of understanding the disease increases, decreasing the harm to mankind. These studies are based on data from the Alzheimer's Disease Neuroimaging Initiative (ADNI), which provides data from multiple modalities. Intuitively, models that integrate data from different modalities outperform their monomodal counterparts.

Gonzalez et al. [1] presented a multimodal ML approach for early diagnosis of AD, which allows an objective comparison of the models used since the dataset and pipeline are the same for all models. Their proposed approach is to use a support vector machine (SVM) and RF on a combination of clinical and neuroimaging data, which would allow a high degree of data diversity while maintaining a suitable degree of bias and variance. For accurate measurements of the performance of the models, the researchers constructed two SVM rating scores for each subject. These scores are added along with the clinical data features into the RF classifier and evaluated using 10k fold cross-validation. The results of the paper showed that using only demographic and clinical data results in a balanced accuracy of 76% with AUC reaching 85%. Ultimately, by combining clinical and neuroimaging features, prediction results improved to a balanced accuracy of 79%, and an AUC of 89%.

Venugopalan et al., presented a research paper [4], which presented a DL approach to predict AD by using integrated multimodal systems relying on data from Magnetic Resonance Imaging (MRI), genetics focusing on single nucleotide polymorphisms and electronic health records, to classify patients into suffering from AD, MCI or Cognitive Normal (CN) where the average healthy individual is CN. The researchers proposed an algorithm where stacked denoising auto-encoders are used to extract features from clinical and genetic data and 3D CNN with MRI data to aid the prediction. Then the extracted features are concatenated into a fully connected layer followed by RF for classification. The results of the internal 10-fold cross-validation showed an accuracy of 88% and a recall of 85%.

El-Sappagh et al. presented an interesting new concept for the implantation in the research paper [5], where a multilayered multimodal system for the detection and prediction of AD was used. The model integrates 11 modalities of the ADNI dataset, making precise decisions along with a set of interpretations for each decision to make the model more robust and accurate. The model has two layers to classify and predicts the target class with minimal errors, the first layer performs multi-class classification for early diagnosis of AD, while the second layer performs binary classification to detect potential progression from MCI to AD within three years of baseline diagnosis. The designed model achieves a cross-validation accuracy of 93.95% in the first layer, while it achieves a cross-validation accuracy of 87.08% in the second layer.

The prediction and progression of AD have been extensively studied, however, research studies on early diagnosis using DL as feature extractors for single modality systems, especially neuroimaging, are less efficient for predicting the progression of AD [6]. Based on the paper [1], We address the challenges of the dataset as it has an unequal distribution of classes in the training dataset and outliers, which affects the accuracy of the classifier. Therefore, in our research paper, we optimize the performance of the prediction of early diagnosis from the multimodal dataset of clinical data and neuroimaging scores that were used in the paper [1] by:

1. Using the Synthetic Minority Oversampling Technique (SMOTE) in the dataset to build larger decision areas containing the points of nearby minority groups.
2. A multi-step framework that uses DL as an optimization technique to extract features of the multimodal dataset and train these features by RF model.

The goal of this paper is to predict the possibility of developing AD given the person is classified as MCI as opposed to CN as people who are in the CN state are hard to predict unless a family history is available. The model will be used to rationalize the state of those suffering from MCI to distinguish the person's nature, and are the closest to CN or AD. If the patient is closer to CN, then the patient is stable (sMCI) however if the patient is closer to AD, then the patient is progressive (pMCI).

2 Materials and Methods

In this section, we first explain the proposed approach as a model, followed by defining the materials to be used, and finally explain our work in steps.

2.1 Proposed Model

The experimental analysis in this work contains four steps (see Fig. 1). The first is to split the data set into test and training data, then balance the training data with SMOTE and create different feature sets. The second step is to extract most of the main features of each feature set using the DNN and experiment with the extracted features using different ML classifiers. The third step is to calculate the different performance metrics for the feature sets and compare each set with these metrics to choose the most appropriate model. Finally, Cross-Validation is applied to estimate the performance of the ML classifiers and to optimize the hyperparameters of each, thus, obtaining the best model which achieves the best accuracy of prediction.

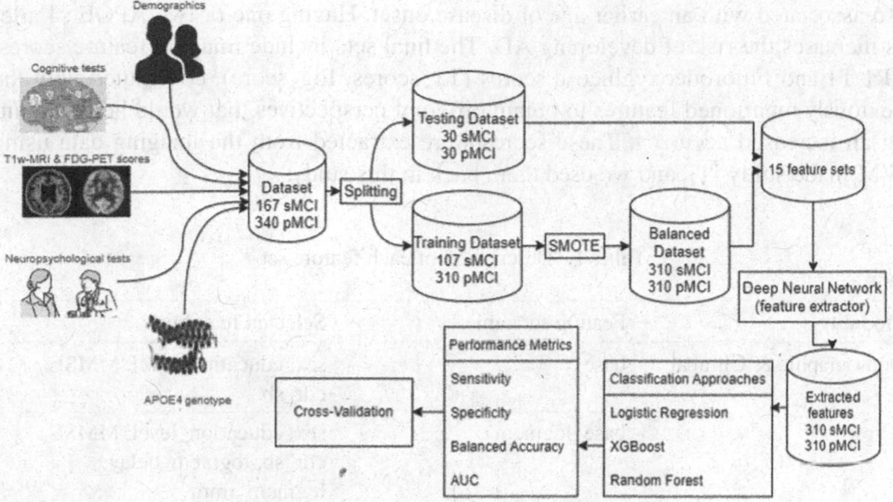

Fig. 1. Overview of the proposed approach.

2.2 Dataset

In previous studies, the data were obtained from ADNI, which is a public dataset. This dataset was released in 2003, with the main goal of measuring the progression of MCI and early AD using a combination of imaging, biological markers, clinical, and neuropsychological assessments. The three main subject classes which are CN (normal), AD and MCI had to have two fundamental tests composed of a mini-mental state examination (MMSE) and clinical dementia rating (CDR) with a range of values that define each class or else they would be ruled out of the data [3, 7]. In this study, we study the performance of the prediction of progression to AD at 36 months, All the data used in the preparation are obtained from ADNI, and the same group of features as in this study [1] was used.

We include 15 different models of the dataset (see Table 1), each of which consists of a base level model which holds demographic details (sex, education_level) plus the MMSE which is a questionnaire that is used to measure cognitive impairment and CDR sum of boxes used to accurately stage the severity of Alzheimer's dementia and MCI. This data is used as the baseline for our model, to classify those who are MCI as either progressive or stable, where pMCI will progress to AD while those who are sMCI will not. In each of the upcoming feature sets, we add a different mix of features with the defined base model. The log memory test which is a standardized assessment of narrative episodic memory is used in the next model. Rey auditory verbal learning test (RAVLT) neuropsychological tool used to assess functions like attention and memory [1]. We also have a series of AD assessment scale cognitive (ADAS) breakdown features that help in the assessment of memory, language, concentration, and praxis at its core (adas_memory, adas_language, adas_concentration, adas_praxis), providing thorough information regarding patient condition. Another critical feature that is taken into consideration is the Apolipoprotein ε4 (APOE4) which increases the risk for AD and is also associated with an earlier age of disease onset. Having one or two APOE ε4 alleles increases the risk of developing AD. The final sets include imaging feature scores, MRI-T1 and Fluorodeoxyglucose scores (T1_scores, fdg_score), convoluted with the previously mentioned features to obtain different perspectives that would help in gaining an increased accuracy. These scores were extracted from the imaging data using SVM in the study [1], and we used them back in this study.

Table 1. Description of each feature set

Modality	Feature set name	Selected features
Demographic & Clinical	base	sex, education_level, MMSE, cdr_sb
	base_logmem	sex, education_level, MMSE, cdr_sb, logmem_delay, logmem_imm
	base_ravlt	sex, education_level, MMSE, cdr_sb, ravlt_immediate
	base_logmem_ravlt	sex, education_level, MMSE, cdr_sb, ravlt_immediate, logmem_delay, logmem_imm
	base_adas	sex, education_level, MMSE, cdr_sb, adas_memory, adas_language, adas_concentration, adas_praxis
	base_ravlt_adas	sex, education_level, MMSE, cdr_sb, adas_memory, adas_language, adas_concentration, adas_praxis, ravlt_immediate
	base_ravlt_apoe	sex, education_level, MMSE, cdr_sb, apoe4, ravlt_immediate
	base_adas_apoe	sex, education_level, MMSE, cdr_sb, apoe4, adas_memory, adas_language, adas_concentration, adas_praxis
	base_ravlt_adas_apoe	sex, education_level, MMSE, cdr_sb, apoe4, adas_memory, adas_language, adas_concentration, adas_praxis, ravlt_immediate

(*continued*)

Table 1. (*continued*)

Modality	Feature set name	Selected features
Demographic, Clinical & imaging	base_t1score	sex, education_level, MMSE, cdr_sb, T1_score
	base_fdgscore	sex, education_level, MMSE, cdr_sb, fdg_score
	base_scores	sex, education_level, MMSE, cdr_sb, T1_score, fdg_score
	base_ravlt_scores	sex, education_level, MMSE, cdr_sb, ravlt_immediate, T1_score, fdg_score
	base_adas_scores	sex, education_level, MMSE, cdr_sb, adas_memory, adas_language, adas_concentration, adas_praxis, T1_score
	base_adas_memtest_scores	sex, education_level, MMSE, cdr_sb, adas_memory, adas_language, adas_concentration, adas_prxis, ravlt_immediate, T1_score, fdg_score

2.3 Data Preprocessing

SMOTE is a method used when the dataset used is imbalanced. SMOTE is most used when the class or target is imbalanced with a severely underrepresented class of data [8]. The algorithm could be separated into multiple steps which can be defined separately to produce the effect of data multiplication. The first step undergone is the under-sampling of the data to trim all the outliers and any possible noise in the minority due to the nature of the algorithm being more complex than merely reiterating or copying and pasting the data back into the data set, for the SMOTE algorithm uses the feature space, or the local area of the minority if graphed. Meaning the feature space will be constructed for the target class in the minority. After the feature space is constructed, the SMOTE algorithm adds new points within the area, mainly near other points in the area using the same protocol as K-nearest neighbors (KNN) [9]. The algorithm can use the Euclidean or the Manhattan distance in constructing the feature space. In our study, we used the Euclidean distance, as it can be used in any space to calculate distance. Since the data points can be represented in any dimension, it is the more viable option. The Euclidean distance is depicted in (1) by taking two or more points to find the squared difference and then calculating the square root of the result.

$$\sqrt{\sum_{i=1}^{k} (x_i - y_i)^2} \quad (1)$$

Although creating virtual data points may seem a source of severe errors, the samples become more general with the increase in data points, preventing bias and variance from occurring. The most crucial step is the increase of data, where the amount of data increase required is determined through the parameters of the SMOTE algorithm. The first parameter is the sampling strategy (ss) where the parameter identifies which class is iterated, whether it is the minority or otherwise. The parameter k is the number of nearest neighbors used to synthesize new data. Furthermore, The out_step parameter determines the step size during calculations in the designated SMOTE algorithm. After setting the parameters (ss = minority, k = 5, out_step = 0.6), the data is inputted into the SMOTE algorithm for synthesizing the new data points.

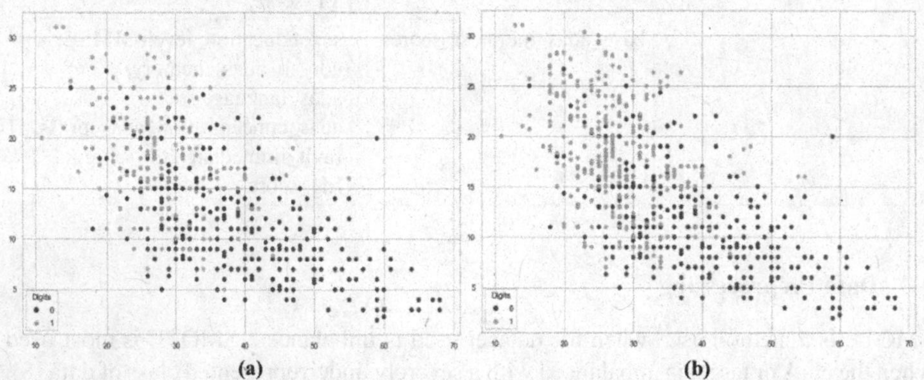

Fig. 2. Scatter plots of two random features of the dataset. (a) and (b) plots illustrate the data points before and after SMOTE is applied respectively.

After splitting the data set, the training data has 310 pMCI and 107 sMCI representing purple and yellow respectively (see Fig. 2), the minority class in the data set is sMCI. As a result of SMOTE, the training data set is balanced to have 310 data points of sMCI.

2.4 Feature Extraction

The process of translating raw data into numerical features that may be processed while keeping the information from the original data set is referred to as feature extraction. Feature extraction can be done manually or automatically through the aid of a Deep Neural Network, which will be more effective and valuable when converting raw data into machine learning algorithms [9].

The DNN is divided into three primary layers, each with its own set of neurons: an input layer, an output layer, and a hidden layer. For the input layer, each node is given the set of characteristics we already obtained from the dataset, and random weights are assigned to the synapses that are vital for attaining the proper outcome in the training phase; weights with a significant positive or negative value will have a substantial influence on the output of the future neuron. These synapses are then linked to neurons in the hidden layer. The hidden layer interprets significant elements from the input data that are predictive of the outputs. A DNN architecture has 3 hidden layers, which have 1,000 neurons, 500 neurons, and 200 neurons respectively and a drop-out layer, however, the output layer has half the number of input features or attributes of the model. As for the optimization algorithm used, the Adam optimizer is used as it is a stochastic gradient descent replacement optimization technique for DL model training, its technique is based on the adaptive approximation of first- and second-order moments [9]. Adam combines the strongest features of the Adaptive Gradient Algorithm and Root Mean Squared Propagation to create an optimization algorithm.

2.5 Classification Approaches

This section explains the basic principle of the classification approaches used in this study. Three different classifiers have been used; Random Forest, Extreme Gradient Boosting, and Logistic Regression.

2.5.1 Random Forest

RF is the set that uses a combination of decision-based trees and data subsets. it is a classification and regression tree or CART that will train on different sets of introductory datasets [10]. For group testing, out-of-bag (OOB) errors are used. The reason OOB was used was due to the RF technique using bagging as a training method. The main principle behind the bagging method is that combining learning models improves the end outcome. In order to obtain a more precise and consistent forecast, RF generates many decision trees and blends them. The main advantage of RF is that it can be applied to both classification and regression problems, which make up the majority of contemporary machine learning systems. The hyperparameters of RF are quite similar to those of a decision tree or a bagging classifier. Fortunately, using the classifier-class of RF eliminates the requirement to combine a decision tree with a bagging classifier.

While the trees are developing, the RF adds more randomness to the model. When dividing a node, it looks for the best feature from a random subset of features rather than the most crucial one. A better model is often produced as a result of the great diversity this causes. As a result, the process for splitting a node in a RF only considers a random subset of the features. By applying random thresholds for each feature in addition to the best available thresholds, the randomness of the trees can even be increased. The RF also makes it very simple to gauge the relative contribution of each feature to the prediction. For data categorization accuracy, each tree receives a vote, and the forest selects the votes with the most classifications.

2.5.2 Extreme Gradient Boosting

Extreme Gradient Boosting or XGBoost is an ensemble learning technique, which means it uses the findings of several models, known as base learners, to create a prediction. Just like in RFs, XGBoost uses decision trees as base learners. Individual decision Trees have high-variance, low-bias models. They are extremely effective at detecting associations in any form of training data, but they struggle to extrapolate well to new data. Furthermore, the trees employed by XGBoost are not standard decision trees [11], CARTs hold real-value scores of whether an instance belongs to a group rather than a single judgement in each leaf node. When the tree has reached its maximum depth, the choice may be made by transforming the scores into categories based on a specific threshold.

2.5.3 Logistic Regression

Logistic Regression (LR) is a ML method that utilizes the sigmoid function to classify the data given to it. LR by nature is a binary classification system where the data is either classified as one or zero, where the two possibilities are arbitrarily assigned. There are other forms of logistic regression, namely Nominal Logistic Regression (NLR) and Ordinal Logistic Regression (OLR). NLR categorizes a data point into one of three or more categories with the categories being unorganized or without a higher priority than another category. OLR categorises the data in a priority or importance level. This is important to note due to the nature of the method as the method can predetermine the possibility earlier if OLR was used instead of Bayesian Linear Regression (BLR), however since our current purposes require an accuracy of 95% or higher, we used BLR, furthermore the processing time of BLR is shorter than the OLR method making it more time efficient as OLR has more categories to fit the data point.

The probability of the outcome will be between the values of 0 and 1 but the outcome will normally collapse either 0 or 1 due to the nature of the sigmoid function. By using this function, we can calculate the probability of an individual developing AD as opposed to remaining in a stable state of mind. Furthermore, the maximum likelihood of an outcome can be calculated by estimating unknown variables and displaying it on a parabolic line.

2.6 Performance Evaluation

In machine intelligence, a viable and trustworthy method of calculating accuracy is the confusion matrix which can produce a receiver operator characteristic (ROC) curve. In such a case the area under the ROC curve is referred to as the AUC. AUC is the best intuitive performance measure and is frequently used to compare different classification algorithms by calculating the probability of ascertaining a given data point in its true class. This is done by measuring the full 2D area under the entire ROC curve through either discrete or continuous methods such as geometric calculations or integration. Since a predetermined resolution threshold is required to report accuracy, the AUC can be used to indicate a more accurate measure of accuracy, which gives an indication of the effectiveness of the chosen model by determining the ratio presented by the ROC graph in a simple and intuitive way which can be identified by any individual [3].

Measures of a test's ability to determine whether a person has or does not have a disease include specificity and sensitivity which are different measurements compared to accuracy allowing different incite in the same information presented by a confusion matrix. Sensitivity relates to the ability of a test to identify a positive result for a person with a disease, it measures the true positive rate of classification making a more interesting value compared to the accuracy (2). In this case, a highly sensitive test means that fewer cases of a disease are missed because there are fewer false negatives.

$$sensitivity = \frac{TruePostive}{TruePostive + Falsenegative} \tag{2}$$

Moreover, another measurement used is the specificity confusion matrix which calculates the classified individuals who do not have a disease as negative, as such it measures the true negative rate of classification (3). The test's high specificity means that there are few false positives. The use of a measurement method with low screening specificity may not be viable, since many people without this condition may test positive and may undergo unnecessary diagnostic procedures.

$$specificity = \frac{Truenegative}{Truenegative + FalsePostive} \tag{3}$$

Lastly, Balanced accuracy is used to measure the efficiency of binary and multi-category classification and is particularly useful when there is an imbalance between classes, meaning that one of the classes appears more frequently than the other. This often occurs in many places where abnormalities and diseases are detected. Balanced accuracy is the arithmetic mean of sensitivity and specificity (4).

$$BalancedAccuracy = \frac{sensitivity + specificity}{2} \tag{4}$$

2.7 Performance Optimization

Based on the previously mentioned performance metrics, the best model is selected, and its accuracy is further optimized through Cross-validation. In our research paper Cross-validation is a necessity in any model where a part of the dataset is used to train the model while the remaining is used to test the model, then the validation will use the complementary subset of the data set and repeat the operation for proper validation. Cross-validation was applied to all trials: results were on average 250 iterations of stratified random splits with 80% of the samples used for training and the remaining 20% for testing. Random Forest classifier was trained with fixed hyperparameters: 100 trees, the tree depth was set at 5 levels, and only the square root of the total number of features is considered when looking for a split [1]. As an evaluation of the classification, we report the results of the mentioned performance metrics and the predicted class for each subject.

3 Results

For each model of the feature set, DNN is applied to perform feature engineering to extract the key features. The number of extracted features for each model is the same as the number of neurons in the output layer, thus reducing the dimensionality of the input data by removing the redundant data. Taking the first model (base) as an example, its dimensions are (620, 4) which means 620 data points of 4 features; thus, the dimensions of the extracted model should be (620, 2). 60% of each model is set as a training set, 20% as a validation set, and the last 20% as a test set. Table 2 shows the validation accuracy of the best epoch of DNN.

Table 2. Validation accuracy and number of extracted features of DNN

	base	base_logmem	base_ravlt	base_logmem_ravlt	base_adas	base_ravlt_adas	base_ravlt_apoe	base_adas_apoe	base_ravlt_adas_apoe	base_t1score	base_fdgscore	base_scores	base_ravlt_scores	base_adas_scores	base_adas_memtest_scores
Val. Acc.	63.4	72	83.5	72.8	76.2	86.7	77.1	93.7	89.5	80.4	83.6	90.8	87.0	91.6	90.6
Ex. feat.	2	3	2	3	4	4	3	4	5	2	2	3	3	4	5

To test our approach, we investigated the performance of different three classifiers, to obtain the best accuracy. The three classifiers are applied to the extracted features and their hyperparameters are optimized using 10-fold cross-validation. For each classifier, a confusion matrix is obtained, and the results are calculated using four metrics; AUC, sensitivity, specificity, and balanced accuracy.

Logistic regression is the first experiment applied and the performance metrics are calculated for each model. As the results show, we obtained the highest balanced accuracy of 88% and an AUC of 94.5% in model base_adas_scores that contains the clinical and neuroimaging features (Table 3).

Table 3. Results of logistic regression

	base	base_logmem	base_ravlt	base_logmem_ravlt	base_adas	base_ravlt_adas	base_ravlt_apoe	base_adas_apoe	base_ravlt_adas_apoe	base_t1score	base_fdgscore	base_scores	base_ravlt_scores	**base_adas_scores**	base_adas_memtest_scores
Specificity	65.1	78.7	70	74	79.2	72.6	74.9	83.9	79.7	76.4	85.3	80.5	74.1	**87.2**	73.4
Sensitivity	69	70.2	87.5	80.7	82.6	89.3	84.4	91.6	88.3	71.1	73.8	76.6	86	**89.4**	93.7
Bal. Acc.	67.1	74.4	78.7	77.3	80.9	80.9	79.6	87.7	84	73.7	79.6	78.5	80	**88.3**	83.6
AUC	75	81.8	88.6	86.7	89.3	90.9	88.7	94.3	90.8	83.2	87.7	87.2	90.6	**94.5**	91.4

For the second experiment, XGBoost is used. The results looked similar to the results of the first experiment (see Table 4). XGBoost showed a balanced accuracy of 78% and an AUC of 95% in the same model (base_adas_scores) of the first experiment. In further analysis, the sensitivity was observed to be higher for each model. A negative result on a test with high sensitivity is useful for ruling out disease. A high sensitivity test is reliable when its result is negative since it rarely misdiagnoses those who have the disease.

Table 4. Results of XGBoost

	base	base_logmem	base_ravlt	base_logmem_ravlt	base_adas	base_ravlt_adas	base_ravlt_apoe	base_adas_apoe	base_ravlt_adas_apoe	base_t1score	base_fdgscore	base_scores	base_ravlt_scores	**base_adas_scores**	base_adas_memtest_scores
Specificity	55.5	69.1	69.7	70.7	76.8	73.3	72	79.9	73.1	67	74.3	71	72.7	**81.9**	79.1
Sensitivity	88	90.5	93.8	94.8	93.9	93.9	93.9	95	93	89.7	92.3	92.3	93.2	**92.8**	95.5
Bal. Acc.	51.6	62.5	62.9	67.8	67.5	69.7	68.6	69.2	71.5	61.6	70.8	75.3	76.9	**78.4**	78.4
AUC	81.2	89.1	91.1	91.2	92.3	91.6	92.4	94.4	92.2	87	91.3	89.9	91.9	**94.9**	92.9

Finally, for the last experiment, RF is used. Random Forest achieved higher results in all the feature sets (see Table 5). The RF algorithm avoids and prevents overfitting by using multiple trees, this gives accurate and precise results. As a result, the experiment showed a balanced accuracy of 92% and AUC of 97% in a different model (base_ravlt_scores), but it has both clinical and neuroimaging features.

Table 5. Results of random forest

	base	base_logmem	base_ravlt	base_logmem_ravlt	base_adas	base_ravlt_adas	base_ravlt_apoe	base_adas_apoe	base_ravlt_adas_apoe	base_t1score	base_fdgscore	base_scores	**base_ravlt_scores**	base_adas_scores	base_adas_memtest_scores
Specificity	74.6	87.1	88.6	90.8	90.2	90.4	90.2	92.6	93.5	76.4	86.3	84.7	**96.5**	84.8	90.5
Sensitivity	68.6	89.4	71.1	79.4	82.7	74.1	76.9	79.8	74.1	74.8	83.6	79.7	**87.2**	85.4	81.5
Bal. Acc.	71.6	88.2	79.9	85.1	86.5	82.2	83.6	86.2	83.8	75.6	85.0	82.2	**91.9**	85.1	86.0
AUC	79.8	94.6	88.8	93.9	93.3	90.3	91.3	92.4	90.8	84.7	92.2	89.3	**97.3**	93.2	92.9

As a comparison of the experiments, apparently, the RF has the highest results among the other two classifiers (see Table 6). It is the appropriate classifier as it is robust to outliers and generalizes the data in an efficient way. As a result, we proved the reliability of the proposed approach and optimized the performance of the prediction of early diagnosis by proposing a novel approach using SMOTE and DL to extract features from multimodal data.

Table 6. AUC comparison of the classifiers

AUC	Clinical	Clinical & neuroimaging
Logistic regression	94.3%	94.5%
XGBoost	94.4%	95.0%
Random Forest	**94.6%**	**97.0%**

4 Conclusion

In this paper, we outperformed the research paper findings [1] for both clinical and the combination of clinical and neuroimaging data, as the below table shows (Table 7).

Table 7. AUC comparison between paper [1] and the proposed approach

AUC	Clinical	Clinical & neuroimaging
Paper [1]	85.0%	89.0%
The proposed approach	**94.6%**	**97.0%**

We proposed a novel diagnostic model for early detection of Alzheimer's disease based on clinical and neuroimaging features. As we tackled the challenges of the dataset, we used SMOTE as preprocessing technique for balancing the dataset, we also used Deep Neural Network as feature extractor for both clinical and neuroimaging features. We tested the reliability of our approach using three classifiers, and they outperformed in their accuracy compared to paper [1], based on both clinical data and combination of clinical and neuroimaging data. The results obtained using this approach can serve as a basis for comparing further approaches in the future.

References

1. Samper-Gonzalez, J., et al.: Reproducible evaluation of methods for predicting progression to Alzheimer's disease from clinical and neuroimaging data. In: SPIE Medical Imaging 2019, San Diego, USA (2019)
2. Alzheimer's Association.: 2016 Alzheimer's disease facts and figures. Alzheimer's Dement. **12**(4), 459–509 (2016)
3. Afzal, S., et al.: Alzheimer disease detection techniques and methods: a review. Int. J. Interact. Multim. Artif. Intell. (In press, 2021)
4. Venugopalan, J., Ton, L., Hassanzadeh, H.R.D., Wang, M.: Multimodal deep learning models for early detection of Alzheimer's disease stage. Sci. Rep. **11**, 3254 (2021)
5. El-Sappagh, S., et al.: A multilayer multimodal detection and prediction model based on explainable artificial intelligence for Alzheimer's disease. Sci. Rep. **11**, 2660 (2021)
6. Sheng, J., Xin, Y., Zhang, Q., et al.: Predictive classification of Alzheimer's disease using brain imaging and genetic data. Sci. Rep. **12**, 2405 (2022)
7. Zhu, Q., et al.: Classification of Alzheimer's disease based on abnormal hippocampal functional connectivity and machine learning. Front. Aging Neurosci. (2022)
8. Fujiwara, K., et al.: Over- and under-sampling approach for extremely imbalanced and small minority data problem in health record analysis. Front. Public Health **8,** 178 (2020)
9. Notley, S., Magdon-Ismail, M.: Examining the use of neural networks for feature extraction: a comparative analysis using deep learning, support vector machines, and K-nearest neighbor classifiers (2018)

10. Alam, M., Rahman, M., Rahman, M.: A Random Forest based predictor for medical data classification using feature ranking, Informat. Med. **15**, 100180(2019)
11. Budholiya, K., Shrivastava, S., Sharma, V.: An optimized XGBoost based diagnostic system for effective prediction of heart disease. J. King Saud Univ. **34**(7), 4514–4523 (2022)

Machine Learning and Optimization

Benchmarking Concept Drift Detectors for Online Machine Learning

Mahmoud Mahgoub[1], Hassan Moharram[1], Passent Elkafrawy[1], and Ahmed Awad[2,3](✉)

[1] Nile University, Giza, Egypt
{m.mahgoub,h.thabet,p.elkafrawy}@nu.edu.eg
[2] University of Tartu, Tartu, Estonia
ahmed.awad@ut.ee
[3] Cairo University, Giza, Egypt

Abstract. Concept drift detection is an essential step to maintain the accuracy of online machine learning. The main task is to detect changes in data distribution that might cause changes in the decision boundaries for a classification algorithm. Upon drift detection, the classification algorithm may reset its model or concurrently grow a new learning model. Over the past fifteen years, several drift detection methods have been proposed. Most of these methods have been implemented within the Massive Online Analysis (MOA). Moreover, a couple of studies have compared the drift detectors. However, such studies have merely focused on comparing the detection accuracy. Moreover, most of these studies are focused on synthetic data sets only. Additionally, these studies do not consider drift detectors not integrated into MOA. Furthermore, None of the studies have considered other metrics like resource consumption and runtime characteristics. These metrics are of utmost importance from an operational point of view.

In this paper, we fill this gap. Namely, this paper evaluates the performance of sixteen different drift detection methods using three different metrics: accuracy, runtime, and memory usage. To guarantee a fair comparison, MOA is used. Fourteen algorithms are implemented in MOA. We integrate two new algorithms (ADWIN++ and SDDM) into MOA.

Keywords: Online machine learning · Concept drifts · Benchmarking

1 Introduction

Nowadays machine learning is considered essential for almost every industry in the world, e.g. healthcare, finance, and manufacturing, to name just a few. Classical machine learning models are designed to act in static environments where data distributions are constant over time. However, with the recent complex systems in real-life, this is not valid anymore. In real applications, generated data distribution is changing over time. We need new techniques to deal with this fact. In addition to that, the massive explosion in the generated data volumes

P. Fournier-Viger et al. (Eds.): MEDI 2022, LNAI 13761, pp. 43–57, 2023.
https://doi.org/10.1007/978-3-031-21595-7_4

due to IoT technology shifts machine learning from being offline to an online and continuous learning task. Thus, rather than learning from static data, classifiers need to learn from data streams. In such learning mode, training and prediction are interweaved [5]. In the meantime, the learning approach cannot keep all the data due to the infinite nature of data streams. Due to the dynamic nature of the data, the latent data distribution learned by online ML algorithms might change over time. If this change goes unnoticed by ML algorithms, the prediction accuracy of the model degrades over time.

The change of the underlying data distribution is known as concept drift [30, 36]. To maintain the prediction accuracy of online ML algorithms, drift detection techniques have been developed [1,2,4,11,18,21,22]. These techniques vary in their underlying detection approach (more details in Sect. 2.1). However, they agree on the input they receive, the classification prediction, and the feedback on that prediction. Upon detection of a drift in the data distribution, the detector raises a flag that is received by the online ML system to take corrective action to restore the prediction accuracy. The latter is out of the scope of the drift detector task.

With the growth of drift detectors, a number of studies over the past decade have compared such drift detectors [3,12,13,15,20]. These studies evaluate drift detection algorithms implemented within MOA [6], the state-of-the-art framework for online ML. The evaluation is mostly focused on drift detection accuracy on synthetic data. Other metrics such as runtime and memory consumption are not addressed. Moreover, detection algorithms not integrated in MOA are not included.

In this paper, we fill this gap in the evaluation and the comparison of the drift detection algorithms. In addition to the fourteen detection algorithms in MOA, we integrate two more algorithms. Namely ADWIN++ [22] and SDDM [21]. Moreover, we cover detection accuracy, detection latency, runtime, and memory consumption. The latter two metrics are of utmost importance from a data engineering and operational points of view.

The rest of this paper is organized as follows. Section 2 briefly describes the background concepts and definitions related to concept drift and discusses the related work. The contribution of the paper is split into two sections. Section 3 describes the benchmark setup. Results and the comparison of the drift detection algorithms are detailed in Sect. 4. Finally, Sect. 5 concludes the paper.

2 Background and Related Work

We start with a background about the different techniques for concept drift detection on data streams. Next, we discuss related work on the comparative evaluation of such techniques.

2.1 Background

Data Streams, Concept Drifts, and Their Types. A data stream can be defined as an unbounded sequence in which the instances have timestamps with various granularity [12,36]. The process that generates the stream can be considered as a random variable X from which the objects $x \in domain(X)$ are drawn. In a classification learning context, a target (or class) variable $y \in domain(Y)$ is available, where Y denotes a random variable over the targets. Thus, the data stream is comprised of instances $(\mathbf{x_1}, y_1)$, $(\mathbf{x_2}, y_2) \ldots (\mathbf{x_t}, y_t)$, where $(\mathbf{x_t}, y_t)$ represents an instance in time t, $\mathbf{x_t}$ represents the vector of feature values and y_t represents the target for that particular instance. In practice, y_t is not necessarily known at time t where the features x_t was observed. y_t is usually known at a later time $t+n$.

As stated in the Bayesian Decision theory [9], the process of classification can be described by the prior probabilities of the targets $P(\mathbf{Y})$, and the target conditional probability density function $P(\mathbf{X}|\mathbf{Y})$. The classification decision is made based on the posterior probabilities of the targets, which can be obtained from:

$$P(\mathbf{Y}|\mathbf{X}) = \frac{P(\mathbf{Y}) \cdot P(\mathbf{X}|\mathbf{Y})}{P(\mathbf{X})} \tag{1}$$

Since $P(\mathbf{Y})$ and $P(\mathbf{X}|\mathbf{Y})$ uniquely determine the joint distribution $P(\mathbf{X}, \mathbf{Y})$, concepts can be defined as the joint distribution $P(\mathbf{X}, \mathbf{Y})$ [12]. A concept at point of time t will be denoted as $P_t(\mathbf{X}, \mathbf{Y})$. In practice, concept drifts occur due to changes in user tastes. For example, changes in trends on Twitter that might make a tweet recommendation obsolete for the user. Mathematically, a concept drift occurs due to the change in the generating random variable X that leads a change in the data distribution. Based on the definition of concept, concept drift between data at point of time t and data at point of time u can formally be defined as a difference in the distributions of the data in these time points:

$$\exists \mathbf{X} : P_t(\mathbf{X}, \mathbf{Y}) \neq P_u(\mathbf{X}, \mathbf{Y}), where\ t, u \in \mathbf{Z} \tag{2}$$

According to [12], the popular patterns of concept drift are the following:

- *Abrupt drift*: when the a learned concept is suddenly replaced by a new concept (Fig. 1a).
- *Gradual drift*: when the change is not abrupt, but goes back and forth between the original and the new concept (Fig. 1b).
- *Incremental drift*: when, as time passes, the probability of sampling from the original concept distribution decreases and the probability of sampling from the new concept increases (Fig. 1c).

Families of Drift Detectors. There are various families of drift detectors. Each family has a different underlying condition to cope with concept drifts. In many cases, the detector utilizes the classification error of an online classifier. In general, there are three families of drift detectors [35]: statistical models, window-based models, and ensemble-based models.

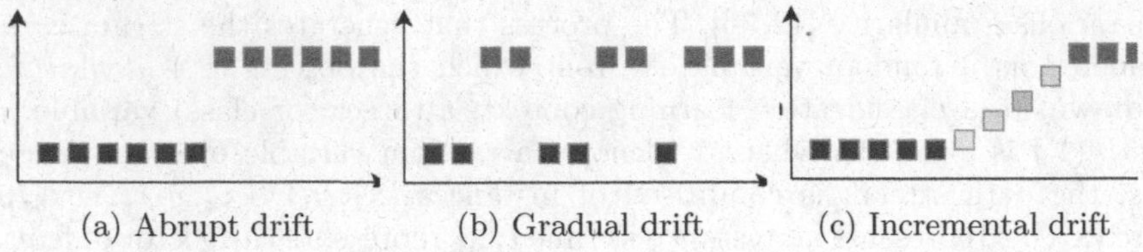

Fig. 1. Patterns of concept drift, X-axis: time, Y-axis: the mean of the data

Statistical detectors are categorized based on the underlying statistical test. The Sequential Probability Ratio Test (SPRT) [33] detects and monitors the change in data using a sequential hypothesis test. Page [24] developed two memory-less models based on SPRT, the Cumulative Sum (CUSUM) and the Page-Hinckley (PH) test. A drawback of the SPRT-based models is that it only depends on two metrics when deciding a concept drift: false alarm and missed detection rates [12]. Another statistical test is the Fisher's Exact test which is employed when the number of errors or correct predictions is small. A model that uses this test is the Fisher Proportions Drift Detector (FPDD) presented by de Lima Cabral and de Barros [18] extending it with the Fisher-based Statistical Drift Detectors (FSDD) and Fisher Test Drift Detector (FTDD). Some models use McDiarmid's inequality for detecting concept drifts such as the McDiarmid Drift Detection Methods (MDDMs) proposed by Pesaranghader et al. [26]. This model uses a weighting scheme represented in a sliding window over prediction results assigning weights to stream elements with higher weights to the most recent ones. A concept drift occurs when there is a significant difference between two weighted means. Three drift detectors are developed after this model depending on the weighting scheme type: MDDM-A model (arithmetic), MDDM-G model (geometric), and MDDME model (Euler).

Some detectors use a base learner (classifier) to classify the future stream elements. The Drift Detection Method (DDM) [11] is the first algorithm to use this concept. Several methods are extended from DDM such as the Early Drift Detection Method (EDDM) [1], and the Reactive Drift Detection Method (RDDM) [2]. The Statistical Drift Detection Method (SDDM) [21], however, does not require feedback from the learner (classifier) to decide about drifts.

Window-Based Detectors. In this family, detectors monitor a sliding window instead of individual stream tuples. The window represents the model's internal memory and varies in size. Some detectors are based on the Very Fast Decision Tree (VFDT) algorithm [8]. The Concept adapting Very Fast Decision Tree (CVFDT) [8] detector is the first to be built on VFDT. It was optimized covering more types of concept drifts through the Efficient Concept-Adapting Very Fast Decision Tree (E-CVFDT) [19] detector.

The Adaptive Window (ADWIN) [4] is considered state-of-the-art in this category. ADWIN almost requires no parameters for its operation except assigning the sensitivity level to change in the data. Moreover, the window expands and shrinks depending on the stream state. By the arrival of a stream element, the internal window is split into two sub-windows covering the whole stream tuples and deciding a drift when a significant difference in the means of the two sub-windows occurs, leading to dropping the oldest elements till no further change is detected. A number of efficient implementations and enhancements have been presented in the literature. Grulich et al. [14] extend ADWIN by presenting three variants: Serial, HalfCut, and Optimistic, making ADWIN more scalable by optimizing its throughput using parallel adaptive windowing techniques. Another optimization is presented in [22] to account for the unbounded growth of the internal window size with steady streams, i.e., streams that have a low frequency of drifts.

Ensemble-Based Detectors. Similar to ensembles of classifiers, an ensemble of drift detectors can be formed to opt for the highest accuracy possible in detecting drifts. Ensemble drift detectors can be divided into two classes: Block-based and Incremental detectors.

The Block-based ensemble detectors processes stream elements in chunks or blocks affected by chunk size. The earliest model to use this concept is the Streaming Ensemble Algorithm (SEA) [32] then it improved by Wang et al. [34] through the Accuracy Weighted Ensemble (AWE) model which replaces the weak-performing classifiers with better-selected ones. Brzeziński and Stefanowski [7] continued the work on AWE presenting the Accuracy Updated Ensemble (AUE) algorithm by improving the memory consumption and accuracy.

Incremental ensemble detectors tend to process stream elements individually in a sequential manner, unlike in chunks. Kolter and Maloof [17] introduced Dynamic Weighted Majority (DWM), the first Incremental ensemble detector. Based on stream state has concept drifts or not, DWM adds or removes the weighted classifiers which are referred as experts. The Learn++ algorithm family is a grouping of Incremental ensemble detectors which uses machine learning to handle concept drifts with imbalanced data [35].

2.2 Related Work

Gama et al. [12] present a comprehensive review on concept drift adaptation. Drift detection is one pillar in adapting to concept drifts. The survey is on a high level discussing the different theories behind drift detection. Wares et al. [35] provide a more recent critical review on concept drift detectors. They conclude their review with several shortcomings in the study of drift detectors. Among these shortcomings are outdated drift detection methods and their comparison and the lack of benchmark setup and real-world data sets for the evaluation-the work we report in this paper addresses these two shortcomings. Lu et al. [20] is another intensive review of concept drift detectors that partially addresses the shortcomings reported in [35]. Namely, the public availability of real-world data sets to evaluate drift detectors.

To the best of our knowledge, the first benchmarking study of drift detectors is presented by Gonçalves et al. [13]. The evaluation covered DDM, EDDM,

ECDD, PHT, STEPD, PL, ADWIN, and DOF. Naive Bayes was used as the base learner. The authors used both synthetic and real-world data sets. However, the synthetic data sets covered only abrupt and gradual drifts. The evaluation of the detectors was focused on accuracy-related aspects. A more recent benchmark is presented by Barros et al. [3]. The evaluation setup largely follows the one in [13]. It includes more detectors such as: HDDM, RDDM, WSTD and SeqDr. Moreover, it uses VFDT as another base learner. The evaluation is concerned with accuracy aspects only.

In this paper, we extend and complement the evaluation done in [3,13] by: – Including more recent drift detectors, namely: SDDM [21] and ADWIN++ [22], – Using synthetic data sets that address all drift types in addition to real-world data sets, – Reporting about operational metrics of runtime, latency and memory consumption in addition to drift detection accuracy measures.

3 Benchmark Setup

In this section, we describe benchmarking, the datasets, the algorithm benchmarked, the parametrization used in the drift detectors, metrics computed, and the evaluation methodology.

3.1 Datasets

Both synthetic and real-world datasets are used for this experiment. The synthetic data are generated using the generator from [14]. This data simulates incremental, gradual, and abrupt concept drifts. Instances in this data represent the prediction made by a Naïve Bayes base learner so they can be applied directly on the concept drift detection methods without the need for a base learner. This helps in focusing on benchmarking the concept drift detection methods themselves without taking into consideration base learners and their possible delays or effects on the experiments. Each dataset in this group consists of two million instances. The incremental dataset (Inc1554) contains 1554 drifts. The gradual dataset (Grad1738) contains 1738 drifts. The abrupt one (Abr1283) contains 1283 drifts.

The second group contains real-world datasets. There are three datasets: Airlines (539383 records), Electricity (45312 records), and INSECTS-Abrupt (balanced) (52848 records). The first two datasets are common in the literature and are publicly available on MOA [6] website. The last dataset is one of many new datasets introduced in [31]. The dataset can be used in benchmarking as its characteristics and pattern are known and avoid the challenges related to the other real-life datasets, i.e. the lack of ground truth about when drift occurred, so we choose to include it in this paper.

3.2 Benchmarked Algorithms

We used all the available algorithms from MOA: DDM [11], EDDM [1], RDDM [2], STEPD [23], SeqDrift1 [25], SeqDrift2 [29], SEED [16], PageHinkleyDM (PH) [24], HDDM_A_Test ($HDDM_A$), HDDM_W_Test ($HDDM_W$) [10],

GeometricMovingAverageDM (GMA) [27], EWMAChartDM (EWMA) [28], CusumDM (CUSUM) [24] and ADWIN [4]. In addition, we have integrated ADWIN++ [22] and SDDM [21] within MOA to have a fair comparison between the different detection methods. ADWIN++ is written in Java so the integration within MOA was straightforward. On the other hand, SDDM is written in Python, so we had to port it to Java and implement the respective interfaces in MOA. However, as SDDM does not require a base learner, the change detector interface of MOA had to be modified to read the full instance data (features and target class). Initially, it supports only reading the target class. Finally, we have defined a new class that implements the change detector interface. The source code for the changes and the benchmark is available on Github[1]

Each drift detection algorithm has a set of hyperparameters that affect detection accuracy as well as other operational aspects. Table 1 lists the chosen hyperparameter values of the respective algorithms that would deliver the highest sensitivity to change in the data. we have chosen those values using grid search tuning. It is worth mentioning that we report only parameters for which we changed their default values. If an algorithm has other parameters not mentioned in the table, they were left with their default values.

3.3 Metrics

In addition to prediction accuracy, we record the runtime and memory usage of each algorithm. Detection accuracy is measured by counting the number of drifts detected by each algorithm. For the synthetic datasets where ground truth drifts are known, we report detection delay which represents the distance, in data instances, from the actual occurrence of the drift to its detection.

To measure the operational metrics, i.e., runtime and memory consumption, we have used the Java Microbenchmark Harness (JMH) toolkit[2]. JMH provides results with 99.9% confidence. JMH has several parameters that can be set to customize the benchmarking: – Warmup: the number of iterations the code runs to warmup. Warmup iterations are very important to avoid variations due to the transient period at the JVM starting. They do not contribute to the measurement results. – Measurement: actual code benchmark execution. The code is run for a number of iterations and the output of these iterations is used to generate the JMH benchmark result. – Benchmark Mode: type of benchmark to be run. Possible types are • Average Time: the average time for executing the code • Throughput: it measures the number of times a code is executed in a certain period of time.

For our benchmarkprovidesve used the following values: – Warmup iterations = 5 – Measurement iterations = 20 – Benchmark Mode = Average Time.

[1] https://github.com/mahmoudmahgoub/moa.

[2] https://github.com/openjdk/jmh.

Table 1. Drift detectors tuned hyperparameters

Algorithm	Parameters	Description	Values
DDM	outcontrolLevel	Change the threshold needed to detect a drift. Smaller values mean the algorithm is more senstive and detects more drifts	1.9
RDDM	driftLevel	Similar to "DDM" outcontrolLevel parameter	1.82
STEPD	alphaDrift	Similar to "DDM" outcontrolLevel parameter	0.045
SeqDrift1	deltaSeqDrift1	Configure the size of the sliding window of the algorithm The size of the window inversely proportional to its value	1
	deltaWarning	A warning level that is raised when the difference between these two estimations is approaching the drift level	1
SeqDrift2	deltaSeqDrift2	Used to calculate epsilon value that determine the drifts	1
SEED	deltaSEED	Used to calculate epsilon value that determine the drifts	1
HDDM_A	driftConfidence	Configure drift level	0.0015
HDDM_W	driftConfidence	Configure drift level	0.0006
ADWIN MOA	deltaAdwin	Similar to "DDM" outcontrolLevel parameter	1
ADWIN++	safe_lim	Fixed minimum ADwin window size limit (buckets)	15
	min_lim	Moving minimum ADwin window size limit (buckets)	51
	max_lim	Fixed maximum ADwin window size limit (buckets)	60
	Theta	Sliding change of min_lim (elements)	70000
	omega	Number of elements to wait after detecting main drift to perform the adaptive buckets dropping	40000
SDDM	driftThreshold	Determine the threshold of drift magnitudates to adress a drift	0.04
	Window size	Configure the window size	500
	Window step	Configure the sliding step of the window	50

To choose those values, we increased the warmup and the measurement iterations value by one unit until the metric values stabilized, i.e. they are not affected by change in warmup and measurement iterations.

Although memory usage can be calculated using different methods, such as using the Java `MemoryMXBean` interface to calculate the memory before and after running the drift detector methods, we preferred to use VisualVM[3], an external profiling tool. VisualVM gives an accurate measurement for memory consumption that are unaffected by the garbage collector calls in the JVM.

4 Results

We report about the metrics discussed in the previous section for the sixteen algorithms using the respective hyperparameter values reported in Table 1. We have conducted the experiments on a computer with an Intel Core i5-8250U processor having 12 GB of RAM and running Windows 11 operating system using Java 8. For measuring runtime and memory consumption, we report the Inc1554 dataset only due to paper length limitation. However, each algorithm followed a similar behavior on the other datasets.

[3] https://visualvm.github.io/.

4.1 Drift Detection Accuracy

Table 2. Accuracy (number of detected drifts)

Datasets	Inc1554	Grad1738	Abr1283	Electricity	Airlines	Insects
ADWIN MOA	1554	1738	1283	2636	5016	373
ADWIN++	1554	1738	1283	2636	5016	373
$HDDM_W$	1478	1490	1509	281	859	2169
RDDM	1221	1149	1310	261	1465	0
STEPD	1298	1345	1245	33206	28004	31132
$HDDM_A$	955	1101	998	481	405	6335
EDDM	629	1741	298	626	688	92
SDDM	644	604	588	570	10778	1047
DDM	493	3190	8785	615	2725	0
SEED	50	68	83	541	1311	86
SeqDrift2	26	47	21	139	544	0
CUSUM	4	11	8	11	101	30
SeqDrift1	1	2	2	62	53	1
EWMA	1	0	0	0	738	0
PH	0	0	0	0	0	0
GMA	0	0	0	0	0	0

Table 2 presents the accuracy results for the sixteen methods. For the synthetic data, ADWIN MOA and ADWIN++ have the highest accuracy as they detect the exact number of the existing drifts in the three synthetic datasets.

RDDM, STEPD, $HDDM_A$, and $HDDM_W$ have relatively good performance when they are used on synthetic data. EDDM has modest accuracy for the incremental and abrupt datasets whereas its accuracy is much higher for the gradual dataset. SDDM accuracy is close ti EDDM. Yet, it does not seem to be affected by the drift type. DDM gives a lot of false positives for the gradual and abrupt datasets. SeqDrift1, SeqDrift2, SEED, PH, GMA, EWMA, and CUSUM have the worst accuracy. The reason can be attributed to their memory-less nature. Thus, they learn the least about the data distribution. This is further evidenced when we discuss memory consumption.

For the real-world datasets, we do not know exactly the number of drifts nor their location in the data. As ADIWN has the highest accuracy in drift detection on the synthetic dataset, we have used it as a reference to compare the accuracy of the other methods on real-world data sets. The methods have very different results compared to ADWIN. For instance, RDDM has poor accuracy on the real-world data set, 10% on Electricity, 30% on Airlines, and 0% on Insects. $HDDM_W$ has a similar behavior as RDDM, except for the Insects dataset. STEPD has a steady huge rate of false positives. Memory-less methods, the last group in

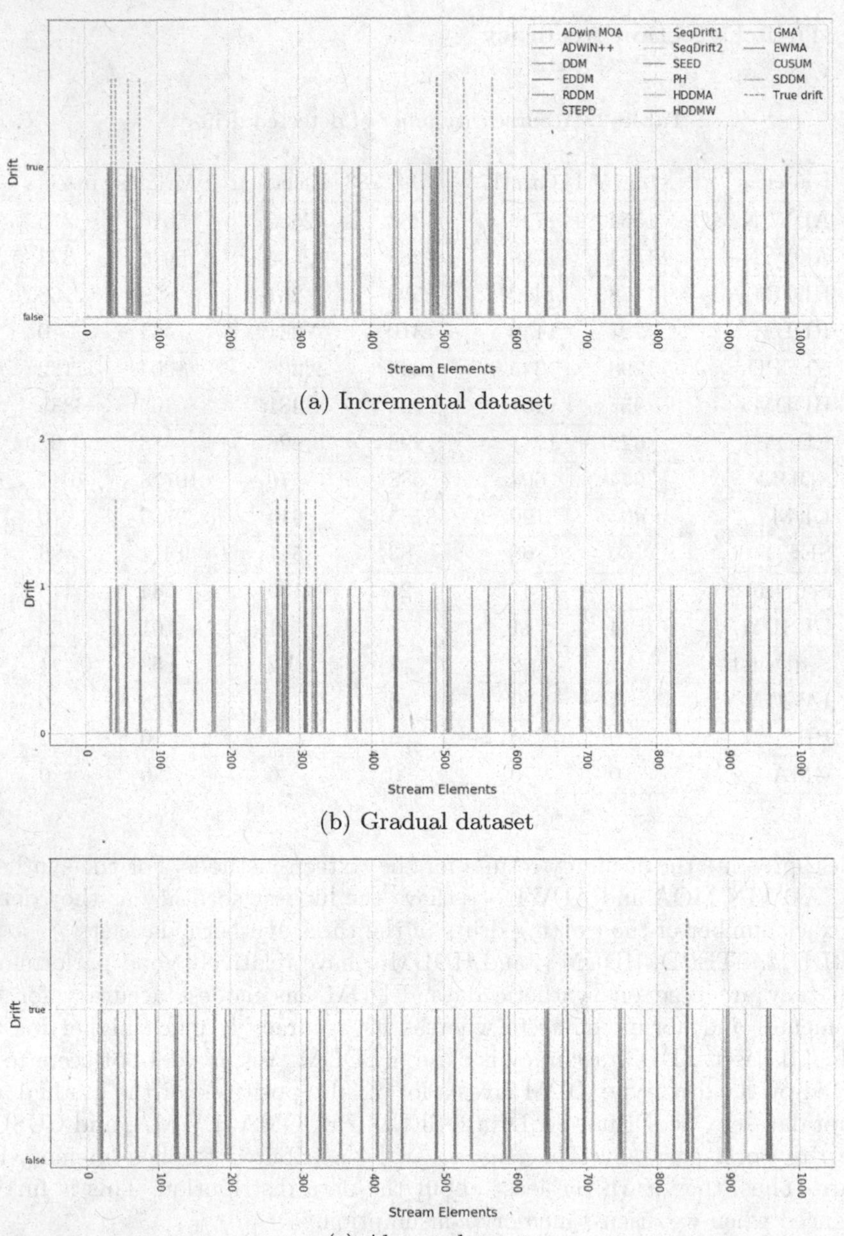

(a) Incremental dataset

(b) Gradual dataset

(c) Abrupt dataset

Fig. 2. Drift detection in the first 1000 instances of the synthetic datasets. X-axis represent the progress of the stream, y-axis shows when a drift is detected.

Table 2, continue to perform poorly on the real-world data set, the best accuracy is 25% reached by SEED on the Airlines dataset.

Figure 2 shows the detected drifts in the first 1000 instances of the synthetic datasets. We can notice that for the algorithms with a good or modest performance, mentioned previously, only ADWIN MOA and ADWIN++ detect them at the correct time, whereas others have detection delays in addition to false positives. The DDM family, including SDDM, depends on some statistical test to monitor the number of errors produced by a model learned on the previous stream items. So, it may be too slow in responding to changes because it may take many observations after the change to detect a drift. The same can be said about memory-less detectors.

On the other hand, ADWIN makes instantiations call of the drift. For instance, upon the arrival of a new stream element, a cut detection is made immediately by splitting the main window into two sliding windows looking for a possible drift. A drift is called when there is a significant difference between the means of these two windows.

4.2 Runtime

Figure 3 shows the average runtime of the different methods. SDDM and ADWIN MOA are the slowest methods. SDDM has a high algorithmic complexity in deciding whether drift is detected due to the internal bucketing and distance measured used [21]. For ADWIN MOA, we have investigated the reason for slowness, and we found that setting the `mintClock` hidden parameter value to 1 is the reason for this slowness. By increasing `mintClock` we get more speed, but accuracy becomes lower. For example, setting `mintClock` to 32 and using the Inc1554 dataset, the execution time drops to 2712.246 ms. However, the number of detected drifts drops to 57. ADWIN++ is better in that sense as it is faster and it preserves its accuracy. In fact, that is the main improvement it brings [22].

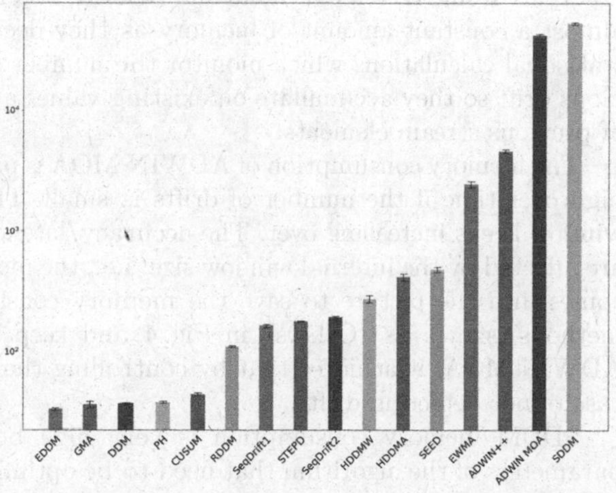

Fig. 3. Average runtime in milliseconds using a logarithmic scale

4.3 Memory Consumption

Figure 4 shows the consumed memory over the time. All the algorithms almost need the same small amount of memory expect for SDDM that consumes almost

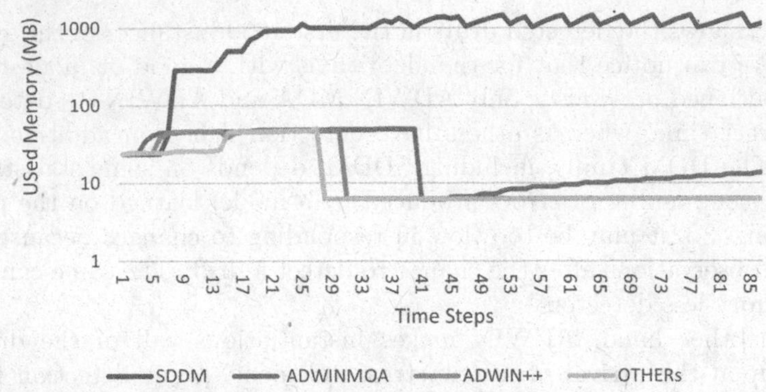

Fig. 4. Consumed memory overtime using a logarithmic scale

about three orders of magnitude higher than the rest of the methods. It almost consumes 1.6 GB.

Apart from ADWIN variants and SDDM, other algorithms tend to consume almost a constant amount of memory as they decide on a drift based on some statistical calculations which monitor the number of errors counted after a previous drift so they accumulate on existing values and no need to keep a window of previous stream elements.

The memory consumption of ADWIN MOA is particularly interesting. It goes high over time if the number of drifts is small. This is because the size of the window keeps increasing over. The accuracy/latency trade-off of ADWIN MOA are affected by the internal window size, i.e., the memory consumed. ADWIN++ comes into the picture to save the memory consumption as low as the other methods, shown as "Others" in Fig. 4 and keep the accuracy and latency of ADWIN MOA. it achieves that by controlling the window size [22] even in the case of not detecting drifts.

SDDM memory consumption is very high because of the many internal parameters of the algorithm that need to be optimized. For example, the bucket size, the distance measure, and the number of features to compute drifts for all affect the accuracy, memory consumption and thus latency of drift detection. The algorithm implementation needs to be optimized. These parameters are orthogonal to those listed in Table 1 for SDDM. The optimization of such parameters is out of scope of the current paper.

5 Conclusion and Future Work

In this paper, we present a comprehensive benchmark for sixteen drift detectors. Fourteen drift detectors are already implemented within MOA. Two other detectors were implemented in Java and integrated into MOA. The benchmark measures and compares detection accuracy on various data sets and two operational metrics, runtime and memory consumption. Overall, ADWIN shows the best accuracy on the synthetic data sets. Namely, ADWIN++ maintains the accuracy of ADWIN while improving runtime and memory consumption. Memory-less detectors are not useful and are not recommended for use in real-life scenarios.

Taking ADWIN as a reference point, we notice a considerable difference in the performance of the other algorithms on real-life data sets. This calls for further investigation on which of those methods delivers the best accuracy. This is the target for future studies. From memory consumption results, more improvements for SDDM implementation are needed.

Acknowledgments. The work of Ahmed Awad is funded by the European Regional Development Funds (Mobilitas Plus Programme grant MOBTT75).

References

1. Baena-Garcıa, M., del Campo-Ávila, J., Fidalgo, R., Bifet, A., Gavalda, R., Morales-Bueno, R.: Early drift detection method. In: Fourth International Workshop on Knowledge Discovery from Data Streams, vol. 6, pp. 77–86 (2006)
2. de Barros, R.S.M., de Lima Cabral, D.R., Gonçalves Jr, P.M.G., de Carvalho Santos, S.G.T.: RDDM: reactive drift detection method. Expert Syst. Appl. **90**, 344–355 (2017)
3. Barros, R.S.M., Santos, S.G.T.C.: A large-scale comparison of concept drift detectors. Inf. Sci. **451–452**, 348–370 (2018)
4. Bifet, A., Gavaldà, R.: Learning from time-changing data with adaptive windowing. In: ICDM, pp. 443–448. SIAM (2007)
5. Bifet, A., Gavaldà, R., Holmes, G., Pfahringer, B.: Machine Learning for Data Streams with Practical Examples in MOA. MIT Press, Cambridge (2018)
6. Bifet, A., Holmes, G., Kirkby, R., Pfahringer, B.: MOA: massive online analysis. J. Mach. Learn. Res. **11**, 1601–1604 (2010)
7. Brzeziński, D., Stefanowski, J.: Accuracy updated ensemble for data streams with concept drift. In: Corchado, E., Kurzyński, M., Woźniak, M. (eds.) HAIS 2011. LNCS (LNAI), vol. 6679, pp. 155–163. Springer, Heidelberg (2011). https://doi.org/10.1007/978-3-642-21222-2_19
8. Domingos, P.M., Hulten, G.: Mining high-speed data streams. In: SIGKDD, pp. 71–80. ACM (2000)
9. Duda, R.O., Hart, P.E., Stork, D.G.: Pattern Classification, 2nd edn. Wiley (2001)
10. Frías-Blanco, I., del Campo-Ávila, J., Ramos-Jiménez, G., Morales-Bueno, R., Ortiz-Díaz, A., Caballero-Mota, Y.: Online and non-parametric drift detection methods based on Hoeffding's bounds. IEEE Trans. Knowl. Data Eng. **27**(3), 810–823 (2015)
11. Gama, J., Medas, P., Castillo, G., Rodrigues, P.: Learning with drift detection. In: Bazzan, A.L.C., Labidi, S. (eds.) SBIA 2004. LNCS (LNAI), vol. 3171, pp. 286–295. Springer, Heidelberg (2004). https://doi.org/10.1007/978-3-540-28645-5_29
12. Gama, J., Zliobaite, I., Bifet, A., Pechenizkiy, M., Bouchachia, A.: A survey on concept drift adaptation. ACM Comput. Surv. **46**(4), 44:1–44:37 (2014)
13. Gonçalves, P.M., de Carvalho Santos, S.G., Barros, R.S., Vieira, D.C.: A comparative study on concept drift detectors. Expert Syst. Appl. **41**(18), 8144–8156 (2014)

14. Grulich, P.M., Saitenmacher, R., Traub, J., Breß, S., Rabl, T., Markl, V.: Scalable detection of concept drifts on data streams with parallel adaptive windowing. In: EDBT, pp. 477–480. OpenProceedings.org (2018)
15. Han, M., Chen, Z., Li, M., Wu, H., Zhang, X.: A survey of active and passive concept drift handling methods. Comput. Intell. **38**(4), 1492–1535 (2022)
16. Huang, D.T.J., Koh, Y.S., Dobbie, G., Pears, R.: Detecting volatility shift in data streams, pp. 863–868 (2014)
17. Kolter, J.Z., Maloof, M.A.: Dynamic weighted majority: a new ensemble method for tracking concept drift. In: ICDM, pp. 123–130. IEEE (2003)
18. de Lima Cabral, D.R., de Barros, R.S.M.: Concept drift detection based on Fisher's Exact test. Inf. Sci. **442**, 220–234 (2018)
19. Liu, G., Cheng, H.R., Qin, Z.G., Liu, Q., Liu, C.X.: E-CVFDT: an improving CVFDT method for concept drift data stream. In: ICCCAS, vol. 1, pp. 315–318. IEEE (2013)
20. Lu, J., Liu, A., Dong, F., Gu, F., Gama, J., Zhang, G.: Learning under concept drift: a review. IEEE TKDE **31**(12), 2346–2363 (2019)
21. Micevska, S., Awad, A., Sakr, S.: SDDM: an interpretable statistical concept drift detection method for data streams. J. Intell. Inf. Syst. **56**(3), 459–484 (2021). https://doi.org/10.1007/s10844-020-00634-5
22. Moharram, H., Awad, A., El-Kafrawy, P.M.: Optimizing ADWIN for steady streams. In: ACM/SIGAPP SAC, pp. 450–459. ACM (2022)
23. Nishida, K., Yamauchi, K.: Detecting concept drift using statistical testing. In: Corruble, V., Takeda, M., Suzuki, E. (eds.) DS 2007. LNCS (LNAI), vol. 4755, pp. 264–269. Springer, Heidelberg (2007). https://doi.org/10.1007/978-3-540-75488-6_27
24. Page, E.S.: Continuous inspection schemes. Biometrika **41**(1/2), 100–115 (1954). https://doi.org/10.1093/biomet/41.1-2.100
25. Pears, R., Sripirakas, S., Koh, Y.S.: Detecting concept change in dynamic data streams. Mach. Learn. **97**, 259–293 (2014). https://doi.org/10.1007/s10994-013-5433-9
26. Pesaranghader, A., Viktor, H.L., Paquet, E.: McDiarmid drift detection methods for evolving data streams. In: IJCNN, pp. 1–9. IEEE (2018)
27. Roberts, S.W.: Control chart tests based on geometric moving averages. Technometrics **1**(3), 239–250 (1959). http://www.jstor.org/stable/1266443
28. Ross, G.J., Adams, N.M., Tasoulis, D.K., Hand, D.J.: Exponentially weighted moving average charts for detecting concept drift. Pattern Recogn. Lett. **33**(2), 191–198 (2012). https://www.sciencedirect.com/science/article/pii/S0167865511002704
29. Sakthithasan, S., Pears, R., Koh, Y.S.: One pass concept change detection for data streams. In: Pei, J., Tseng, V.S., Cao, L., Motoda, H., Xu, G. (eds.) PAKDD 2013. LNCS (LNAI), vol. 7819, pp. 461–472. Springer, Heidelberg (2013). https://doi.org/10.1007/978-3-642-37456-2_39
30. Sobolewski, P., Wozniak, M.: Enhancing concept drift detection with simulated recurrence. In: Pechenizkiy, M., Wojciechowski, M. (eds.) New Trends in Databases and Information Systems. AISC, vol. 185, pp. 153–162. Springer, Heidelberg (2012). https://doi.org/10.1007/978-3-642-32518-2_15
31. Souza, V.M.A., dos Reis, D.M., Maletzke, A.G., Batista, G.E.A.P.A.: Challenges in benchmarking stream learning algorithms with real-world data. Data Min. Knowl. Discov. **34**(6), 1805–1858 (2020). https://doi.org/10.1007/s10618-020-00698-5
32. Street, W.N., Kim, Y.: A streaming ensemble algorithm (SEA) for large-scale classification. In: SIGKDD, pp. 377–382. ACM (2001)

33. Wald, A.: Sequential Analysis. Courier Corporation (1973)
34. Wang, H., Fan, W., Yu, P.S., Han, J.: Mining concept-drifting data streams using ensemble classifiers. In: SIGKDD, pp. 226–235. ACM (2003)
35. Wares, S., Isaacs, J., Elyan, E.: Data stream mining: methods and challenges for handling concept drift. SN Appl. Sci. **1**(11), 1–19 (2019). https://doi.org/10.1007/s42452-019-1433-0
36. Webb, G.I., Lee, L.K., Petitjean, F., Goethals, B.: Understanding concept drift. CoRR abs/1704.00362 (2017)

Computational Microarray Gene Selection Model Using Metaheuristic Optimization Algorithm for Imbalanced Microarrays Based on Bagging and Boosting Techniques

Rana Hossam Elden[1(✉)], Vidan Fathi Ghoneim[1], Marwa M. A. Hadhoud[1], and Walid Al-Atabany[2,1]

[1] Biomedical Engineering Department, Faculty of Engineering, Helwan University, Cairo, Egypt
{ranahossamelden,vidanfathighoneim, marwa_hadhoud}@h-eng.helwan.edu.eg

[2] Information Technology and Computer Science School, Nile University, Giza, Egypt
w.al-atabany@nu.edu.eg

Abstract. Genomic microarray databases encompass complex high dimensional gene expression samples. Imbalanced microarray datasets refer to uneven distribution of genomic samples among different contributed classes which can negatively affect the classification performance. Therefore, gene selection from imbalanced microarray dataset can give rise to misleading, and inconsistent nominated genes that would alter the classification performance. Such unsatisfactory classification performance is due to the skewed distribution of the samples across the microarrays toward the majority class. In this paper, we propose a modified version of Emperor Penguin Optimization (EPO) algorithm combined with Random Forest (RF) of Bagging and Boosting Classification named by EPO-RF to select the most informative genes based on classification accuracy using imbalanced microarray datasets. The modified version of EPO was built to be based on decision trees that takes in consideration the criterion of tree splitting weights to handle the imbalanced microarray datasets. Average gene expression binary values are used as a preliminary step for exploring disease trajectories with the aid of metaheuristic optimization feature selection algorithms. Results show that the proposed model revealed its superiority compared to well-known established metaheuristic optimization algorithms, e.g., Harris Hawks Optimization (HHO), Grey Wolf Optimization (GWO), Salp Swarm Optimization (SSO), Particle Swarm Optimization (PSO), and Genetic Algorithms (GA's) using several pediatric sepsis microarray datasets for patients who admitted to the Intensive Care Unit (ICU) for the first 24 h.

Keywords: Gene selection · Imbalanced microarray · Metaheuristic · Oversampling · Random Forest

1 Introduction

Affymetrix microarray datasets represent a powerful analysis tool used for determination of disease-relevant gene through the analysis of mRNA expression profile of thousands

P. Fournier-Viger et al. (Eds.): MEDI 2022, LNAI 13761, pp. 58–71, 2023.
https://doi.org/10.1007/978-3-031-21595-7_5

of genes. Unfortunately, not all assorted microarray genes are expressed in all tissues needed to be removed as they aren't related to the state of the disease and represent irrelevant and redundant that mislead the machine learning algorithms [1, 2]. Therefore, relevant informative gene selection is a matter of concern needed to enhance the classification performance of the diseases trajectories [3, 4]. Classification imbalanced microarray datasets poses a challenge for machine learning predictive modeling as the distribution of samples across the assorted classes is biased or skewed [5]. Therefore, recently, considerable attention has been paid for tackling the imbalanced datasets. Tang et al. [5] proposed granular Support Vector Machines repetitive cost-sensitive learning undersampling algorithm (GSVM-RU) that minimize the negative impact of information loss while maximizing the positive effect of data filtering within the undersampling process using less number of support vectors. While Krawczyk et al. [6] introduced an effective ensemble of cost-sensitive Decision Trees (DT) for imbalanced classification. Sáez et al. [7] suggested Synthetic Minority Oversampling Technique (SMOTE) combined with Iterative-Partitioning Filter (IPF) used for balancing the biased datasets. The study was based on synthetizing new samples not correlated with noisy and borderline samples. Xiao et al. [8] analyzed the effectiveness of a novel class-specific cost regulation extreme learning machine (CCR-ELM) that based on determination of class-specific regulation cost for handling misclassification of each class in the performance index to be not sensitive to the dispersion degree of the utilized dataset. Lopez-Garcia et al. [9] suggest a hybrid metaheuristic feature selection algorithm named by Genetic Algorithm (GA) with a cross entropy (CE) based on ensembles called Adaptive Splitting and Selection (AdaSS) that partitions the feature space into clusters. AdaSS establishes a different classifier for each partition through adjusting the weights of the different base classifiers using the discriminant function of the collective decision-making method. Krawczyk et al. [10] confirmed the robustness of boosting strategy combined with evolutionary undersampling in handling imbalanced datasets using an enhanced ensemble classifier named EUSBoost. Aljarah et al. [11] introduced whale optimization algorithm (WOA) as a novel metaheuristic optimization algorithm trained with multilayer perceptron (MLP) neural networks classifier. WOA was utilized to determine the optimal values for weights and biases to minimize the mean square error (MSE) fitness function of MLP to overcome the problem of imbalanced datasets. Whereas Aljarah et al. [12] implemented a machine learning algorithm based on radial basis function (RBF) neural networks using Biogeography-Based Optimizer. Likewise, Roshan and Asadi [13] implemented an ensemble of bagging classifiers with evolutionary undersampling techniques. On the other hand, some chaotic metaheuristic approaches and machine learning algorithms applied on these problems [14–17] but classification performance is a matter of concern.

In spite of many studies in the field of handling imbalanced datasets classification, they were restricted to undersampling the datasets which results in information loss. Therefore, the present study deploys a modified version of a novel metaheuristic algorithm known as Emperor Penguin Optimization (EPO) for training a supervised Random Forest (RF) classifier of bagging and boosting ensembles to overcome the problem of imbalanced dataset classification. The results were assessed using discovery imbalanced sepsis microarray datasets from the same platform GPL570 by conducting statistical analysis of the experimental results through the calculation of the classification accuracy.

2 Methodology

The preprocessed microarray dataset which encompasses the average gene expression values undergoes binary conversion to be adaptable for EPO algorithm. Since imbalanced microarray dataset can alter the classification performance based on the improper gene selection, an oversampling strategy was applied for synthesizing new sampling to adjust the class distribution of the utilized dataset. Thereafter, the most informative genes were extracted from the preprocessed dataset using Emperor Penguin Optimization algorithm based on the best classification performance achieved by the internal embedded RF classifier within EPO algorithm. The performance of the selected genes in differentiating were further evaluated using supervised machine learning algorithms to confirm the robustness of the proposed modified EPO model. Figure 1 depicts the outline of the proposed architecture of the modified version of EPO algorithm for gene selection from imbalanced datasets.

2.1 Emperor Penguin Optimization (EPO)

A novel nature based metaheuristic optimization algorithm known as Emperor Penguin Optimization (EPO) [18] was applied to nominate the most informative genes used for the early predication of the disease trajectories. EPO imitates the huddling behavior of Aptenodytes Forsteri emperor penguins for warmth and protection. The mathematical formulation model of EPO encompasses: Identification the huddle boundary, compute temperature around the huddle, determining the distance, and find an effective mover.

Two dimensional L-shape polygon plane was suggested as the huddle plane in which emperor penguins position themselves randomly which exemplify the assorted microarray genes. Each microarray gene at least is surrounded by two gene neighbors which chosen randomly. After that, each microarray gene tries to update its position randomly towards the position of central microarray gene situated in the center of L-shaped polygon region which is characterized by the highest effective fitness rate during the current iteration as shown in Fig. 2. The reason of huddle behavior of emperor penguins to conserve energy and maximize the temperature in the huddle for surviving against cold. In order to find the huddle boundary, the wind velocity around it which is much greater than the movement velocity of penguins determined using the following equation:

$$\Psi = \nabla \Phi \tag{1}$$

where, Φ represents the wind velocity, and Ψ determines the gradient of the wind velocity.

The temperate profile of the huddle is mathematical modeled using the following equation:

$$T' = \left(T - \frac{Max_{iteration}}{x - Max_{iteration}} \right) T = \begin{cases} 0, \textit{ if } R > 1 \\ 1, \textit{ if } R < 1 \end{cases} \tag{2}$$

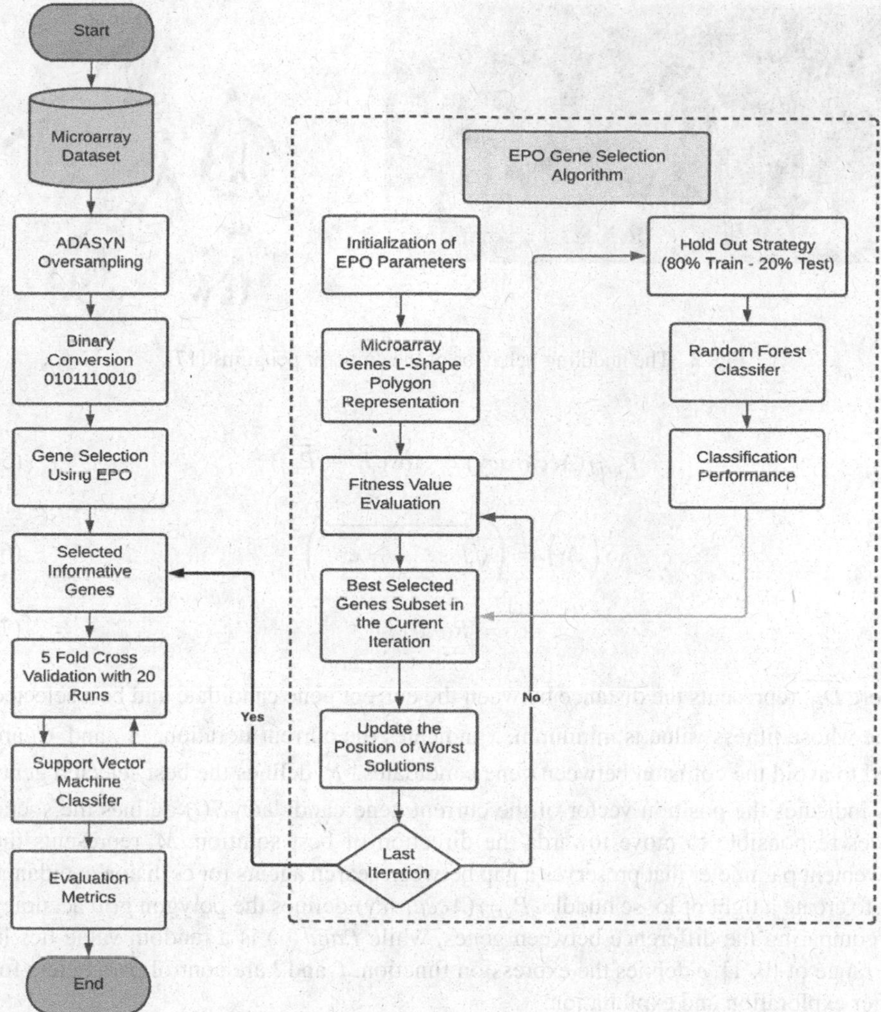

Fig. 1. System architecture of the modified version of EPO algorithm for imbalanced microarray gene selection.

where, T represents the temperature, R is the radius of the polygon, x depicts the current iteration, and $Max_{iteration}$ defines the maximum number of iterations.

Thereafter, we will calculate the distances between assorted genes which means that genes update their positions according to the best emperor penguin position which is mathematically defined as follows:

$$\overrightarrow{D_{ep}} = Abs\left(S\left(\overrightarrow{A}\right) \cdot \overrightarrow{P}(x) - \overrightarrow{C} \cdot \overrightarrow{P_{ep}}(x)\right) \tag{3}$$

$$\overrightarrow{A} = \left(M * \left(T' + P_{grid}(Accuracy)\right) * Rand()\right) - T' \tag{4}$$

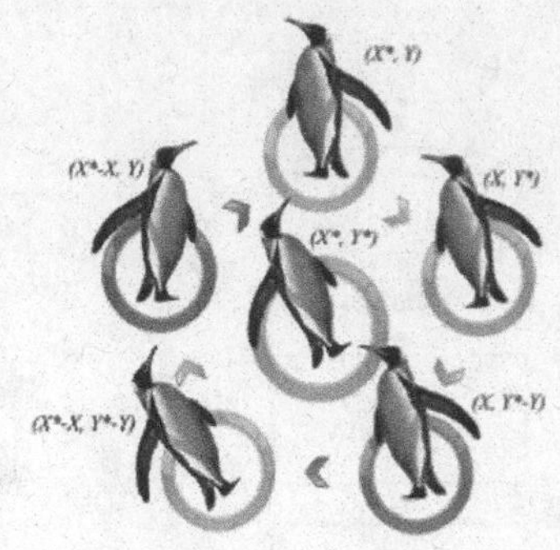

Fig. 2. The huddling behavior of the emperor penguins [17].

$$P_{grid}(Accuracy) = Abs(\overrightarrow{P} - \overrightarrow{P_{eq}}) \tag{5}$$

$$S\left(\overrightarrow{A}\right) = \left(\sqrt{f \cdot e^{-\frac{x}{l}} - e^{-x}}\right)^2 \tag{6}$$

$$\overrightarrow{C} = Rand(\ ) \tag{7}$$

where $\overrightarrow{D_{ep}}$ represents the distance between the current gene candidate and best selected gene whose fitness value is minimum, x indicates the current iteration. $\overrightarrow{A}$, and $\overrightarrow{C}$ are used to avoid the collision between gene candidates. $\overrightarrow{P}$ defines the best selected gene, $\overrightarrow{P_{eq}}$ indicates the position vector of the current gene candidate. $S(\ )$ defines the social forces responsible to move towards the direction of best solution. M represents the movement parameter that preserves a gap between search agents for collision avoidance not to create a tight or loose huddle. $P_{grid}(Accuracy)$ defines the polygon grid accuracy by comparing the difference between genes, while $Rand(\ )$ is a random value lies in the range of [0, 1]. e defines the expression function. f and l are control parameters for better exploration and exploitation.

Meanwhile, the positions of emperor penguins are updated according to the best obtained optimal solution as follows:

$$\overrightarrow{P_{eq}}(x+1) = \overrightarrow{P}(x) - \overrightarrow{A} \cdot \overrightarrow{D_{ep}} \tag{8}$$

where $\overrightarrow{P_{eq}}(x+1)$ represents updated position of gene candidate.

2.2 Adaptive Synthetic Sampling Approach (ADASYN)

Unfortunately, imbalanced microarray datasets pose a challenge for classification predictive model due to the skewed distribution of assorted samples toward the majority class. Resampling techniques represent the widely adopted technique for handling imbalanced datasets. It encompasses undersampling algorithms for removing samples from the majority class and oversampling algorithms for synthetizing more samples for the minority class. Despite, undersampling represents a main core of resampling strategies for the imbalanced datasets but involves removing random records from the majority class, which can cause loss of information. Therefore, we have resorted to the over-ssampling strategies. Two common oversampling techniques are used to handle the imbalanced microarrays which they are; Synthetic Minority Oversampling Technique (SMOTE) [19], which is based on synthesizing new data points for the minority class by randomly selecting a data point at first and finding out its K nearest neighbors and the Adaptive Synthetic Sampling method (ADASYN) [20], which is a modified version of SMOTE featured by its dependence on the density distribution of the minority class that analyzes the samples of the minority class found in spaces dominated by the majority class that are difficult to classify, this in order to generate samples in the lower density areas of the minority class. Hence, ADASYN has been adopted in our proposed algorithm to handle the microarray dataset to be adaptable for gene selection using EPO algorithm. The mathematical model of ADASYN is as follows:

Suppose the training dataset contains N samples $\{X_i, Y_i\}$, i = 1, 2, 3, ..., N, where X_i represents the microarray samples and Y_i is the class labels related to the assorted samples. We need to determine the degree of class imbalance:

$$\begin{gathered} d = m_{\min} / m_{maj} \\ m_{\min} \le m_{maj} \\ m_{\min} + m_{maj} = N \end{gathered} \tag{9}$$

where $m_{\min}$, and m_{maj} represent the minority and majority class samples, respectively. After that, the total number of the new synthetic data that need to be generated for the minority class to handle unbalancing are needed to be determined using the following equation:

$$G = (m_{min} - m_{maj}) \times \beta \tag{10}$$

where β is a real number whose ϵ[0, 1] value used to specify the desired balance level after generation of the synthetic data. For each microarray samples in the minority class, K nearest neighbor's samples based on the Euclidean distance in the search space are detected as follows:

$$r_i = \frac{\Delta_i}{k}, i = 1, 2, \ldots\ldots, m_{min} \tag{11}$$

where Δi is the number of examples in the K nearest neighbors of the present samples from the majority class. And then, we need to normalize the values of r_i over the minority samples as follows:

$$\widehat{r_i} = \frac{r_i}{\sum_{i=1}^{m_{min}} r_i} \tag{12}$$

After determining the number of K-nearest neighbors for each randomly selected sample X_i. The number of synthetic data samples for each minority example are needed to be determined from which a balanced density distribution can be achieved using the following equation:

$$g_i = \widehat{r_i} \times G \tag{13}$$

Finally, the synthetic data samples are generated using the following mathematical model:

$$s_i = X_i + (X_{zi} - X_i) \times \lambda \tag{14}$$

where X_{zi} represents the K-nearest neighbors samples from X_i, while λ is a random real number $\epsilon[0, 1]$.

2.3 Random Forest Classifier (RF)

Random Forest classifier (RF) [21] is one of the most popular supervised machine learning that was built algorithm based on an ensemble of DT trained through bagging or bootstrap aggregating. The classification performance of RF depends on the majority voting of the ensemble trees, therefore increasing the number of trees increases the precision of the classification outcome and reduces the overfitting of microarray datasets. RF utilizes two ensemble techniques; Bagging or Boosting algorithms. Bagging technique is based on generating a different training subset from training samples with replacement, and the final output is based on majority voting while Boosting combines weak learners to get strong learners through creating sequential models.

2.4 EPO for Feature Selection

The modified version of EPO for gene selection from the imbalanced microarray datasets is assembled based on three successive stages:

Initialization
The initial population of the microarray genes are represented in the suggested L-shape polygon using in which each data point represents the average gene expression level value. Before evaluation the fitness value of each gene candidate, we have performed binary conversion using the sigmoidal transfer function as follows the same as in [22]:

$$\begin{gathered} Trans.Fun(V(x)) = \frac{1}{1+e^{-V(x)}} \\ \overrightarrow{P_{eq}}(x+1) = \begin{cases} 1 & Rand < T(Y(x+1)) \\ 0 & Rand \geq T(Y(x+1)) \end{cases} \end{gathered} \tag{15}$$

where $V(x)$ represents the velocity at iteration of the gene candidate. Using Eq. (15), if $\overrightarrow{P_{eq}}(x+1)$ is 1, this means that such gene candidate are highly selected and otherwise will not selected.

Updating Solutions
After the binary conversion of the gene candidate, the fitness value is evaluated to determine the best solution. This step is terminated after the maximum number of iterations reached.

Supervised Classification
The best selected gene candidates have been randomly divided using hold-out strategy in which 80% are used for the training set and 20% are used for testing. RF classifier trained

using bagging and boosting algorithms is used the main supervised machine learning algorithm for evaluating the effectiveness of EPO algorithm in imbalances microarrays gene selection. The classification performance was assessed through determining accuracy, sensitivity, and specificity of the classification phase [23].

3 Experimental Results and Discussion

To confirm the validity of the proposed EPO-RF, 2 sepsis microarray datasets validated using MATLAB program (R2021b) on computing environment with Intel® Core™ i5 (2.50 GHz) CPU with RAM 8GB and operating system Microsoft Windows 7. The optimal ensemble model of the RF classifier has been selected based on achieving the minimum number of selected genes and highest accuracy performance. And then, a comparative study was performed to validate the robustness of the proposed model versus five well-known established metaheuristic optimization algorithms, e.g., Harris Hawks Optimization (HHO) [24], Grey Wolf Optimization (GWO) [25], Salp Swarm Optimization (SSO) [26], Particle Swarm Optimization (PSO) [27], and Genetic Algorithms (GA's) [28]. The description of the utilized microarray datasets in terms of number of samples, and data categories are declared in Table 1. The microarray dataset with the accession number (GSE66099) has been used to benchmark the performance of the selected ensemble of the proposed algorithm in differentiating the subtypes of sepsis to be further used as the main ensemble of EPO-RF algorithm. Table 2 outlines the performance of the proposed modified algorithm in terms of the number of selected genes and Accuracy (ACC.) using SVM with RBF executed for 20 runs of 5-fold cross validation with the maximum number of iteration was 100. Results stated that the attribute bagging named by Random Subspace represents the optimal ensemble technique in dealing such imbalanced datasets as it enhances the performance of the machine learning algorithm by avoiding the overfitting of the dataset through eliminating the correlation between estimators in an ensemble by training them on random data samples of features with replacement instead of the whole entire feature set. For further confirmation the outstanding performance of the proposed algorithm, a comparative study was conducted versus the aforementioned well-known metaheuristic optimization algorithms. The results of the competitive algorithms with the parameters settings are listed in Table 3 that confirms the robustness of the selected EPO-RF algorithm with less number of informative selected genes.

Table 1. List of the utilized microarray datasets with their clinical description.

Dataset	Population	Samples	CO	SIRS	SP	SPS
GSE66099	Pediatric	276	47	30	18	181
GSE13904		139	18	22	32	67

Table 2. The performance of the ensemble techniques of EPO-RF in gene selection.

Dataset	Ensemble method	Ensemble problem support	Type	ACC
GSE66099	**Subspace**	**Random subspace**	Bagging	**98.67**
	AdaBoostM1	Adaptive boosting	Boosting	97.80
	GentleBoost	Gentle adaptive boosting	Boosting	97.30
	LogitBoost	Adaptive logistic regression	Boosting	96.80
	Bag	Bagging	Bagging	97.09
	LpBoost	Linear programming boosting	Boosting	96.86
	RobustBoost	Robust boosting	Boosting	97.03
	RusBoost	Random undersampling boosting	Boosting	97.49
	TotalBoost	Totally corrective boosting	Boosting	97.05

Table 3. Performance evaluation and parameters setting of competitor metaheuristic optimization algorithms using microarray dataset GSE66099.

Method	Selected genes	Parameters	GSE66099
Proposed method	**236**	**M = 2, f = 3, l = 1**	**98.67**
HHO	226	–	78.45
GWO	3223	–	78.16
SSO	10403	L = 1	74.52
PSO	10148	Cognitive factor = 2 Social factor = 2 Inertia weight = 0.9	78.13
GA's	9822	Crossover rate = 0.08 Mutation rate = 0.01	78.05

To emphasize that the proposed model is adaptable to other microarray datasets, the performance of gene selection was evaluated through the analysis of the aforementioned microarrays that listed in Table 4. From the table, it's evident that the proposed algorithm achieved the highest classification performance and showed the exploitation capability of the proposed algorithm.

In terms of average statistical evaluation metrics for the overall classification model, Table 5 depicts the average accuracy performance of the competitive metaheuristic optimization algorithms using SVM classifier implemented in the same experimental environment. By analyzing the aforementioned table, we can conclude that the proposed model outperforms other suggested optimizers in 24 classification models out of 25 ones. To summarize, Fig. 3 and Fig. 4 depict the average confusion matrices for the EPO-RF and other compared algorithms. Figures allow us to determine the best optimizer which maximized the accuracy.

Table 4. Statistical evaluation results of the proposed algorithm using SVM classifier with 20 runs of fivefold cross-validation procedure. Acc., Accuracy; Spec., Specificity; Sens., Sensitivity; Prec., Precision.

Dataset	Classification model	Class	Acc.	Sens.	Spec.	Prec.
GSE66099	CO vs. SIRS vs. SP vs. SPS	CO	98.67	98.86	99.98	99.94
		SIRS		91.13	99.70	99.11
		SP		99.73	99.63	99.88
		SPS		99.11	98.60	95.78
GSE13904		CO	91.84	100.00	96.46	88.76
		SIRS		100.00	98.00	93.84
		SP		97.85	95.23	84.75
		SPS		77.29	100.00	100.00

Table 5. Average accuracy performance of the competitive optimization algorithms using SVM classifier with 20 runs of fivefold cross-validation procedure.

Dataset	EPO-RF	PSO	SSO	GWO	HHO	GA's
GSE66099	98.67	78.13	74.52	78.16	78.45	78.05
GSE13904	91.84	57.36	77.03	57.94	55.74	57.11

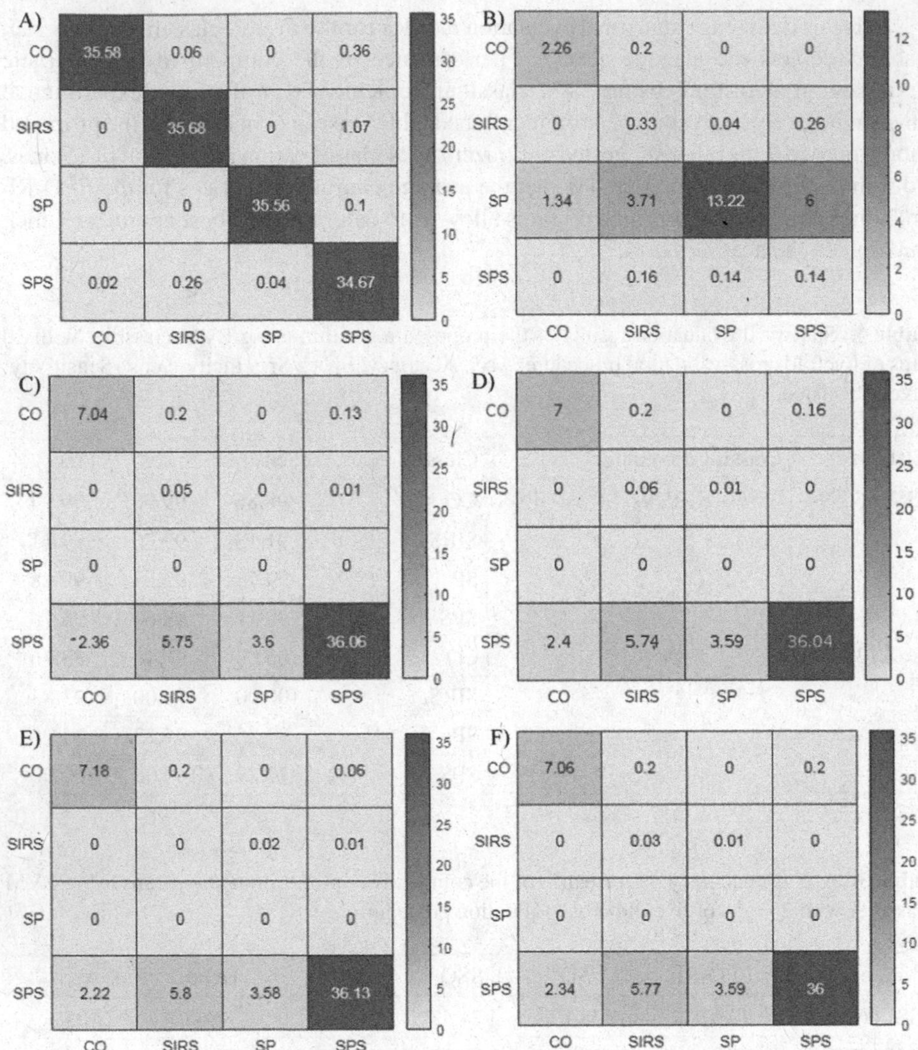

Fig. 3. (A–F) The average confusion matrix of the EPO-RF algorithm in comparison with other metaheuristic algorithms for GSE66099 A) EPO, B) PSO, C) GWO, D) SSA, E) HHO, and F) GA's.

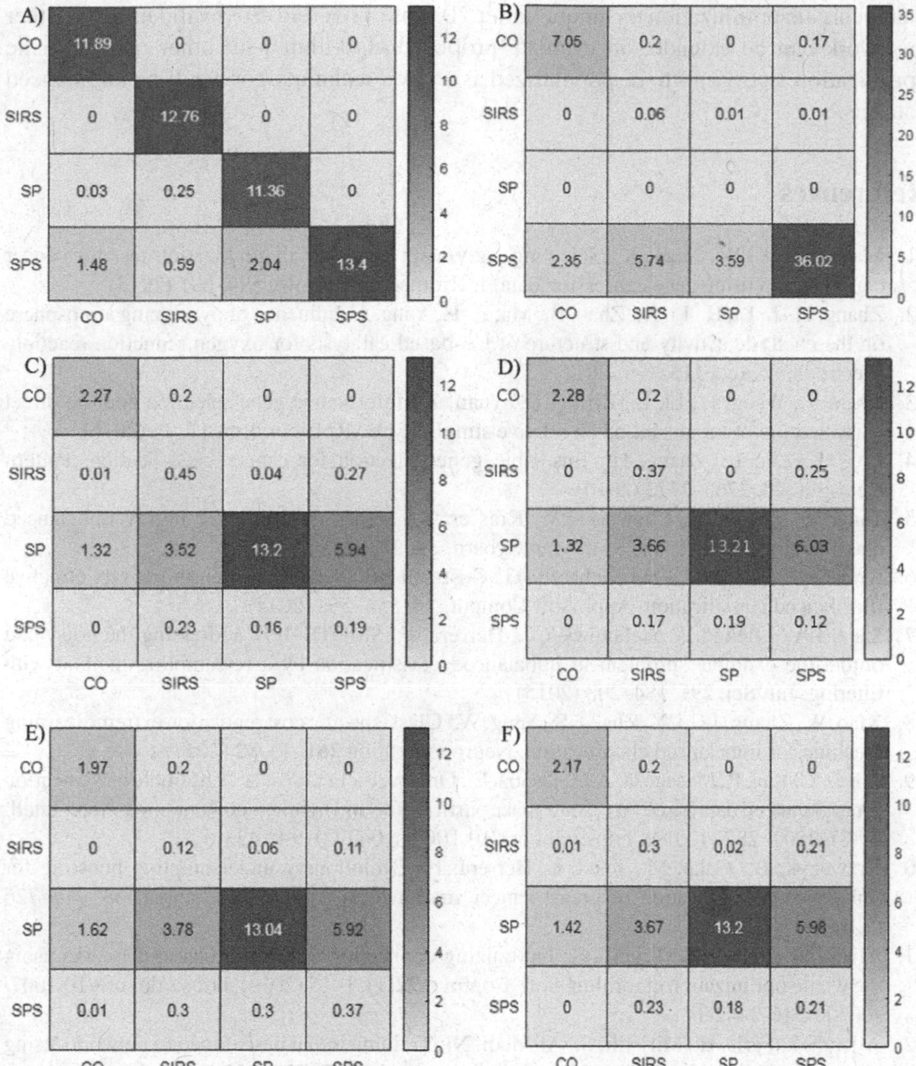

Fig. 4. (A–F) The average confusion matrix of the EPO-RF algorithm in comparison with other metaheuristic algorithms for GSE13904 A) EPO, B) PSO, C) GWO, D) SSA, E) HHO, and F) GA's.

4 Conclusion

A modified version of a nature inspired metaheuristic optimization algorithm named by EPO-RF has been suggested for informative gene selected from imbalanced microarray datasets. The performance of the proposed algorithm was conducted using two real world microarray datasets. The average experimental results confirmed the robustness of EPO-RF in comparison to existing PSO, SSO, GWO, HHO, and GA's well known metaheuristic optimization techniques after 20 runs of fivefold cross validations. Further this work can be extended to evaluate proposed algorithm with other metaheuristic optimization algorithms to be popularized as a novel technique for handling imbalanced datasets.

References

1. Mao, Z., Cai, W., Shao, X.: Selecting significant genes by randomization test for cancer classification using gene expression data. J. Biomed. Inform. **46**, 594–601 (2013)
2. Zhang, H.-J., Li, H., Li, X., Zhao, B., Ma, Z.-F., Yang, J.: Influence of pyrolyzing atmosphere on the catalytic activity and structure of Co-based catalysts for oxygen reduction reaction. Electrochim. Acta **115**, 1–9 (2014)
3. Chen, Y., Wang, L., Li, L., Zhang, H., Yuan, Z.: Informative gene selection and the direct classification of tumors based on relative simplicity. BMC Bioinform. **17**, 44 (2016)
4. Liu, H., Liu, L., Zhang, H.: Ensemble gene selection for cancer classification. Pattern Recognit. **43**, 2763–2772 (2010)
5. Tang, Y., Zhang, Y., Chawla, N.V., Krasser, S.: SVMs modeling for highly imbalanced classification. IEEE Trans. Syst. Man Cybern. **39**, 281–288 (2009)
6. Krawczyk, B., Woźniak, M., Schaefer, G.: Cost-sensitive decision tree ensembles for effective imbalanced classification. Appl. Soft Comput. **14**, 554–562 (2014)
7. Sáez, J.A., Luengo, J., Stefanowski, J., Herrera, F.: SMOTE–IPF: addressing the noisy and borderline examples problem in imbalanced classification by a re-sampling method with filtering. Inf. Sci. **291**, 184–203 (2015)
8. Xiao, W., Zhang, J., Li, Y., Zhang, S., Yang, W.: Class-specific cost regulation extreme learning machine for imbalanced classification. Neurocomputing **261**, 70–82 (2017)
9. Lopez-Garcia, P., Masegosa, A.D., Osaba, E., Onieva, E., Perallos, A.: Ensemble classification for imbalanced data based on feature space partitioning and hybrid metaheuristics. Appl. Intell. **49**(8), 2807–2822 (2019). https://doi.org/10.1007/s10489-019-01423-6
10. Krawczyk, B., Galar, M., Jeleń, Ł, Herrera, F.: Evolutionary undersampling boosting for imbalanced classification of breast cancer malignancy. Appl. Soft Comput. **38**, 714–726 (2016)
11. Aljarah, I., Faris, H., Mirjalili, S.: Optimizing connection weights in neural networks using the whale optimization algorithm. Soft. Comput. **22**(1), 1–15 (2016). https://doi.org/10.1007/s00500-016-2442-1
12. Aljarah, I., Faris, H., Mirjalili, S., Al-Madi, N.: Training radial basis function networks using biogeography-based optimizer. Neural Comput. Appl. **29**(7), 529–553 (2016). https://doi.org/10.1007/s00521-016-2559-2
13. Roshan, S.E., Asadi, S.: Improvement of bagging performance for classification of imbalanced datasets using evolutionary multi-objective optimization. Eng. Appl. Artif. Intell. **87**, 103319 (2020)
14. Hashim, F., Mabrouk, M.S., Al-Atabany, W.: GWOMF: Grey Wolf Optimization for motif finding. In: 2017 13th International Computer Engineering Conference (ICENCO), pp. 141–146 (2017)
15. Elden, R.H., Ghoneim, V.F., Al-Atabany, W.: A computer aided diagnosis system for the early detection of neurodegenerative diseases using linear and non-linear analysis. In: 2018 IEEE 4th Middle East Conference on Biomedical Engineering (MECBME), pp. 116–121 (2018)
16. Elden, R.H., Ghoneim, V.F., Hadhoud, M.M.A., Al-Atabany, W.: Studying genes related to the survival rate of pediatric septic shock. In: 2021 3rd Novel Intelligent and Leading Emerging Sciences Conference (NILES), pp. 93–96 (2021)

17. Abdelnaby, M., Alfonse, M., Roushdy, M.: A hybrid mutual information-LASSO-genetic algorithm selection approach for classifying breast cancer (2021)
18. Dhiman, G., Kumar, V.: Emperor penguin optimizer: a bio-inspired algorithm for engineering problems. Knowl. Based Syst. **159**, 20–50 (2018)
19. Chawla, N., Bowyer, K., Hall, L., Kegelmeyer, W.: SMOTE: synthetic minority over-sampling technique. J. Artif. Intell. Res. **16**, 321–357 (2002)
20. Haibo, H., Yang, B., Garcia, E.A., Shutao, L.: ADASYN: adaptive synthetic sampling approach for imbalanced learning. In: 2008 IEEE International Joint Conference on Neural Networks (IEEE World Congress on Computational Intelligence), pp. 1322–1328 (2008)
21. Breiman, L.: Random forests. Mach. Learn. **45**, 5–32 (2001)
22. Dhiman, G., et al.: BEPO: a novel binary emperor penguin optimizer for automatic feature selection. Knowl. Based Syst. **211**, 106560 (2021)
23. Prince John, R., Lewall David, B.: Sensitivity, specificity, and predictive accuracy as measures of efficacy of diagnostic tests. Ann. Saudi Med. **1**, 13–18 (1981)
24. Heidari, A.A., Mirjalili, S., Faris, H., Aljarah, I., Mafarja, M., Chen, H.: Harris hawks optimization: algorithm and applications. Future Gener. Comput. Syst. **97**, 849–872 (2019)
25. Mirjalili, S., Mirjalili, S.M., Lewis, A.: Grey wolf optimizer. Adv. Eng. Softw. **69**, 46–61 (2014)
26. Mirjalili, S., Gandomi, A.H., Mirjalili, S.Z., Saremi, S., Faris, H., Mirjalili, S.M.: Salp swarm algorithm: a bio-inspired optimizer for engineering design problems. Adv. Eng. Softw. **114**, 163–191 (2017)
27. Kennedy, J., Eberhart, R.: Particle swarm optimization. In: Proceedings of ICNN 1995 - International Conference on Neural Networks, vol. 1944, pp. 1942–1948 (1998)
28. Whitley, D.: A genetic algorithm tutorial. Stat. Comput. **4**, 65–85 (1994)

Fuzzing-Based Grammar Inference

Hannes Sochor(✉), Flavio Ferrarotti, and Daniela Kaufmann

Software Competence Center Hagenberg GmbH (SCCH), Hagenberg, Austria
{hannes.sochor,flavio.ferrarotti,daniela.kaufmann}@scch.at

Abstract. In this paper we propose and suggest a novel approach for grammar inference that is based on grammar-based fuzzing. While executing a target program with random inputs, our method identifies the program input language as a human-readable context-free grammar. Our strategy, which integrates machine learning techniques with program analysis of call trees, uses a far smaller set of seed inputs than earlier work. As a further contribution we also combine the processes of grammar inference and grammar-based fuzzing to incorporate random sample information into our inference technique. Our evaluation shows that our technique is effective in practice and that the input languages of tested recursive-descending parser are correctly inferred.

Keywords: Fuzzing · Grammar-based fuzzing · Software testing · Grammar inference · Grammar learning · Program analysis

1 Introduction

Software testing is one of the most important phases of the software lifecycle. This includes testing not only for functional correctness but also for safety and security. Finding bugs and security vulnerabilities presents a difficult task when facing complex software architectures. An integral part of software testing is fuzzing software with more or less random input and tracking how the software reacts. Having knowledge about the input structure of the software under test enables the fuzzer to generate more targeted inputs which significantly increases the chance to uncover bugs and vulnerabilities by reaching deeper program states.

As such the most successful fuzzers all come with some sort of model that describes the input structure of the target program. One of the most promising methods poses grammar-based fuzzing, where inputs are generated based on a context-free grammar which fully covers the so-called input language of a program. This makes it possible for the grammar-based fuzzer to produce inputs that are valid or near-valid, considerably raising its success rate. Although grammar-based fuzzing is a very successful method, in most cases such a precise description of the input language is not available.

The automation of learning input languages for a program, in our instance in the form of a synthesized context-free grammar, is still an issue in current grammar-based fuzzing techniques and is not completely resolved yet. With these

P. Fournier-Viger et al. (Eds.): MEDI 2022, LNAI 13761, pp. 72–86, 2023.
https://doi.org/10.1007/978-3-031-21595-7_6

capabilities, we would be able to apply grammar-based fuzzing to a wider range of problems. Additionally it would be possible to utilize the inferred model for additional security analysis, such as comparing the inferred grammars of different implementations to determine whether they are equivalent.

While some current grammar-based fuzzing tools can be used more broadly but suffer from mistakes as a result, others are connected to specific programming paradigms or languages. In addition, state of the art grammar inference tools heavily depend on a good starting set of seed inputs to be able to correctly infer a grammar [8,10]. However, a good set of inputs is frequently not available, which leaves much room for improvement. Especially in the setting of security analysis that includes grammar-based fuzzing, a proper set of seeds is vital, as the inferred grammar has to be as accurate as possible.

In this paper we propose a novel automated method for grammar-based fuzzing, which automatically learns the grammar that is later used for fuzzing. In our technique we start from an incomplete seed grammar that is extracted from a small set of seed inputs. While fuzzing the target program, we actively learn and continuously enhance this grammar. These improvements in the grammar are based on information that we gain while executing our target program with randomly generated inputs and observing the response of the program. Our method makes use of a machine learning algorithm in combination with program analysis, more precisely, by extracting the call tree of some executed inputs. Our approach is generic and may be utilized regardless of the programming language of the target program because the learning process we use is black-box and the extraction of call trees is not based on a particular programming language. In addition, the input set needed for our approach is significantly smaller than in state of the art grammar inference tools [8,10]. The fundamental disadvantage of our approach is that it is restricted to recursive top-down parsers, which account for up to 80% of all parsers in use today [12]. Most grammatical inference tools also share this restriction, according to [8,10].

Our experimental results show that our fuzzing-based grammar inference method enables us to learn a context-free grammar from tested recursive top-down parsers with the maximum possible accuracy in every case that we considered. Our technique accomplishes this in a relatively quick time while employing a simple program analysis technique. We further generate a human-readable grammar that may be applied to further security analysis.

The paper is structured as follows: We provide background information on formal definitions and learnability in language theory in Sect. 2. Our main contribution can be found in Sect. 3 where we present our method in detail, followed by an experimental evaluation of our approach in Sect. 4 as well as an in-depth discussion on related work in Sect. 5. We give our conclusion in Sect. 6.

2 Preliminaries

In this section we introduce the necessary notation and theory for the rest of the paper. We assume the reader is familiar with basic concepts of language theory. An excellent reference for that is the classical book by Hopcroft and Ullman [9].

Let Σ be an *alphabet*, i.e., a finite set of symbols. A finite sequence of symbols taken from Σ is called a *word* or *string* over Σ. The free monoid of Σ, i.e., the set of all (finite) strings over Σ plus the empty string λ, is denoted as Σ^* and known as the *Kleene star* of Σ. If $v, w \in \Sigma^*$, then $vw \in \Sigma^*$ is the *concatenation* of v and w and $|vw| = |v| + |w|$ is its length. If $u = vw$, then v is a *prefix* of u and w is a *suffix*. A *language* is any subset of Σ^*.

A *grammar* is formally defined as a 4-tuple $G = (N, \Sigma, P, S)$, where N and Σ are finite disjoint sets of *nonterminal* and *terminal symbols* respectively, $S \in N$ is the *start symbol* and P is a finite set of *production rules*, each of the form:

$$(\Sigma \cup N)^* N (\Sigma \cup N)^* \rightarrow (\Sigma \cup N)^*.$$

We say G *derives* (or equivalently *produces*) a string y from a string x in one step, denoted $x \Rightarrow y$, iff there are $u, v, p, q \in (\Sigma \cup N)^*$ such that $x = upv$, $p \rightarrow q \in P$ and $y = uqv$. We write $x \Rightarrow^* y$ if y can be derived in zero or more steps from x, i.e., $\Rightarrow^*$ denotes the reflexive and transitive closure of the relation $\Rightarrow$.

The language of G, denoted as $\mathcal{L}(G)$, is the set of all strings in Σ^* that can be derived in a finite number of steps from the start symbol S. In symbols,

$$\mathcal{L}(G) = \{w \in \Sigma^* \mid S \Rightarrow^* w\}$$

In this work, Σ always denotes the "input" alphabet (e.g., the set of ASCII characters) of a given executable (binary) program p. The set of *valid inputs of p* is defined as the subset of Σ^* formed by all well formed inputs for p. In symbols:

$$validInputs(p) = \{w \in \Sigma^* \mid w \text{ is a well formed input for } p\}$$

The definition of a well formed input for a given program p depends on the application at hand. In our setting we only need to assume that it is possible to determine whether a given input string w is well formed or not for a program p by simply running p with input w.

As usual, we assume that $validInputs(p)$ is a context-free language. Consequently, there is a context-free grammar G_p such that $\mathcal{L}(G_p) = validInputs(p)$. Recall that a grammar is context-free if its production rules are of the form $A \rightarrow \alpha$ with A a single nonterminal symbol and α a possibly empty string of terminals and/or nonterminals.

Our main contribution in this paper is a novel algorithm that takes as input a program p and a finite (small) subset I of $validInputs(p)$, and infers a grammar G_p such that $\mathcal{L}(G_p)$ approximates $validInputs(p)$. I is usually called *seed input*. We say "approximates" since it is *not* decidable in our setting (see [5]) whether $\mathcal{L}(G_p) = validInputs(p)$. To evaluate how well $\mathcal{L}(G_p)$ approximates $validInputs(p)$, we measure the precision and recall of $\mathcal{L}(G_p)$ w.r.t. $validInputs(p)$ as in [2], among others.

In our setting we first fix a procedure to calculate the probability distribution of a language, starting from its corresponding grammar. Following [2] we use random sampling of strings. Let $G = (N, \Sigma, P, S)$ be a context-free grammar.

As a first step, G is converted to a *probabilistic context-free grammar* by assigning a *discrete distribution* $\mathcal{D}_A$ to each nonterminal $A \in N$. As usual, $\mathcal{D}_A$ is of size $|P_A|$, where P_A is the subset of productions in P of the form $A \to \alpha$. Here, we assume that $\mathcal{D}_A$ is *uniform*. We can then *randomly sample a string x from the language* $\mathcal{L}(G, A) = \{w_i \in \Sigma \mid A \Rightarrow^* w_i\}$, denoted $x \sim \mathcal{P}_{\mathcal{L}(G,A)}$, as follows:

- Using $\mathcal{D}_A$ select randomly a production $A \to A_1 \cdots A_k \in P_A$.
- For $i = 1, \ldots, k$, recursively sample $x_i \sim \mathcal{P}_{\mathcal{L}(G,A_i)}$ if $A_i \in N$; otherwise let $x_i = A_i$.
- Return $x = x_1 \cdots x_k$.

The *probability distribution* $\mathcal{P}_{\mathcal{L}(G)}$ *of the language* $\mathcal{L}(G)$ is simply defined as the probability $\mathcal{P}_{\mathcal{L}(G,S)}$ induced by sampling strings in the probabilistic version of G defined above.

We can now measure the quality of a learned (or inferred) language $\mathcal{L}'$ with respect to the target language $\mathcal{L}$ in terms of precision and recall.

- The *precision* of $\mathcal{L}'$ w.r.t. $\mathcal{L}$, denoted *precision*$(\mathcal{L}', \mathcal{L})$, is defined as the probability that a randomly sampled string $w \sim \mathcal{P}_{\mathcal{L}'}$ belongs to $\mathcal{L}$. In symbols, $\Pr_{w \sim \mathcal{P}_{\mathcal{L}'}}[w \in \mathcal{L}]$.
- Conversely, the *recall* of $\mathcal{L}'$ w.r.t. $\mathcal{L}$, denoted *recall*$(\mathcal{L}', \mathcal{L})$, is defined as $\Pr_{w \sim \mathcal{P}_{\mathcal{L}}}[w \in \mathcal{L}']$.

We say that $\mathcal{L}'$ is a good approximation to $\mathcal{L}$ if it has both, high precision and high recall. Note that, a language $\mathcal{L}' = \{w\}$, where $w \in \mathcal{L}$, has perfect precision, but most likely has also very low recall. On the other hand, $\mathcal{L}' = \Sigma^*$ has perfect recall, but probably low precision.

It is well known that there are effective algorithms that can infer a finite automaton (and hence also a regular grammar or regular expression) to recognize regular languages $\mathcal{L}$. These algorithms need however a "teacher" that can answer *membership* and *equivalence queries* w.r.t. $\mathcal{L}$. The first and most well known algorithm of this kind was introduced by Dana Angluin [1] and is known as L*.

The *membership query* inquires as to whether a provided string is a part of the target language. Since it pertains to determining if a particular input to a program p belongs to *validInputs*(p), this can obviously be addressed in our context. The *equivalence query* tests if the target language exactly matches the language that a given automaton (or grammar in our example) recognizes. Otherwise, the "teacher" ought to be able to offer a counterexample.

We leverage the approach NL* presented in [4] as part of our strategy (i.e. as a subroutine) in our algorithm to infer context free grammars from program input samples. This approach is based on L* and helps learning regular languages efficiently, although it learns residual-finite state machines as opposed to deterministic finite automata. The assumption is the same as for L*: a "teacher" who is able to respond to membership and equivalence questions. However, because we are unable to answer the equivalency question in our environment, we must instead rely on statistical sampling to look for counterexamples.

Algorithm 1: Fuzzing-based Grammar Inference Algorithm

Input : Seed inputs $I \subseteq \text{validInputs}(p)$, set of terminals Σ and program p
Output: Inferred grammar G'_p

```
G_s := findSeedGrammar(I, p);
N_s := G_s.getNonTerminals();
G'_p := G_s.clone();
for A ∈ N_s do
    c := null;
    repeat
        M := runNL*(p, A, Σ, N_s, c);
        c := searchForCounterexample(M, p, A, G_s, 1000, 10)   ▷ See Algorithm 2;
    until c = null;
    α := M.toRegularExpression();
    G'_p.add("A → α")
return G'_p
```

3 Algorithm

The goal of our approach is to infer a context-free grammar G'_p given a program p, a set of terminal symbols Σ as well as some valid seed inputs I. Ideally the language $L(G'_p)$ produced by our inferred grammar G'_p should be able to produce the input language *validInputs*(p) of p such that $L(G'_p) = \mathit{validInputs}(p)$. To achieve our goal, we apply the following steps:

1. **Seed Grammar Extraction:** First we extract a seed grammar G_s from p using the valid seed inputs in I.
2. **Seed Grammar Expansion:** Next we continuously expand the rules of G_s. to achieve a better coverage of *validInputs*(p). We do this by utilizing the NL* algorithm. While learning the rules of G'_p, we apply grammar-based fuzzing to find counterexamples needed during the learning process.
3. **Grammar-based fuzzing:** At some point we have inferred a grammar G'_p where finding a counterexample is hard because we have, or nearly have, identified *validInputs*(p). We can run our grammar-based fuzzer indefinitely at this point until we uncover another counterexample, if one exists.

In this section we first give a formal description of our algorithm, followed by an example to better illustrate the learning process. Finally, we briefly discuss how our algorithm may be applied in a grammar-based fuzzing setting.

3.1 Learning Context-Free Grammars

Assume we have a program p. We want to learn a grammar G_p such that $L(G_p)$ approximates *validInputs*(p) as well as possible in terms of precision and recall (see preliminaries). As usual, the set of terminal symbols Σ of the target grammar G_p, or equivalently the set of characters accepted by p, is assumed to be known. We further assume an initial (finite) subset I of *validInputs*(p). Our

Algorithm 2: Adapted Equivalence Query

```
Function: searchForCounterexample(M, p, A, G_s, n, m)
Input   : Automaton M, program p, Seed Grammar G_s, NonTerminal A, Set
          of parse Trees T and maximum number n and m of trials and
          mutations per trial, respectively.
Output  : Counterexample string if found. Otherwise, null.
G := M.toGrammar();
for (i := 0; i < n; i++) do
    w := G.generateString();
    if ¬membershipQuery(G_s, w, A, T, p) return w        ▷ See Algorithm 3;
    for (j := 0; j < m; j++) do
        w' := w.applyMutation();
        if w' ∉ L(G) ∧ membershipQuery(G_s, w', A, T, p) return w';
return null
```

Algorithm 3: Adapted Membership Query

```
Function: membershipQuery(G, w, A, T, p)
Input   : Grammar G = (N, Σ, P, S), w ∈ (N ∪ Σ)*, A ∈ N \ {S}, set T of
          parse trees, program p.
Output  : true if A → w is deemed to be a good candidate for extending the
          productions of G. Otherwise false.
if ¬∃Tx(T ∈ T ∧ x ∈ nodes(T) ∧ label(x) = A) return false;
s := deriveString(w, G)          ▷ Derives s ∈ Σ* from w using G ;
T_A := parseTree(A, w, s, G)     ▷ Using left-most derivation and A → w ;
for each T ∈ T do
    for each x ∈ nodes(T) with label(x) = A do
        T.replaceSubTree(x, T_A)              ▷ Subtree rooted at node x;
    input := T.toString();
    if input ∉ validInputs(p) return false;
    else
        if p.parseTree(input) = T return true else return false;
```

strategy is based on extracting a seed grammar from p using I, and then expanding it using grammar-based fuzzing of p until we obtain a good approximation of $validInputs(p)$. Grammar-based fuzzing of p means that we execute p with randomly sampled inputs produced from a given grammar. The concrete strategy is described in Algorithm 1.

The seed grammar extraction is done by the function findSeedGrammar(I, p) (line 1 in Algorithm 1). The set of non-terminals N_s of the seed grammar G_s is formed by the names of all functions called by executing p with inputs from I (line 2 in Algorithm 1). The set of productions P_s of G_s is obtained as follows. For each $u \in I$,

1. Extract from p and u a parse tree T_u for u, where the internal nodes of T_u are labelled by non-terminals in N_s and the leaves are labelled by terminals in Σ.

2. For each internal node e of T_u with children $c_1, \ldots, c_n$ ordered from left to right, add the rule $A \to L_1 \ldots L_n$ to P_s, where $A, L_1, \ldots, L_n$ are the labels of the nodes $e, c_1, \ldots, c_n$, respectively.

The extraction of the parse trees T_u simply requires to track the function calls during an execution of p with input u. This can be done in multiple ways. In our case, we use the dynamic symbolic execution framework in our eknows platform [13]. A perfectly good alternative is to use dynamic tainting as in [8]. We omit here the technical details of this process as they are well known.

By construction, $I \subseteq L(G_s)$. Furthermore, if we assume that p uses a recursive descent parsing technique where each procedure/function implements one of the nonterminals of a grammar G_p such that $L(G_p) = \mathit{validInputs}(p)$, then $L(G_s) \subseteq \mathit{validInputs}(p)$. Moreover, for every nonterminal symbol A of G_s, we have that $\mathcal{L}(G_s, A) = \{w \in \Sigma \mid A \Rightarrow^* w\}$ is non empty, i.e., for every nonterminal A of the seed grammar there is at least one finite derivation that starts with A and produces a word in Σ^*.

Next the algorithm tries to augment the set of productions in G_s, i.e., in the seed grammar, to better approximate $\mathit{validInputs}(p)$ (line 4–11 in Algorithm 1). It proceeds by considering, for every non-terminal symbol $A \in N_s$, a new rule of the form $A \to \alpha$, where α is a regular expression such that $\mathcal{L}(\alpha) \subseteq (N_s \cup \Sigma)^*$. We search for a suitable α using the NL* algorithm [4] couple with the procedure described in Algorithm 3 and 2 to answer the necessary membership and equivalence queries, respectively. At the end of this process, we obtain α by translating (following the well known standard procedure) the automaton returned by NL* into an equivalent regular expression (line 10).

Each time NL* needs to answer a membership query for a string $w \in N_S \cup \Sigma$, it calls the procedure described in Algorithm 3 with the following parameters:

- Seed grammar G_s.
- String w.
- Non-terminal A.
- Set $\mathcal{T}$ of parse trees such that $T_u \in \mathcal{T}$ iff $u \in I$ and T_u is the tree extracted in the previous stage (i.e., during the inference process of the seed grammar G_s) by tracking the function calls in the run of p with input u.
- Program p.

The function *deriveString* in line 2 of Algorithm 3 produces a string $s \in \Sigma^*$ starting from w by applying the rules of G with a grammar-based fuzzing technique. Note that the function $\mathit{parseTree}(A, w, s, G)$ in line 3 returns the parse tree corresponding to the left-most derivation $A \Rightarrow w \Rightarrow^* s$ of G. On the other hand, $p.\mathit{parseTree}(\mathit{input})$ in line 10 returns the parse tree obtained by tracking the function calls in the run of p with input, using the procedure explained earlier. The remaining parts of this algorithm are self-explanatory.

Whenever NL* needs answer to the equivalence query for an automaton M, we apply the heuristic described in Algorithm 2. Notice, that we do not have in this context a properly defined regular language that NL* needs to learn. Instead, we search for a counterexample string in $\mathcal{L}(M)$ that does not satisfy

```
  parse → expr
   expr → term | term+term | term−term
   term → factor | factor∗factor | factor/factor
 factor → 1 | 2 | 3 | (expr)
```

Listing 1. Target Grammar

the (adapted) membership query expressed by Algorithm 3, or vice versa. The search for a counterexample is performed until one is found or a maximum number of trials n has been reached. In each trial, the algorithm first derives a string $w \in (N_s \cup \Sigma)$ by applying grammar-based fuzzing with a grammar G equivalent to the automaton M (line 3). If w does not satisfy the (adapted) membership query, the algorithm returns w as a counterexample (line 4). Otherwise, it generates a mutation w' of w (line 6). If w' is not in the language recognized by G (or equivalently M) but satisfies the (adapted) membership query, then the algorithm returns w' as a counterexample (line 7). Otherwise, it tries different mutations of w up to a maximum number m. For the cases considered in the experiments reported in this paper, $n = 1000$ and $m = 10$ gives us optimal results and good performance.

3.2 Example

We will provide an example run of our algorithm to give a better understanding of the formal definition above. We will use the example grammar provided in Listing 1 as G_p. Assume we have a parser p where $validInput(p) = L(G_p)$. We know the set of terminals $\Sigma = \{1, 2, 3, (,), +, -, *, /\}$ as well as the initial set of seed inputs $I = \{``1"\}$. We start extracting a seed grammar G_s by tracking p while executing $u \in I$. This returns a parse tree T_u. Figure 1 displays T_u as well as the result of transforming T_u to G_s.

Next we want to expand the rules of G_s. At this point we will use the expansion process of the rule `factor` → 1 as a showcase example. To expand the rule, we want to learn `factor` → $\alpha_{\texttt{factor}}$. First we have to identify which symbols are used by $\alpha_{\texttt{factor}}$. We do this by using static analysis to identify which functions are called by `factor` and add the according non-terminals as well as Σ. In our case the set of symbols is $\{\Sigma \cup \texttt{expr}\}$. Now we can apply NL* to learn $\alpha_{\texttt{factor}}$. To do so, we have to answer both membership-queries as well as equivalence-queries. First we will give some examples on how membership-queries are answered while learning the rules `expr` → $\alpha_{\texttt{expr}}$ as well as `factor` → $\alpha_{\texttt{factor}}$:

1. $w \in \mathcal{L}(\alpha_{\texttt{expr}})$ for $w =$ `term term`: We start by replacing the children of `expr` in T_u with w. Next we replace left non-terminals in the tree by applying the rules of G_s. This process is illustrated in Fig. 2. Finally, we transform the newly built parse tree to $w' = 11$ and execute p with w'. As $w' \notin validInputs(p)$, we return $w \notin \mathcal{L}(\alpha_{\texttt{expr}})$.
2. $w \in \mathcal{L}(\alpha_{\texttt{expr}})$ for $w =$ "`term+term`": Again, we start by replacing the children of `expr` in T_u with w. Next we replace left non-terminals in the tree by

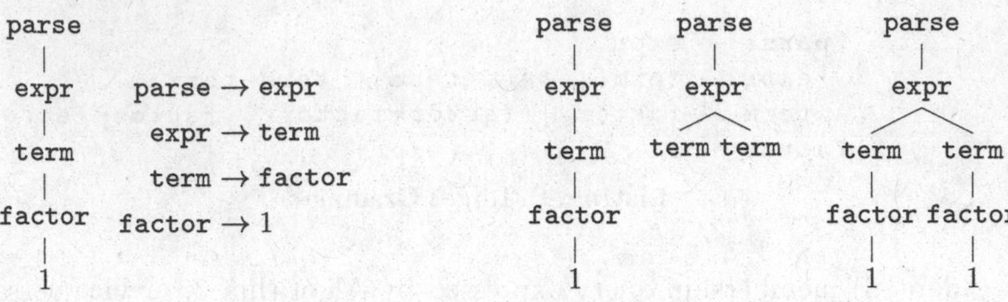

Fig. 1. From parse tree to grammar

Fig. 2. Parse tree evolution of query 1

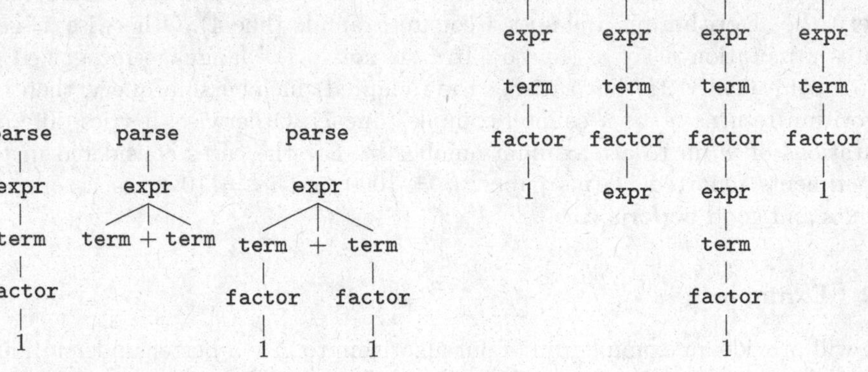

Fig. 3. Parse tree evolution of query 2

Fig. 4. Parse tree evolution of query 3

applying the rules of G_s as illustrated in Fig. 3. When we execute p with input $w' = 1+1$ we see that $w' \in \mathit{validInputs}(p)$. Now we check the call tree of the execution and see that it is equivalent to the parse tree of w', so we return $w \in \mathcal{L}(\alpha_{\texttt{expr}})$.

3. $w \in \mathcal{L}(\alpha_{\texttt{factor}})$ for $w =$ "expr": Again, we replace the children of `expr` in T_u with w and resolve left non-terminals with G_s. This leads to the trees shown in Fig. 4. Executing p with input $w' = 1$ shows that $w' \in \mathit{validInputs}(p)$. Note that the call tree of executing w' (see the most right tree in Fig. 4) is not equivalent to the parse tree of w' so we return $w \notin \mathcal{L}(\alpha_{\texttt{factor}})$.

Next we provide an example on how we will answer an equivalence-query while learning the rule $\texttt{expr} \rightarrow \alpha$. We convert the given automaton M to a regular expression $\alpha = \texttt{term} \mid \texttt{term} + \texttt{term}$. Next we build a new grammar G_α containing a single rule of the form $S \rightarrow \texttt{term} \mid \texttt{term} + \texttt{term}$ with S being the start symbol. We continue by searching for a counterexample by generating new words using a grammar-based fuzzer using G_α. Assume we find a counterexample $c = \texttt{term} - \texttt{term}$ by applying a mutation to a generated word $\texttt{term} + \texttt{term}$. We return c and continue to answer membership-queries until the next equivalence-query is performed. In our case, the next automaton would translate to the regular expression $\alpha = \texttt{term} \mid \texttt{term}(-\texttt{term} \mid +\texttt{term} \mid \epsilon)$. As this already is

```
parse → expr
  expr → term | term ( −term | +term | ϵ)
  term → factor | factor ( /factor | ∗factor | ϵ)
factor → 1 | 2 | 3 | (expr)
```

Listing 2. Learned Grammar

Table 1. Experiment results

Target	Precision	Recall	MQ	EQ	Time
ExprParser	1.0	1.0	6 898	7	14 s
MailParser	1.0	1.0	8 482	6	1 m 42 s
HelloParser	1.0	1.0	7 168	2	1 m 09 s
AdvExprParser	1.0	1.0	22 984	8	1 m 47 s
JsonParser	1.0	1.0	35 058	35	3 m 23 s

the correct α, we will not be able to find another counterexample and stop fuzzing when the maximum specified amount of tries is reached. Finally, we add $\texttt{expr} \rightarrow \texttt{term} \mid \texttt{term}(-\texttt{term} \mid +\texttt{term} \mid \epsilon)$ to G'_p. We repeat the process described above for every non-terminal in G_s. The resulting grammar G'_p is shown in Listing 2. This grammar is equivalent to G_p given in Fig. 1.

3.3 Application in Grammar-Based Fuzzing

While learning a rule $A \rightarrow \alpha$ we systematically explore a small part of p, more specifically we explore the function which is contributed to the non-terminal A. When an arbitrary membership-query is executed, we guide the parser to the exact part of the code where the query is parsed by executing the query in a known context. This has the effect that, while performing an equivalence query, we can effectively fuzz exactly that part of the code until we find a counterexample. Additionally, a positive side effect is that the mutations we insert will most likely explore border-line cases within the context of A which reduces the search space effectively. As the learning of a rule $A \rightarrow \alpha$ is completely independent from learning another, we may learn all the rules simultaneously. This enables us to run one grammar-based fuzzer targeting each function of p separately for as long as we need, pausing fuzzing only when we are able to fine-tune the used grammar by a deterministic search with NL*.

4 Experiments and Evaluation

In this section we evaluate the performance of our fuzzing-based grammar inference method. Given a single input word, we apply our technique to different parsers and calculate the precision and recall of the inferred grammar. For each test run we start with a given target grammar G_p and automatically

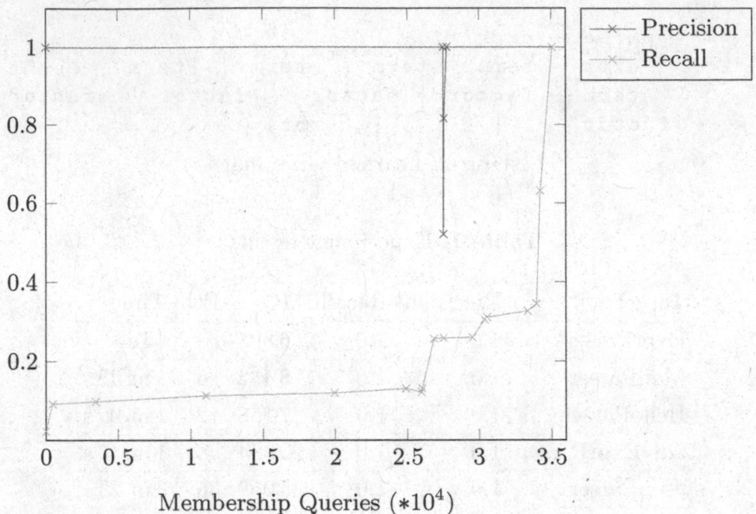

Fig. 5. Json Learning process

generate a parser p that accepts our target grammar G_p using the compiler-compiler Coco/R[1]. We apply our fuzzing-based grammar inference algorithm on this setup. If a specified maximum amount of allowed membership queries has been reached without finding a counterexample, execution is stopped and the current state of the inferred grammar G'_p is returned. Finally, we calculate $precision(L(G'_p), L(G_p))$ and $recall(L(G'_p), L(G_p))$ by randomly sampling 1 000 words each.

Table 1 displays the results, with the first column indicating the targeted parser. The second and third columns show the resulting precision and recall of our extracted grammar, followed by the total amount of unique membership-queries (MQ) performed, the total number of equivalence-queries (EQ) performed, and the time elapsed. In contrast to previous approaches that rely on a good number of representative seed inputs to perform appropriately (cf. Sect. 5), we have used sets I with no more than two seed inputs each. Our results in Table 1 show that we are able to recover a grammar that perfectly matches the target grammar for the 1 000 inputs examined. This shows that our approach is indeed robust to recover context-free grammars from recursive top-down parsers.

In the following we provide more details for the experiment "JsonParser". The target grammar is given in Listing 3. Listing 4 shows our learned grammar. Additionally Fig. 5 shows a more detailed analysis of the learning process for "JsonParser". We calculate $precision(L(G'_p), L(G_p))$ and $recall(L(G'_p), L(G_p))$ every time an equivalence query is performed, where G'_p is the current state of the inferred grammar. Again, we use 1 000 randomly sampled words each. As can be seen in Fig. 5, precision stays at 1 most of the time during the learning process, due to the fact that $L(G_s) \subseteq L(G_p)$ (see Sect. 3.1).

[1] https://ssw.jku.at/Research/Projects/Coco/.

Initial Input: “{‘a’. $1e - 0$,‘’ : [$true, true$]}”

```
      Json → Element.
   Element → Ws Value Ws.
        Ws → " ".
     Value → Object|Array|
             String|Number|
             "true"|"false"|"null".
    Object → "{" Ws [String Ws ":"
             Element["," Members]] "}".
   Members → Member ["," Members].
    Member → Ws String Ws ":" Element.
    Array  → "[" Ws [Value Ws
             [ "," Elements ]] "]".
  Elements → Element [ "," Elements ].
    String → "'" Characters "'".
Characters → ϵ|Character Characters.
 Character → "a"|"b"|"c".
    Number → Integer Fraction
             Exponent.
   Integer → ["−"] ("0"|
             Onenine [Digits]).
    Digits → Digit [Digits].
     Digit → "0"|Onenine.
   Onenine → "1"|"2"|"3".
  Fraction → ϵ|"." Digits.
  Exponent → ϵ|"E" Sign Digits|
             "e" Sign Digits.
      Sign → ϵ|"+"|"−".
```

Listing 3. Json Target Grammar

```
      Json → Element.
   Element → Ws Value Ws.
        Ws → " ".
     Value → Object|Array|
             String|Number|
             "true"|"false"|"null".
    Object → "{" (Ws "}"|Ws ("}"|
             String Ws ":" (Element
             "}"| Element ("}"|","
             Members "}")))).
   Members → Member|
             Member ("," Members|ϵ).
    Member → Ws String Ws ":" Element.
     Array → "[" (Ws "]"|Ws ("]"|
             Value (Ws "]"|Ws ("]"|
             "," Elements "]")))).
  Elements → Element|
             Element ("," Elements|ϵ).
    String → "'" Characters "'".
Characters → ϵ|Character Characters.
 Character → "a"|"b"|"c".
    Number → Integer Fraction
             Exponent.
   Integer → "0"|Onenine|
             Onenine (Digits|ϵ)|
             "−" (Onenine|Onenine
             (Digits|ϵ)|"0").
    Digits → Digit|Digit (Digits|ϵ).
     Digit → "0"|Onenine.
   Onenine → "1"|"2"|"3".
  Fraction → ϵ|"." Digits.
  Exponent → ϵ|"E" Sign Digits|
             "e" Sign Digits.
      Sign → ϵ|"+"|"−".
```

Listing 4. Json Learned Grammar

If only a portion of the desired language is accepted by the rule at the measurement point, precision remains at 1. Precision may gradually drop as you learn more rules over time. This could occur when attempting to identify the correct body of a rule, in particular when the rule accepts a superset of the wanted language. These inaccuracies are automatically fixed when a counterexample is found. For example the drop in precision in Fig. 5 occurs while learning a rule which consumes integers. The learned automaton accepts words containing preceding zeros as well as words containing more than one “-” at the beginning. Both are not accepted by the parser. As such, the precision of the learned grammar was lowered to 0.5. After the drop in precision, first the issue with multiple “-” symbols is fixed by providing a counterexample. This raises precision to 0.8. Finally, after providing another counterexample and consequently disallowing preceding zeros, the rule is learned correctly and precision increases back to 1.

In terms of recall, we see a consistent increase over time as the learnt grammar is expanded, and as a result, the learned language grows significantly. When a rule that consumes terminals is learned, the boost in recall is often greater. For example, the final spike in recall occurs while learning the rules for parsing digits and mathematical symbols.

We must remark that we have rarely used optimizations in our implementation, which leaves a lot of room for improvement. Possible performance improvements include (i) using hash-tables to store previously seen membership-queries instead of a plain-text list, (ii) replacing the early-parser used to determine whether a grammar produces a given word with something more efficient, (iii) using paralellization to speed up fuzzing, and (iv) to simultaneously learn the different rules of the seed grammar.

5 Related Work

Extracting context-free grammars for grammar-based fuzzing is not a new idea. Several methods exist for grammar learning which try to recover a context-free grammar by means of membership-queries from a black-box, such as by beginning with a modestly sized input language and then generalizing it to better fit a target language [2,15]. Another approach synthesizes a grammar-like structure during fuzzing [3]. However, this grammar-like structure has a few shortcomings, e.g., multiple nestings that are typical in real-world systems are not represented accurately [8]. Other methods use advanced learning techniques to derive the input language like neural networks [7] or Markov models [6]. Although black-box learning is generally promising, it suffers from inaccuracies and incompleteness of learned grammars. It is shown in [1] that learning a context-free language from a black box cannot be done in polynomial time. As a result, all pure black-box methods must give up part of the accuracy and precision of the learnt grammars.

Due to limitations with black-box approaches there exist several white-box methods to recover a grammar. If full access to the source code of a program is given, described methods fall under the category of grammar inference. Known methods for inferring a context-free language using program analysis include AUTOGRAM [10] and MIMID [8]. Unlike its predecessor AUTOGRAM, which relies on data flow, MIMID uses dynamic control flow to extract a human readable grammar. Finally, [11] describes how a grammar can be recovered using parse-trees of inputs, which is then improved with metrics-guided grammar refactoring. All of the aforementioned grammar inference methods share the same flaw: They all primarily rely on a predetermined set of inputs from which a grammar is derived that corresponds to this precise set of inputs. If some parts of a program are not covered by the initial set of inputs, the resulting grammar will also not cover these parts. However there exist some methods that attempt to automatically generate valid input for a given program, such as symbolic execution [14].

6 Conclusion

Our main contribution is a novel approach for grammar inference that combines machine learning, grammar-based fuzzing and program analysis. Our approach, in contrast to other efforts, reduces reliance on a good set of seed inputs while keeping other advantageous features of grammar inference techniques. This reduction in the original input set causes us to perform more membership queries since we need to uncover paths that we lose by randomly sampling the input set. We exchanged some of the benefits of complex program analysis, such as dynamic symbolic execution, for less complex program analysis, such as call tree extraction, to speed up the execution because our approach was designed with grammar-based fuzzing in mind. This was done in order to process such a vast number of inputs effectively. We can cease grammar inference and resume fuzzing the target program using the inferred grammar whenever we are certain that we have a good enough approximation of our input language. Despite the trade-offs outlined above, our preliminary findings show that we can still learn the target input language accurately in a reasonable amount of time, especially for more complicated input languages like JSON.

In the future, we aim to improve our learning technique by looking into ways to learn context-free grammars from any program without being restricted by recursive top-down parsers. Furthermore, we want to enhance the current implementation with a range of performance optimizations so that we may utilize it to uncover security problems in a real-world scenario.

References

1. Angluin, D., Kharitonov, M.: When won't membership queries help? J. Comput. Syst. Sci. **50**(2), 336–355 (1995)
2. Bastani, O., Sharma, R., Aiken, A., Liang, P.: Synthesizing program input grammars. In: PLDI, pp. 95–110. ACM (2017)
3. Blazytko, T., et al.: GRIMOIRE: synthesizing structure while fuzzing. In: USENIX Security Symposium, pp. 1985–2002. USENIX Association (2019)
4. Bollig, B., Habermehl, P., Kern, C., Leucker, M.: Angluin-style learning of NFA. In: IJCAI, pp. 1004–1009 (2009)
5. Gold, E.: Language identification in the limit. Inf. Control **10**(5), 447–474 (1967)
6. Gascon, H., Wressnegger, C., Yamaguchi, F., Arp, D., Rieck, K.: PULSAR: stateful black-box fuzzing of proprietary network protocols. In: Thuraisingham, B., Wang, X.F., Yegneswaran, V. (eds.) SecureComm 2015. LNICST, vol. 164, pp. 330–347. Springer, Cham (2015). https://doi.org/10.1007/978-3-319-28865-9_18
7. Godefroid, P., Peleg, H., Singh, R.: Learn& fuzz: machine learning for input fuzzing. In: ASE, pp. 50–59. IEEE Computer Society (2017)
8. Gopinath, R., Mathis, B., Zeller, A.: Inferring input grammars from dynamic control flow. CoRR abs/1912.05937 (2019)
9. Hopcroft, J.E., Ullman, J.D.: Introduction to Automata Theory. Languages and Computation. Addison-Wesley, Boston (1979)
10. Höschele, M., Zeller, A.: Mining input grammars with AUTOGRAM. In: ICSE (Companion Volume), pp. 31–34. IEEE Computer Society (2017)
11. Kraft, N., Duffy, E., Malloy, B.: Grammar recovery from parse trees and metrics-guided grammar refactoring. IEEE Trans. Softw. Eng. **35**(6), 780–794 (2009)
12. Mathis, B., Gopinath, R., Mera, M., Kampmann, A., Höschele, M., Zeller, A.: Parser-directed fuzzing. In: PLDI, pp. 548–560. ACM (2019)

13. Moser, M., Pichler, J.: eknows: platform for multi-language reverse engineering and documentation generation. In: 2021 IEEE International Conference on Software Maintenance and Evolution (ICSME), pp. 559–568 (2021)
14. Moser, M., Pichler, J., Pointner, A.: Towards attribute grammar mining by symbolic execution. In: SANER, pp. 811–815. IEEE (2022)
15. Wu, Z., et al.: REINAM: reinforcement learning for input-grammar inference. In: ESEC/SIGSOFT FSE, pp. 488–498. ACM (2019)

Natural Language Processing

In the Identification of Arabic Dialects: A Loss Function Ensemble Learning Based-Approach

Salma Jamal[1(✉)], Salma Khaled[1], Aly M. Kassem[2], Ayaalla Eltabey[1], Alaa Osama[1], Samah Mohamed[1], and Mustafa A. Elattar[1]

[1] School of Information Technology and Computer Science, Nile University, Giza, Egypt
sagamal@nu.edu.eg

[2] School of Computer Science, University of Windsor, Windsor, Canada

Abstract. The automation of a system to accurately identify Arabic dialects many natural language processing tasks, including sentiment analysis, medical chatbots, Arabic speech recognition, machine translation, etc., will greatly benefit because it's useful to understand the text's dialect before performing different tasks to it. Different Arabic-speaking nations have adopted various dialects and writing systems. Most of the Arab countries understand modern standard Arabic (MSA), which is the native language of all other Arabic dialects. In this paper we propose a method for identifying Arabic dialects Using the Arabic Online Commentary dataset (AOC), which includes three Arabic dialects-Gulf, Levantine, and egyptian-alongside MSA. Our approach includes two ensemble learning strategies using two BERT-based models and different loss functions such as focal loss, dice loss, and weighted cross-entropy loss. The first strategy is between the two proposed models using the loss function that performed best on the models, and the other is between the same model but using different loss functions, which resulted in 83.3%, 80.1%, 85.8%, 81.45%, Precision, Recall, Accuracy and Macro-F1 on the test set respectively.

Keywords: Arabic dialect identification · Imbalanced dataset · Ensemble loss functions

1 Introduction

More than 2 billion people use Arabic as their liturgical language, making it the sixth or seventh most widely spoken language in the world. It is a member of the Semitic language family. One of the hardest languages to learn is typically Arabic. First and foremost, Arabic has a 28-symbol alphabet that can change meaning depending on where it appears in a word. Additionally, Arabic is read from right to left, which is completely counter-intuitive to how most westerners read. The letters and the diacritics, which alter the sound values of the letters to which they

P. Fournier-Viger et al. (Eds.): MEDI 2022, LNAI 13761, pp. 89–101, 2023.
https://doi.org/10.1007/978-3-031-21595-7_7

are attached, make it a two-dimensional language. Additionally, a word's meaning might vary depending on its diacritics. All printed materials are written in modern standard Arabic (MSA), which is the official language of all Arab nations. However, each Arabic-speaking country learns a particular dialect of Arabic as its native tongue, which is defined as the linguistic traits of a particular community based on the region, such as Egyptian, Maghrebi, Gulf, etc. Many applications, including sentiment analysis, Arabic speech recognition, E-health chatbots, Machine Translation, etc., can benefit if we can automate a system to detect Arabic dialects accurately. Due to the task's significance in many domains, researchers have become interested in working on it. However, it is challenging as it's more difficult than simply understanding a particular language.

In this paper, we propose a method for automatically identifying Arabic dialects. We used the Arabic Online Commentary Dataset (AOC) [25], which was produced by gathering a significant amount of reader comments on the articles from three Arabic online newspapers. Alongside MSA, it includes three Arabic dialects Gulf, Levantine, and Egyptian, which are representative of the nations where the three newspapers are published. Given that MSA is the most prevalent label in the dataset this produces a class-imbalance problem, and we focused on finding a solution to this problem in our method. We have two alternative ensemble learning strategies in our pipeline, but before we get there, we will discuss the model details.

We employed Bert-based models [6], which are transformer-based models [15] that only use the encoder part of the transformer as opposed to the original transformer's encoder and decoder architecture. The Transformers attention mechanism, which finds contextual relationships between words (or subwords) in a text, is used by the model which shows significant improvements in the results. Additionally, the state-of-the-art results demonstrated that the language model can understand the context and flow of language more deeply and perform better on numerous NLP tasks.

As previously stated, our dataset had a class-imbalance problem that needed to be addressed, thus following the literature [19] instead of utilizing the standard Cross-Entropy loss, which is inappropriate for dealing with this kind of datasets, we used various loss functions. We fine-tuned the MARBERT-V2 [2] pre-trained BERT model using the weighted cross-entropy loss, Focal loss, and Dice loss on our dataset because they have been shown to improve the results in state-of-the-art research when working with an imbalanced dataset [19].

The findings demonstrated that some losses increased precision but not recall, and vice versa. Therefore, we used ensemble techniques between the same model but with different loss functions which was proposed in [12] as the initial ensemble learning strategy in our pipeline, and it enhanced the results. Then, we applied the same methodology to AraBERTv0.2-Twitter [5], a different Arabic pre-trained BERT based-model, and the outcome was the same: employing an ensemble of different loss functions with the same model enhances the results. Furthermore, instead of applying an ensemble with the same model and different loss functions, the second ensemble strategy in our pipeline was to use ensemble techniques between two different models with the loss function that yielded the best results on that model and it enhanced the model performance.

The rest of the paper is organized as follows. A review of earlier Arabic dialect identification literature is in Sect. 2. The proposed dataset is described in Sect. 3. The methods and pipeline are in Sect. 4. The results and evaluation are discussed in Sect. 5. Finally, we conclude in Sect. 6.

2 Related Work

A lot of work has gone into developing a method for reliably identifying Arabic dialects, although it is more difficult than just identifying a particular language because Arabic Dialects shares a lot of vocabulary. In recent years, the problem received a lot of attention.

Abdelali et al. [1] discovered that a significant source of errors in their method was caused by the naturally occurring overlap between dialects from nearby countries. They created a dataset of 540 k tweets that contained a significant amount of dialectical Arabic tweets by applying filtering techniques to remove tweets that are primarily written in MSA. Employing two models, a SVM classifier and a Transformer model (mBERT and AraBERT), their strategy achieved a macro-averaged F1-score of 60.6%.

A survey on deep learning techniques for processing Arabic dialectal data as well as an overview of the identification of Arabic dialects in text and speech was conducted by Shoufan et al. [23].

Salameh et al. [21] proposed a method for classifying Arabic dialects using a dataset that included 25 Arabic dialects from certain Arab cities, in addition to Modern Standard Arabic. They experimented several Multinomial Naive Bayes (MNB) models, and their strategy was able to achieve an accuracy of 67.9% for sentences with an average length of 7 words.

Malmasi et al. [17] proposed a method to identify a set of four regional Arabic dialects (Egyptian, Gulf, Levantine, and North African) and Modern Standard Arabic (MSA) in a transcribed speech corpus that has a total of 7,619 sentences in the training set. They achieved an F1-score of 51.0% by employing and ensemble learning technique between different SVM models but with different feature types.

Using six different deep learning techniques, including Convolution LSTM and attention-based bidirectional recurrent neural networks, Elaraby et al. [10] benchmarked the AOC dataset [25]. They tested the models in a variety of scenarios, including binary and multi-way classification. Using different embeddings, they attained accuracy of 87.65% on the binary task (MSA vs. dialects), 87.4% on the multi-classification task (Egyptian vs. Gulf vs. Levantine), and 82.45% on the 4-way task.

As a solution to the VarDial Evaluation Campaign's Arabic Dialect Identification task, Mohamed Ali [4] proposed three different character-level convolutional neural network models with the same architecture aside from the first layer to solve the task. MSA, Egyptian, Gulf, Levantine, and North African dialects are among the five included in the dataset. The first model used a one-hot character representation, the second model used an embedding layer before the convolution layer, and the third model used a recurrent layer before the convolution layer, which produced the best results 57.6% F1-score.

Obeid et al. proposed ADIDA [20], an automated method for identifying Arabic dialects. The algorithm outputs a vector of probabilities showing the possibility that a sentence entered is from one of 25 dialects and MSA. They employed the Multinomial Naive Bayes (MNB) classifier that Salameh et al. [21] proposed using the MADAR corpus. They achieved a 67.9% accuracy.

The Nuanced Arabic Dialect Identification (NADI) 2020 [3] shared task, which was divided into two sub-tasks of country-level identification and province-level identification, was the subject of a pipeline proposed by El Mekki et al. [9]. For sub-task one, their pipeline consisted of a voting ensemble learning approach with two models: the first model is AraBERT [5] with a softmax classifier, and the second model is TF-IDF with word and character n-grams to represent tweets. They achieved a 25.99% F1-score for the first sub-task, placing them second. They also applied a hierarchical classification strategy for the second sub-task, fine-tuning Arabert for each country to forecast its provinces after applying the country-level identification, and they were ranked first with an F1-score of 6.39%.

Because the self attention mechanism in the transformer models [24] captures the long range dependencies, Lin et al. [15] assumed that using a transformed-based Arabic Dialect identification system will improve the result rather than if we used CNN-based system. However, they used the self attention mechanism with down-sampling to reduce the computational complexity. They evaluated their technique on the ADI17 dataset [22], which performed better than CNN-based models with an accuracy of 86.29%.

Issa et al. [13], proposed a solution to the country-level Identification sub-task in the second Nuanced Arabic Dialect Identification (NADI) [3]. They applied two models and assessed the results. A pre-trained CBOW word embedding was utilised with an LSTM as the first model, while linguistic features were used in the second model as low-dimensional feature embeddings that were fed via a simple feed-forward network. Their F1-scores were 22.10% for the first model and 18.60% for the second, demonstrating that the use of language features did not improve the performance.

The results of this study show that most studies relied on the standard weighted cross-entropy loss rather than conducting further research to solve the issue of data imbalance that most Arabic datasets suffer. They also concentrated on using each dialect's unique linguistic characteristics in their pipeline. Following the literature [19], in order to more effectively handle the issue of data imbalance, the goal of this research is to overcome prior constraints by employing and evaluating different loss functions on the Arabic dialects identification task.

3 Dataset

We used a subset of the Arabic Online Commentary dataset (AOC) [25]. As stated earlier, it was generated by gathering comments from three Arabic news papers publications, each of which represented one of the dialects as follows:

- Al-Ghad → leventine.
- Al-Riyadh → Gulf.
- Al-Youm Al-Sabe' → Egyptian.

The dataset was annotated using an annotation interface where random phrases were displayed to the annotator, who had to make two statements about it: first, its dialectal content, such as whether it is mostly dialect, does not have any dialect, or is it mixed; and second, its dialect type, such as Leventine, Egyptian, Gulf, or MSA. As expected, each newspaper contained more comments in the local dialect of the nation in which it was published; yet, MSA predominated in all of them, creating a problem of class imbalance that we concentrated on resolving in our methods. The dataset contains 108,173 K comments. Figure 1 illustrates the dataset labels distribution.

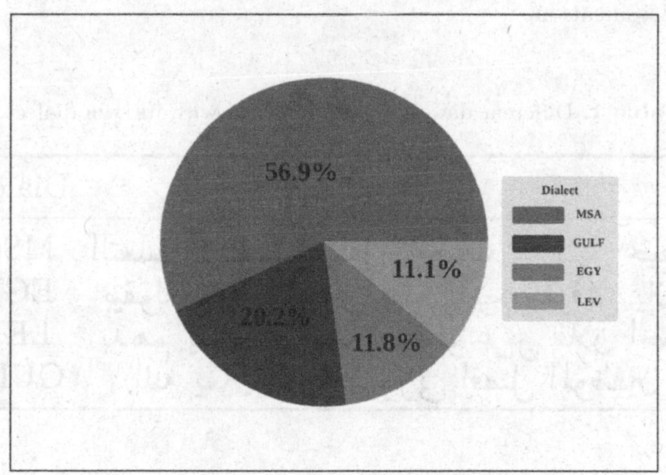

Fig. 1. Distribution of the dataset labels

4 Methods

In this section, we'll go over the steps and techniques for the proposed method, covering everything from data prepossessing and cleaning to the pre-trained models overviews and ensemble learning techniques to the prediction output.

4.1 Data Preprocessing

Following is a summary of the steps we took in this stage, which was to analyse the text to make sure it was properly formatted and devoid of unnecessary characters. Sample texts from the dataset are shown in Table 1 along with their dialect.

- Arabic Stopwords were eliminated in order to draw attention to the text's most crucial information.
- In order to normalize the data, we substituted:
  - أ آ إ → ا
  - ى ئ → ي
  - ة → ه
  - گ → ك
  - ؤ → و
- We eliminated the punctuation and numerals because they're not useful for dialect classification.

Table 1. Different dataset examples, each with its own dialect.

Text	Dialect
التعليقات السابقه اقل توصف بها انها سخيفه	MSA
بيقولك مره صعيدي طلع الهرم معرفش ينزل	EGY
بدهم يتنافسو عشان يقولو عرس فلان احلي	LEV
الله يعين عاطلين ورى افضل الموظفين	GULF

4.2 Pre-trained Models

In this section, we will discuss the Transformer-based (BERT) pre-trained models that we used to tackle the problem. These models showed promising results when tested against the state-of-the-art literature for a number of reasons. First, BERT learns the contextual relationships between the words in a phrase (Transformer attention mechanism). In addition, the model reads the full sequence of text at once, enabling it to understand the context of a word based on all of its surroundings. This property of the model makes it a bidirectional or nondirectional model. Two pre-trained BERT models were applied:

MARBERT-V2. One of the three models proposed in [2], but with a longer sequence length of 512 tokens. These Transformers Language models were pre-trained on 1 billion Arabic tweets, which were created by randomly selecting tweets from a large dataset of approximately 6 billion tweets made up of 15.6 billion tokens, in order to improve transfer learning on most Arabic dialects alongside MSA. The model was trained without the next sentence prediction components but has the same network architecture as the BERT Base (masked language model).

AraBERTv0.2-Twitter. An extension of AraBERT [5] which is another Arabic BERT language model that was trained on a combination of manually scraped articles from Arabic news website, two large Arabic corpora-the 1.5 billion words Arabic Corpus [8] with articles from 8 different countries and the 3.5 million articles from the Open Source International Arabic News Corpus [26] with news sources from 24 Arab countries. Additionally, AraBERTv0.2-Twitter is a refined version of AraBERT that was pre-trained on 60 million Arabic Tweets with a maximum word length of 64 using the masked language model method. The model's performance was assessed using three Arabic natural language understating tasks: sentiment analysis, named entity recognition, and question answering, the results were promising.

4.3 Loss Functions

As we previously mentioned, the proposed dataset has a class imbalance problem, following the literature [19], employing different loss functions instead of using the standard cross-entropy loss helps improves the model performance. Three different loss functions were employed:

Weighted Cross-Entropy loss (WCE): The standard cross-entropy loss inherits bias toward the majority class when making predictions, and the model becomes more confident in predicting the majority class. In addition, it fails to pay more attention to hard examples: the ones where the model repeatedly makes significant errors on, because it cannot differentiate between the hard and easy examples. Consequently, we used the Weighted cross-entropy loss to try to overcome those issues. Although it can handle the issue of class imbalance, it is unable to differentiate between easy and hard examples.

Focal Loss (FL): Is proposed in [14]. it can deal with dataset class imbalance, by applying a modulating term (scaling factor) to the cross entropy loss $(1-p_t)^{\gamma}$, which decays to zero as confidence in the correct class increases, it focuses on the examples that the model gets wrong (hard examples) rather than the ones that are easily classified. In the case of the misclassified examples, the pi is small, making the modulating factor approximately or very close to 1, the loss function is therefore unaffected, acting as a Cross-Entropy loss. Additionally, this automatically down-weights the importance of easy samples during training and

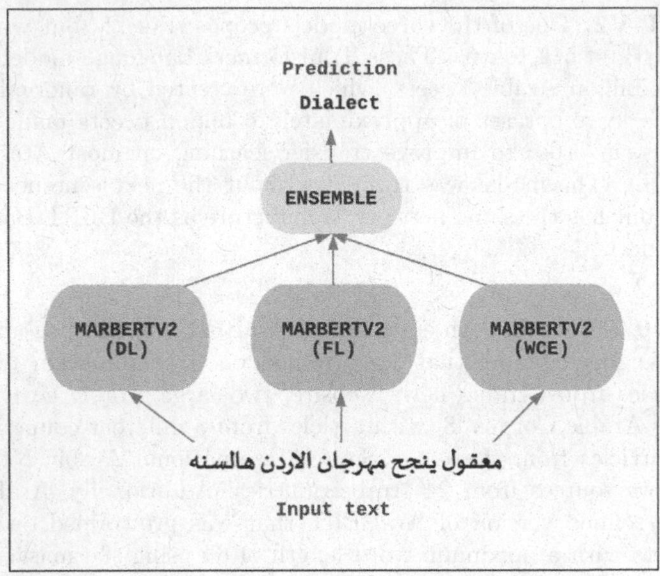

Fig. 2. The proposed ensemble learning model

focuses the model on difficult ones. We used the focal loss with label smoothing parameter to address the issue of overconfidence.

Dice Loss (DL): The Dice coefficient [7], also known as the harmonic mean of sensitivity and precision, is used to balance the two because they have different denominators and true positives in the numerator. In addition, the denominator of the dice loss equation is changed to its square form to speed up convergence as proposed in [18].

4.4 Proposed Ensemble Learning Model

Ensemble learning techniques enhance the performance of models, due to its ability to achieve better outcomes than any single contributing model. Additionally, it improves the robustness of the model by decreasing the spread or dispersion of the results and predictions. Combining the prediction of different models improves the outcome as the models that contribute to the ensemble make different errors on the samples. Utilizing the same model with a different loss functions was one of the two ensemble learning strategies we employed in our pipeline. The other strategy involved using different models with the loss function that achieved the best results on that model. The final proposed model illustrated in Fig. 2 was an ensemble between the same BERT-based model MARBERT-V2 but with different loss functions Focal loss (FL), Dice loss (DL) and weighted Cross-Entropy loss (WCE).

5 Results and Discussion

This section will outline all the experiments we conducted employing the proposed pre-trained models, different loss functions, and ensemble methodologies.

5.1 Experimental Results

Different Pre-trained Models. We fine-tuned the proposed BERT-based models with each of the different loss functions we previously described to see which one would perform best. As indicated in Table 2, we noticed that each loss function may yield a good recall but a poor precision, and vice versa.

Table 2. Results of fine-tuning the models using different loss functions

Model	Loss function	PR–RR	ACC–F1
MARBERT-V2	WCE	80.2–81.0	83.7–79.8
	FL	**83.9–78.8**	**85.7–81.0**
	DL	82.5–78.2	83.4–80.1
AraBERTv0.2-T	WCE	79.4–78.7	84.2–78.4
	FL	**82.1–77.6**	**85.8–79.4**
	DL	80.7–75.5	84.4–77.7

Ensemble Learning Methodologies. In an effort to improve the model performance on the proposed dataset as shown in Table 3, we applied two ensemble learning strategies as follows:

I- Different Loss Functions with the Same Model: As previously mentioned, the same model can achieve better results in recall but not in precision, and vice versa using different loss functions. Therefore, following the literature [12] we first tried an ensemble of MARBERT-V2 with the different proposed loss functions, weighted Cross-Entropy, Focal loss, and Dice loss, and it achieved the best results in our pipeline. Additionally, we tried the same strategy with AraBERTv02-twitter to test that it truly works, and the results were the same.

II- Different Models: The second ensemble learning strategy was between the two proposed pre-trained BERT models, MARBERT-V2 and AraBERTv02-twitter, the Focal loss was used as the loss function as it achieved the best results on both models.

Table 3. Results of applying the two ensemble learning methodologies which shows that ensemble learning using MARBERT-V2 model with different loss function achieved the highest MACRO-F1 score.

Model	Loss function	PR–RR	ACC–F1
Ensemble (MARBERT-V2)	**FL, WCE, DL**	**83.3–80.1**	**85.8–81.45**
Ensemble (AraBERTv0.2-T)	FL, WCE, DL	81.9–78.5	84.5–79.80
Ensemble(MARBERT-V2, AraBERTv0.2-T)	FL	84.4–79.1	85.7–81.30

5.2 Discussion

In this section will discuss the results of implementing different models, loss functions, performance metrics, ensemble learning strategies and the result of the proposed model compared to the previous models.

Performance Metrics: Different metrics, such as Precision and Recall, were utilised to assess the performance of the pipeline; however, since Accuracy is is an inappropriate metric to utilise given the imbalance class-distribution in the dataset, we focused on the Macro-F1 score.

Loss Function: In both models, Focal Loss outperformed the other loss functions in the Macro-F1 score, achieving 81.0% with MARABERTV2 and 79.4% with AraBERTv02-t. This is due to the fact that Focal Loss down-weights the easy examples so that, despite their huge number, their contribution to the overall loss is minimal, hence addressing the issue of class imbalance better. As shown in Table 4, the proposed ensemble learning model combined with different loss functions such as Focal-loss and Dice-Loss outperformed the previous models with a significant margin even with including the four dialects.

BERT-Based Models: MARBERT-V2 surpassed AraBERTv02-t in terms of Macro-F1 score when fine-tuned on the dataset with all of the proposed loss functions, possibly because the model pre-training data was similar to the proposed dataset as well as the fact that the model was pre-trained on a longer sequence length than AraBERTv02-t and the proposed dataset has a long sequence length thus, allowing it to capture the meaningful contextual meanings.

Ensemble Learning Strategies: Employing ensemble learning technique between the two proposed pre-trained models enhanced the results, because each model was pre-trained on a different dataset, it may have produced different embedding for the same text and, consequently, different error. Furthermore, using ensemble learning with the same model but with different loss functions improved the performance because each loss function learns differently and generates errors differently. As a result, using ensemble learning will improve the average prediction performance as well as, it will lower the variance element of the prediction errors generated by the contributing models of the ensemble.

Table 4. Results Of Different Systems On AOC Dataset

System	Loss function	Included dialectics	Accuracy
Elaraby and Abdul-Mageed [10]	N/A	MSA, EGY, LEV, GLF	82.45%
Zaidan and Callison-Burch [25]	N/A	MSA, EGY, LEV, GLF	69.40%
Elfardy and Diab [11]	N/A	MSA, EGY	85.50%
Lulu and Elnagar [16]	N/A	EGY, LEV, GLF	71.40%
Proposed Ensemble Model	FL,WCE,DL	MSA, EGY, LEV, GLF	85.80%

6 Conclusion and Future Works

In this paper, we proposed a method to automatically identify Arabic dialects using the Arabic online commentary dataset (AOC). Instead of utilizing the standard Cross-Entropy loss, we used different loss functions to address the problem of data imbalance as proposed in [19]. Additionally, we used two ensemble learning strategies to enhance the model performance. Our final proposed ensemble BERT-based model achieved 83.3%, 80.1%, 85.8%, 81.45%, Precision, Recall, Accuracy and Macro-F1 on the test set respectively. In the future work, we intend to evaluate the effectiveness of our pipeline using different datasets as well as the Country-level dialect identification problem as it is more complex.

Acknowledgments. This research is supported by the Vector Scholarship in Artificial Intelligence, provided through the Vector Institute.

References

1. Abdelali, A., Mubarak, H., Samih, Y., Hassan, S., Darwish, K.: QADI: arabic dialect identification in the wild. In: Proceedings of the Sixth Arabic Natural Language Processing Workshop, pp. 1–10 (2021)
2. Abdul-Mageed, M., Elmadany, A., Nagoudi, E.M.B.: Arbert & marbert: deep bidirectional transformers for arabic. arXiv preprint arXiv:2101.01785 (2020)
3. Abdul-Mageed, M., Zhang, C., Bouamor, H., Habash, N.: Nadi 2020: The first nuanced Arabic dialect identification shared task. In: Proceedings of the Fifth Arabic Natural Language Processing Workshop, pp. 97–110 (2020)
4. Ali, M.: Character level convolutional neural network for Arabic dialect identification. In: Proceedings of the Fifth Workshop on NLP for Similar Languages, Varieties and Dialects (VarDial 2018), pp. 122–127 (2018)
5. Antoun, W., Baly, F., Hajj, H.: Arabert: transformer-based model for Arabic language understanding. arXiv preprint arXiv:2003.00104 (2020)
6. Devlin, J., Chang, M.W., Lee, K., Toutanova, K.: BERT: pre-training of deep bidirectional transformers for language understanding. arXiv preprint arXiv:1810.04805 (2018)
7. Dice, L.R.: Measures of the amount of ecologic association between species. Ecology **26**(3), 297–302 (1945)
8. El-Khair, I.A.: 1.5 billion words Arabic corpus. arXiv preprint arXiv:1611.04033 (2016)

9. El Mekki, A., Alami, A., Alami, H., Khoumsi, A., Berrada, I.: Weighted combination of BERT and n-gram features for nuanced Arabic dialect identification. In: Proceedings of the Fifth Arabic Natural Language Processing Workshop, pp. 268–274 (2020)
10. Elaraby, M., Abdul-Mageed, M.: Deep models for Arabic dialect identification on benchmarked data. In: Proceedings of the Fifth Workshop on NLP for Similar Languages, Varieties and Dialects (VarDial 2018), pp. 263–274 (2018)
11. Elfardy, H., Diab, M.: Sentence level dialect identification in Arabic. In: Proceedings of the 51st Annual Meeting of the Association for Computational Linguistics (Volume 2: Short Papers), pp. 456–461 (2013)
12. Hajiabadi, H., Molla-Aliod, D., Monsefi, R., Yazdi, H.S.: Combination of loss functions for deep text classification. Int. J. Mach. Learn. Cybern. **11**(4), 751–761 (2020)
13. Issa, E., AlShakhori, M., Al-Bahrani, R., Hahn-Powell, G.: Country-level Arabic dialect identification using RNNs with and without linguistic features. In: Proceedings of the Sixth Arabic Natural Language Processing Workshop, pp. 276–281 (2021)
14. Lin, T.Y., Goyal, P., Girshick, R., He, K., Dollár, P.: Focal loss for dense object detection. In: Proceedings of the IEEE International Conference on Computer Vision, pp. 2980–2988 (2017)
15. Lin, W., Madhavi, M., Das, R.K., Li, H.: Transformer-based Arabic dialect identification. In: 2020 International Conference on Asian Language Processing (IALP), pp. 192–196. IEEE (2020)
16. Lulu, L., Elnagar, A.: Automatic Arabic dialect classification using deep learning models. Proc. Comput. Sci. **142**, 262–269 (2018)
17. Malmasi, S., Zampieri, M.: Arabic dialect identification in speech transcripts. In: Proceedings of the Third Workshop on NLP for Similar Languages, Varieties and Dialects (VarDial3), pp. 106–113 (2016)
18. Milletari, F., Navab, N., Ahmadi, S.A.: V-net: fully convolutional neural networks for volumetric medical image segmentation. In: 2016 Fourth International Conference on 3D Vision (3DV), pp. 565–571. IEEE (2016)
19. Mostafa, A., Mohamed, O., Ashraf, A.: GOF at Arabic hate speech 2022: breaking the loss function convention for data-imbalanced Arabic offensive text detection. In: Proceedings of the 5th Workshop on Open-Source Arabic Corpora and Processing Tools with Shared Tasks on Qur'an QA and Fine-Grained Hate Speech Detection, pp. 167–175. European Language Resources Association, Marseille, France, June 2022. http://www.lrec-conf.org/proceedings/lrec2022/workshops/OSACT/pdf/2022.osact-1.21.pdf
20. Obeid, O., Salameh, M., Bouamor, H., Habash, N.: ADIDA: automatic dialect identification for Arabic. In: Proceedings of the 2019 Conference of the North American Chapter of the Association for Computational Linguistics (Demonstrations), pp. 6–11 (2019)
21. Salameh, M., Bouamor, H., Habash, N.: Fine-grained Arabic dialect identification. In: Proceedings of the 27th International Conference on Computational Linguistics, pp. 1332–1344 (2018)
22. Shon, S., Ali, A., Samih, Y., Mubarak, H., Glass, J.: Adi17: a fine-grained Arabic dialect identification dataset. In: IEEE International Conference on Acoustics, Speech and Signal Processing (ICASSP), pp. 8244–8248 (2020)
23. Shoufan, A., Alameri, S.: Natural language processing for dialectical Arabic: a survey. In: Proceedings of the Second Workshop on Arabic Natural Language Processing, pp. 36–48 (2015)

24. Vaswani, A., et al.: Attention is all you need. In: Advances in Neural Information Processing Systems, vol. 30 (2017)
25. Zaidan, O.F., Callison-Burch, C.: Arabic dialect identification. Comput. Linguist. **40**(1), 171–202 (2014)
26. Zeroual, I., Goldhahn, D., Eckart, T., Lakhouaja, A.: OSIAN: open source international Arabic news corpus-preparation and integration into the Clarin-infrastructure. In: Proceedings of the Fourth Arabic Natural Language Processing Workshop, pp. 175–182 (2019)

Emotion Recognition System for Arabic Speech: Case Study Egyptian Accent

Mai El Seknedy[1(✉)] and Sahar Ali Fawzi[1,2]

[1] Nile University, Giza, Egypt
{Mai.Magdy,sfawzi}@nu.edu.eg
[2] Faculty of Engineering, Cairo University, Giza, Egypt
saharfawzi@eng1.cu.edu.eg

Abstract. Speech Emotion Recognition (SER) systems are widely regarded as essential human-computer interface applications. Extracting emotional content from voice signals enhances the communication between humans and machines. Despite the rapid advancement of Speech Emotion Recognition systems for several languages, there is still a gap in SER research for the Arabic language. The goal of this research is to build an Arabic-based SER system using a feature set that has both high performance and low computational cost. Two novel feature sets were created using a mix of spectral and prosodic features, which were evaluated on the Arabic corpus (EYASE) constructed from a drama series. EYASE is the Egyptian Arabic Semi-natural Emotion speech dataset that consists of 579 utterances representing happy, sad, angry, and neutral emotions, uttered by 3 male and 3 female professional actors. To verify the emotions' recognition results, surveys were conducted by Arabic and non-Arabic speakers to analyze the dataset constituents. The survey results show that recognition of anger, sadness, and happiness are sometimes misclassified as neutral. Machine learning classifiers Multi-Layer Perceptron, Support Vector Machine, Random Forest, Logistic Regression, and Ensemble learning were applied. For valence (happy/angry) emotions classification, Ensemble learning showed best results of 87.59% using the 2 proposed feature sets. Featureset-2 had the highest recognition accuracy with all classifiers. For multi-emotions classification, Support Vector Machine had the highest recognition accuracy of 64% using featureset-2 and benchmarked Interspeech feature sets. The computational cost of featureset-2 was the lowest for all classifiers, either for training or testing.

Keywords: Speech emotion recognition · Arabic corpus · Mel frequency cepstral coefficients · Prosodic features · Mel-spectrogram features

1 Introduction

Emotion expression is an important component of human communication as it helps in transferring feelings and offering feedback. Recently, high interest in speech emotion recognition systems (SER) evolved. Speech emotion recognition systems attempt to detect desired emotions using voice signals regardless of semantic content [1]. Advances

P. Fournier-Viger et al. (Eds.): MEDI 2022, LNAI 13761, pp. 102–115, 2023.
https://doi.org/10.1007/978-3-031-21595-7_8

in artificial intelligence technology have an impact on human-machine interacting applications as SER systems. Nowadays, SER is taking a significant role in our digital world as being adopted in many applications like call centers [2], computer gaming [3], online e-learning [4], autonomous vehicles, police criminal investigations, and medical diagnosis as in psychological diseases analysis [5]. SER proved to have a high impact to the medical sector as represented by the authors of research [6]. They proved a psychophysiological alternative mode for analyzing emotions through speech using SER systems for patients suffering from Autism Spectrum Disorder (ASD) patients who usually suffer from defects in facial expressions.

The number of speech emotion databases is growing for different languages as English, German, French, Italian Spanish, Urdu, Mandarin, Turkish, Japanese, Hindi, and Korean [7–10]. Although Arabic is the fifth most spoken language, with 274 million speakers worldwide, the available Arabic datasets with the different Arabic dialects are very limited [11]. An Arabic language semi-natural emotion speech dataset (EYASE) with an Egyptian dialect have been developed by Abdel-Hamid et al. [12].

The objective of this research is to develop SER model for identifying the Egyptian Arabic speaker's emotional state. The proposed SER model uses novel feature set combinations that provide high classification accuracies and low computational time. Different supervised machine learning models is included for classification.

This paper is organized as follows: a literature review with advances in SER is presented in Sect. 2, the methodology is introduced in Sect. 3 including the dataset used, features extraction, the machine learning models and the experimental setup, Sect. 4 presents the results of the experiments and finally, Sect. 5 introduces the conclusion and future work.

2 Literature Review

SER systems undergo enormous evolution over the past decade. The system methodologies depend on the dataset used. Dataset types can be acted, elicited and non-acted [7]. Furthermore, the datasets differ based on the speaker's gender, age [8].

Features used to describe the utterances are the key players in SER systems. The set of features related to the emotions description can greatly enhance the recognition success rate. Features of different domains are used in SER systems. Prosodic features which describe the intonation and rhythm of the speech signal such as pitch, intensity, and fundamental frequency F0 [7, 8, 13]. Spectral features represented by Mel spectrograms and Mel-frequency cepstral coefficients (MFCC) are widely used in SER [14]. Advanced techniques as deep neural networks depend on MFCCs and spectrogram images to train the system using CNN and LSTM [15]. Furthermore, there are linear prediction coefficients (LPC) based features and Voice Quality Features as Jitter and Shimmer. Selected sets of features for precise emotion recognition depends on the corpus used, the language and the classification algorithm [10].

INTERSPEECH 2009 Emotion Challenge feature set (IS09) [16] and INTERSPEECH 2010 Paralinguistic Challenge Paralinguistic Challenge feature set (IS10) [17] are considered as benchmark for many SER systems. A feature set IS09-10 generated from combining IS09 and IS10 features, was introduced by Klaylat et al. [10], results in improvement in some cases.

A wide variety of machine learning algorithms are used for classification in SER domain as Hidden Markov models (HMM), Gaussian Mixture Models (GMM), Support Vector Machine (SVM), tree-based models as Random Forest (RF), K-Nearest Neighbor (K-NN), Logistic regression (LR), and recently Artificial Neural Networks (ANN). Advantages and limitations of these algorithms are surveyed by El Ayadi et al. [8] and Koolagudi et al. [18]. Recent research focuses on ensemble learning and majority voting combining the advantages of different classifiers to create a model capable of enhancing the prediction results [19].

Artificial neural networks are being explored in SER. Literatures showed that deep neural networks need more optimization to be included in the speech emotion recognition field. No significant improvement was reported using the features sets as input to the deep neural networks [20].

3 Methodology

The SER system includes three basic building phases, as shown in Fig. 1. First phase focuses on the chosen Arabic dataset (EYASE). Phase two comprises the construction of the two proposed feature sets, features' normalization, and features selection analysis. The third phase focuses on classification, and it employs five machine learning models: MLP, SVM, Random Forest, Logistic Regression, and Ensemble Learning.

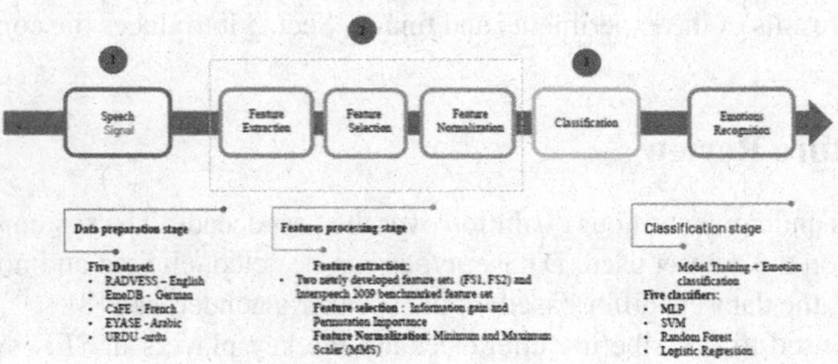

Fig. 1. The proposes SER system phases

3.1 Datasets

EYASE: Egyptian Arabic Semi-natural Emotion speech database is created from a drama series and consists of 3 male and 3 female aged 22–45 years old actors with 12 to 22 years of professional experience. It includes four basic emotions: angry, happy, neutral, and sad. A total of 579 wav files with a sampling rate of 44.1 kHz [12].

3.2 Features Extraction

Different features have been extracted and used in SER systems, but there is no generic or precise set of features that can be used for the best results [21]. Features' sets generally includes different prosodic and spectral features [7].

Two new features sets were developed based on previous work experiences. Featureset-1 consists of MFCCs, Mel-spectrogram, spectral contrast, root mean square, the tonal centroid features, chroma features using 2 statistical functions. An enhanced featureset-2 included more prosodic features as F0, zero-crossing rate and reduced MFCCs coefficients to the first 14 and 8 Mel spectrogram filters. The statistical functions were increased to 6 functions. Python library Librosa [22] was used to implement features extraction algorithms. Features sets details are displayed in Table 1. OpenSmile toolkit [23] was used to generate benchmarked features sets IS09, IS10 and IS09-10 to compare their performance on EYASE dataset.

Table 1. Shows the set of features used during our research, the statistical functions used, and the tools used in feature extraction.

Feature	Description	Statistical functions	Tool
Featureset-1 Total features: 194	40 MFCC, 128 Mel-spectrogram, 12 Chroma, Tonnetz, 8 Contrast, and RMS	Mean and standard deviation	Librosa (python library)
Featureset-2 Total features: 122	14 MFCC, 8 Mel-spectrogram, RMS, 12 Chroma, Tonnetz, 8 Contrast, Zero crossing rate, Fundamental frequency (F0), Pitch Contour, and Signal's low-frequency band mean energy (SLFME)	Min, max, standard deviation, mean, range, and percentile (25, 50, 75,90)	Librosa + pYAAPT (pitch tracking algorithm in python)
IS09 (INTERSPEECH 2009 Emotion Challenge feature set) **[79]** **Total features: 384**	RMS, 12 MFCC, ZCR, Voicing probability, and F0	Min, max, standard deviation, mean, range, maxPos, minPos, linregc1, linregc2, linregerrQ, skewness, and kurtosis	Opensmile tool

(continued)

Table 1. (*continued*)

Feature	Description	Statistical functions	Tool
IS10 (INTERSPEECH 2010 Paralinguistic Challenge feature set) **[80]** **Total features: 1582**	Pcm loudness, 14 MFCC, Log Mel Freq Band, lsp Freq, F0 Env, Voicing final unclipped, JitterLocal, JitterDDP, Shimmer local, and Shimmer DDP	Min, max, standard deviation, mean, range, maxPos, minPos, linregc1, linregc2, linregerrQ, skewness, kurtosis, quartiles (1–3), iqr (1–3), percentile, and pctlrange0–1	Opensmile tool
IS09-10 [9] **Total features: 816**	RMS, 14 MFCC, ZCR, Pcm loudness, Log Mel Freq Band, lsp Freq, F0 Env, Voicing final unclipped, JitterLocal, JitterDDP, Shimmer local, and Shimmer DDP	Min, max, standard deviation, mean, range, maxPos, minPos, linregc1, linregc2, linregerrQ, skewness, kurtosis, quartiles (1–3), iqr (1–3), percentile, and pctlrange0–1	Opensmile tool

3.3 Features Importance

Choosing the most suitable features to detect emotions from the speech signal is considered a key player in enhancing SER performance. Two algorithms were applied to rank features according to their impact on classification model used.

1. **Information Gain:** It uses entropy (randomness of data) to measure the information gained with each iteration or tree split of the machine learning model used [24].
2. **Permutation Importance:** It estimates and ranks feature importance based on the impact of each feature on the trained model predictions [25]. Table 2 represents the top 10 features for each classifier.

Table 2. Features importance for used models

Classifier	Features
MLP	Spectral contrast (1), Chroma (4), RMS(1), MFCC(2), F0(1), Pitch Contour (1)
SVM	Chroma (2), MFCC (3), Spectral contrast (3), RMS(2)
RF	Mel spectrogram (6), MFCC (2), RMS (1), Chroma (1)
LR	Chroma (4), Spectral contrast (1), MFCC (3), Pitch Contour (1), RMS(1)

Results show that, MFCC is the most dominant feature across all the classifiers. Mel-Spectrogram features are highest with random forest which supports Information gain results.

3.4 Feature Scaling

Machine learning algorithms have better performance when features are normalized. This reduces the effect of speakers' variabilities, different languages and recordings conditions on the recognition process. Normalization techniques includes Standard Scaler, Minimum and Maximum Scaler (MMS) and Maximum Absolute Scaler (MAS) [26, 27]. MMS is used in this work to normalize features to a 0-1 range applying Eq. 1. Both MMS and Standard scaler were applied and by comparing results, MMS showed better results.

$$X_{scaled} = (X - \min)/(\max - \min) \quad (1)$$

where X is the input features, min is the features' minimum value and max is the features' maximum value.

Shapiro-Wilk test was performed on featureset-2 to accept or reject the null hypothesis (H0) that the data had a normal distribution using the estimated p-value [10]. The majority of the 122 features rejected the H0 as the p-values were less than the accepted confidence 0.05. Histograms' visualization concluded that the data is more skewed than normally distributed as in Fig. 2.

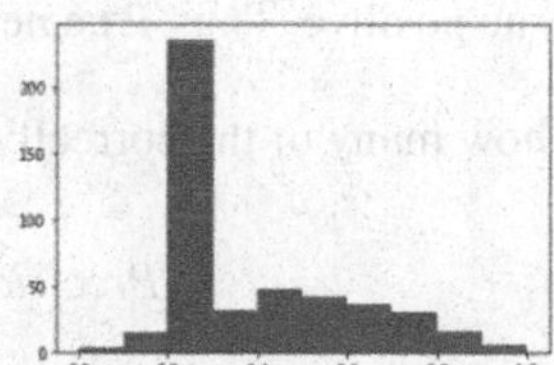

Fig. 2. Represents samples of the features histograms (ZCR and MFCC)

3.5 Machine Learning Models

Five classification techniques were considered based on their performance in previous literature. SVM proved to have good performance with multi-dimensional data and accurate results [7, 27, 28]. Random forest, Logistic regression, and Multi-Layer Perceptron (MLP) classifiers are widely used in SER systems [10, 27, 29]. Finally, Ensemble learning is performed integrating the four mentioned classifiers using the majority voting technique. GridSearchCV method was used to fine-tune classifier's parameters (Table 3).

Table 3. Hyper parameters tuning for classifiers

Model	Hyper parameters
Multi-layer perceptron	alpha = 0.0001, batch_size = 5, solver = 'adam', hidden_layer_sizes = (400), learning_rate = constant, max_iter = 300
SVM	kernel = 'rbf', C = 10, gamma = 0.001, decision_function_shape = 'ovr'
Random forest	criterion = 'entropy', n_estimators = 500,max_depth = 20
Logistic regression	C = 1.0, solver = 'lbfgs', penalty = 'l2', max_iter = 1000

3.6 Evaluation Metrics

10-fold cross-validation was applied to ensure statistical stability and generalization of the model. Where, in 10-fold cross-validation, the database is randomly partitioned into 10 equal size subsamples. Of the 10 subsamples, 1 subsample which is 10% of the database is considered as the testing data to validate the classification model, and the remaining 9 subsamples are used as training data. The reported accuracy is the average of the 10 folds tests. We used 4 evaluation metrics during our experiments. Accuracy: where it gives an overall measure of the percentage of correctly classified instances.

$$Accuracy = \frac{Tp + Tn}{Tp + Tn + Fp + Fn} \tag{2}$$

where, Tp is True positive, Tn is True negative, Fp is False positive, Fn is False negative

Precision: is how many of the correctly predicted emotional classes were positive.

$$Precision = \frac{Tp}{Tp + Fp} \tag{3}$$

Recall: is how many of the actual positive emotional classes were correctly predicted.

$$Recall = \frac{Tp}{Tp + Fn} \tag{4}$$

Confusion Matrix: is another way to analyze how many samples were miss-classified by the model by giving a comparison between actual and predicted labels.

4 Results and Discussions

This section presents the findings and discussions of the Arabic SER experiments. First, the results of the Arabic dataset analysis survey are introduced then the SER performance with different classifiers discussed. In addition, a comparison with the previous research conducted by Abdel-Hamid et al. [12].

4.1 Corpus Survey Analysis

To validate the recently released Arabic dataset AYASE, 2 surveys were conducted on a sample of utterances representing the different emotions. The participants of the first survey were 64 Arabic speakers while 20 non-Arabic speakers participated in the second one. Figure 3 shows the surveys' results which emphasizes the ambiguity of defining the happy emotion for most participants. Furthermore, eliminating the semantic sense of the utterances in second survey, led to identify 'happy' as neutral for all the sample utterances used and angry emotions were also misidentified as neutral.

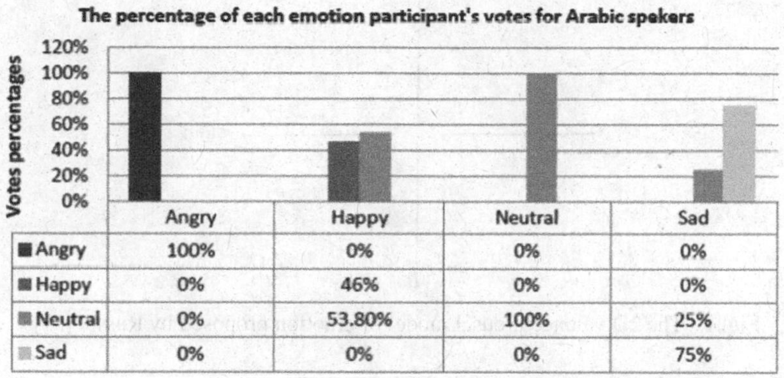

	Angry	Happy	Neutral	Sad
Angry	100%	0%	0%	0%
Happy	0%	46%	0%	0%
Neutral	0%	53.80%	100%	25%
Sad	0%	0%	0%	75%

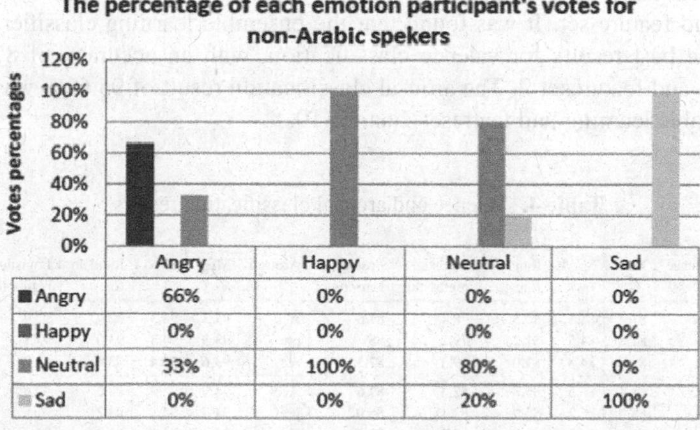

	Angry	Happy	Neutral	Sad
Angry	66%	0%	0%	0%
Happy	0%	0%	0%	0%
Neutral	33%	100%	80%	0%
Sad	0%	0%	20%	100%

Fig. 3. Survey Analysis results, the percentage of each emotion participant's votes for Arabic and non-Arabic speakers

4.2 Speech Emotion Recognition Experiments

The recognition performance was analyzed using MLP, SVM, Random Forest, Logistic Regression, and ensemble learning classifiers. Evaluation criteria were done using 10 folds to capture the generalization model performance as well as the accuracy, precision, and recall. Three experiments were performed: Valence-Arousal classification, Anger emotional classification, and Multi emotion classification.

1. **Valence-Arousal classification**

Valence and arousal are the two main dimensions defining emotions, as shown in Fig. 4 [30, 31].

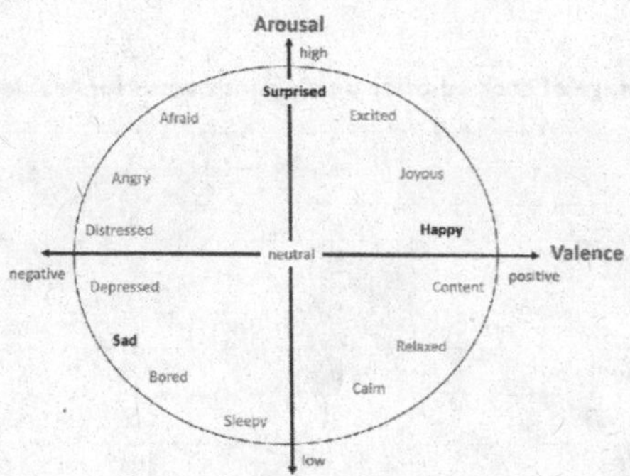

Fig. 4. The 2D valence-arousal model of emotion proposed by Russel [31]

The valence-arousal recognition results are represented in Table 4, showing each classifier and feature set. It was found that the ensemble learning classifier approach achieved the best results for valence classification, with an accuracy of 87.59% for featureset-1 and featureset-2. The arousal classification result of 95.62% was obtained using ensemble learning and featreset-2 and IS10.

Table 4. Valence and arousal classification results

Feature-set	Valence	MLP	SVM	Random forest	Logistic regression	Ensemble learning	Arousal	MLP	SVM	Random forest	Logistic regression	Ensemble learning
Featureset-1	Acc	81.3	84.1	80.5	78.4	**87.6**	Acc	93.3	**94.3**	92.9	92.9	93.9
	pre	81.8	84.5	81.2	79.2	**88**	pre	93.3	**94.3**	93.0	93.1	94.0
	rec	82.2	84.8	81.0	79.3	**87.5**	rec	93.6	**94.4**	93.4	93.2	94.2
Featureset-2	Acc	85.1	86.5	85.8	82.3	**87.6**	Acc	94.6	95.3	93.6	92.6	**95.6**
	Pre	85.0	86.2	85.9	82.1	**88.0**	pre	94.7	95.3	93.7	92.7	**95.8**
	rec	85.0	86.3	85.8	82.3	**87.5**	rec	94.4	95.2	93.7	92.6	**95.4**
IS10	Acc	79.1	**84.4**	79.8	82.2	83.7	Acc	95.0	95	93.3	95.6	**95.6**
	pre	79.2	**84.2**	80.0	81.8	83.8	pre	95.4	95.4	93.4	96	**96**
	rec	78.7	**84.7**	79.8	82.2	83.5	rec	94.7	94.7	93.0	95.4	**95.4**
IS-09	Acc	82.3	**83.7**	82.3	82.0	83.4	Acc	91.6	**92.3**	89.6	92.0	**92.2**
	Pre	82.3	83.3	82.4	81.6	83.0	pre	92.6	**92.6**	89.5	92.3	**92.8**
	rec	82.8	84.2	**91.8**	82.5	83.4	rec	92.3	**92.3**	89.5	91.8	**92.0**
IS-09-10	Acc	78.7	**82.3**	82.0	78.4	81.2	Acc	94.6	94	94.0	**95.3**	95.0
	Pre	79.0	**82.3**	81.6	78.6	81.5	pre	94.6	93.6	93.8	**95.1**	94.8
	rec	79.2	**82.9**	82.0	79.0	81.5	rec	94.8	93.7	94.0	**95.5**	95.1

2. Anger Detection

The most essential need in emotion detection applications is anger emotion recognition. Anger detection is one of the most significant emotions to detect since it is commonly utilized in contact centers and retail businesses to measure client happiness. As well as in the medical field, such as recognizing if a patient is in an angry state based on his voice signal. The Anger classification results are shown in Table 5, showing that SVM and IS09-10 had the greatest accuracy of 91.33%, MLP and featureset-2 had the same accuracy of 91%, and feature-set surpassed other features using ensemble learning approach with an accuracy of 90.00%. Across all of the featuresets, SVM has the best average accuracy. Angry classification rate ranges vary from 84% to 91%.

Table 5. Anger classification results

Feature-set	Anger	MLP	SVM	Random forest	Logistic regressin	Ensemble learning
Featureset-1	Acc	86.3	86.3	84.3	86.3	86
	pre	87.1	86.9	85	87.3	86.9
	rec	83	86.8	85.2	86.8	86.5
Featureset-2	Acc	**91**	**90.7**	88	82.3	**90**
	Pre	91	90.7	88.9	82.1	**90.5**
	rec	91.7	91.2	88.8	82.3	**90.5**
IS10	Acc	87.3	89	86.7	89.2	88.7
	pre	87	89	86.3	81.8	83.8
	rec	87.7	89.4	87	82.2	83.5
IS-09	Acc	86	83.7	82.3	86.0	88.4
	Pre	86	83.3	82.4	81.6	83.0
	rec	86	84.2	86.8	82.5	83.4
IS-09-10	Acc	89.3	**91.3**	86.7	88.4	89.2
	Pre	89.4	**91.2**	87.6	88.6	81.5
	rec	89.4	**91.7**	82.0	89.0	81.5

3. Multi emotion classification

Multi-emotion classification performance was tested by incorporating the four emotions: angry, happy, neutral, and sad.

In Table 6. The greatest accuracy was determined to be 65% when utilizing featureset-1 and the ensemble learning approach. In addition, SVM outperformed other classifiers in terms of average accuracy, except ensemble learning approach and overall feature sets, where it attained an accuracy of 64% for featureset-2, 64.6% for IS09, 64% for IS10, and 63% for IS09-10.

From previous results, it was concluded that featureset-2 results are generally higher than the other feature sets (IS09, IS10, and IS09-10). Featureset-2 superseded other

Table 6. Multi-emotion classification (Angry/Happy/Neutral/Sad)

Feature-set	MLP	SVM	Random forest	Logistic regressin	Ensemble learning
Featureset-1	62.4	50.6	62.3	62.9	**65**
	63	52.3	61.3	62.6	64.6
	62	50	61.4	62.3	64
Featureset-2	64.6	64	62.5	61.3	64
	64.7	62.7	60	60.3	62.8
	64.9	63	61	60.4	63
IS10	61	64.6	59	61.5	63
	60	64.5	57	60.6	62
	60	63.7	58	60.7	62
IS-09	59	64	60	60.5	62,7
	58	63.5	58.3	59.6	61.5
	58.8	63.7	59	60	62
IS-09-10	59	63	60.8	59	61.5
	58	62.5	58.6	58.5	60.7
	58	62.6	59.5	58.7	60.8

features using MLP where it achieved a performance improvement of 4% compared to IS09, 5.36% to IS10 and IS09-10. Ensemble learning showed good results in featureset-1 and featureset-2 compared to other classifiers with accuracies 65.11% and 64.08% respectively whereas in IS09, IS10 and IS09-10. From the recognition results, featureset-2 gives the best performance overall classifiers either showing enhancement as in MLP, random forest and logistic regression or nearly similar results as in SVM and ensemble learning. Overall performance concerning performance accuracy, precise featureset size, and running time, featureset-2 proved to be the best choice.

4. Arabic SER - Previous work comparison

Table 7 introduces a comparison between the proposed models using featureset-2 and previous results by Abdel-Hamid who introduced the Egyptian dataset "EYASE" [12]. Using the same dataset ensures measuring the models' performance. She used a feature set composing of prosodic, spectral and wavelet features of a total of 49 features, as well as using linear SVM classifier and KNN for classification.

The arousal classification of the proposed model achieved 1% enhancement using SVM classifier. An enhancement of 2.2% was achieved when comparing KNN results.

For valence classification, we achieved nearly the same result in the case of both SER systems using SVM and an enhancement of 1.39% when ensemble learning was used technique versus their SVM SER. Leading to improvement of 4% when compared their results using knn versus our model using either SVM or ensemble learning. For Anger classification, Lamia [8] used many features combined for anger detection ending up with the best result of 91% using Prosodic, LTAS, and Wavelet features and the lowest result of 81% using MFCC and Formants features. Compared with our model, which achieved 91% using MLP and featureset-2 and 90% using ensemble learning and SVM. For Multi classification, Lamia achieved 66.8% using SVM and 61.7% using Knn compared with our model results of 64.61% using SVM and 64.07% using ensemble learning. So, we were able to achieve enhancement over their Knn model with 3% but they superseded by 2% for SVM classifier. The justification here is that the LTAS and wavelet features are very effective with the Arabic language in the case of multi-emotion classification. That was concluded by Lamia as well when exploring feature importance in multi-classification as LTAS took a high rank among other features. That's why it was concluded that in Multi-classification the absence of LTAS differs a bit in the performance but still, our target during research is to have a baseline featureset for cross-corpus not just Arabic and to not be computationally expensive.

Table 7. Comparison between our work using featureset-2 and previous SER research work

Paper	Emotion category	Classifier	Accuracy	Evaluation criteria
Abdel-Hamid [1]	Arousal	SVM	94.3%	**10-K folds**
		KNN	93.3%	
	Valence	SVM	86.2%	
		KNN	83.7%	
	Multi classification	SVM	66.8%	
		KNN	61.7%	
	Anger classification	SVM	91.00%	
Proposed models	Arousal	Ensemble-learning	95.62%	**10-K folds**
	Valence	SVM	86.50%	
		Ensemble-learning	87.59%	
	Multi classification	Ensemble-learning	64.07%	
		MLP	64.61%	
	Anger classification	MLP	91.00%	
		Ensemble-learning/SVM	90.00%	

5 Conclusion

Different speech feature sets were used to train five different machine learning models. The correlation between each feature and the classifier was investigated, as well as which feature has the greatest impact on each classifier. It was found that MFCC is one of the most dominant features across the four classifiers. In comparison to the previous SER, our model improved Arabic results by 1–2%. SVM showed best classification results in many cases. MLP is a highly promising classifier that verifies the current research trend of neural networks and their different forms. Furthermore, Ensemble learning was effective and was highly sensitive to the overall other 4 models predication rates, reflecting multiple classifier point of view. The new state-of-the-art featureset-2 created and implemented as a mix of spectral and prosodic features, outperformed previous benchmarked feature sets such as Interspeech feature sets IS09, IS10, and IS09-10. Furthermore, featureset-2 has the lowest computational time to train the models compared to other feature sets.

In the future, there's a lot of opportunity for supplementing the model with additional multilingual data and expanding the input dataset as much as possible. Apply novel methods such as Convolutional neural network (CNN), LSTM, and transfer learning, as well as deep learning methods. Furthermore, consider feeding the voice stream straight into the neural network model without first extracting speech characteristics, which might speed up the process. Implementing the transfer learning approach by considering spectrogram images as an input feature to the CNN model.

References

1. Likitha, M.S., Gupta, S.R.R., Hasitha, K., Raju, A.U.: Speech based human emotion recognition using MFCC. In: 2017 International Conference on Wireless Communications, Signal Processing and Networking (WiSPNET), pp. 2257–2260 (2017). https://doi.org/10.1109/WiSPNET.2017.8300161
2. Blumentals, E., Salimbajevs, A.: Emotion recognition in real-world support call center data for latvian language. In: CEUR Workshop Proceedings, vol. 3124 (2022)
3. Stankova, M., Mihova, P., Kamenski, T., Mehandjiiska, K.: Emotional understanding skills training using educational computer game in children with autism spectrum disorder (ASD) - case study. In: 2021 44th International Convention on Information, Communication and Electronic Technology, MIPRO 2021 – Proceedings, pp. 672–677 (2021). https://doi.org/10.23919/MIPRO52101.2021.9596882
4. Du, Y., Crespo, R.G., Martínez, O.S.: Human emotion recognition for enhanced performance evaluation in e-learning. Prog. Artif. Intell. 1–13 (2022). https://doi.org/10.1007/S13748-022-00278-2
5. Roberts, L.: Understanding the Mel Spectrogram (2020). https://medium.com/analyticsvidhya/understanding-the-mel-spectrogram-fca2afa2ce53
6. Rashidan, M.A., et al.: Technology-assisted emotion recognition for autism spectrum disorder (ASD) children: a systematic literature review. IEEE Access **9**, 33638–33653 (2021)
7. Akçay, M.B., Oğuz, K.: Speech emotion recognition: emotional models, databases, features, preprocessing methods, supporting modalities, and classifiers. Speech Commun. **116**, 56–76 (2020)
8. El Ayadi, M., Kamel, M.S., Karray, F.: Survey on speech emotion recognition: features, classification schemes, and databases. Pattern Recognit. **44**(3), 572–587 (2011)
9. Mori, S., et al.: Emotional speech synthesis using subspace constraints in prosody. In: 2006 IEEE International Conference on Multimedia and Expo, pp. 1093–1096 (2006)
10. Klaylat, S., Osman, Z., Hamandi, L., Zantout, R.: Emotion recognition in Arabic speech. Analog Integr. Circ. Sig. Process. **96**(2), 337–351 (2018)
11. Szmigiera, M.: The most spoken languages worldwide 2021. https://www.statista.com/statistics/266808/the-most-spoken-languages-worldwide/
12. Abdel-Hamid, L.: Egyptian Arabic speech emotion recognition using prosodic, spectral and wavelet features. Speech Commun. **122**, 19–30 (2020)

13. Mirsamadi, S., Barsoum, E., Zhang, C.: Automatic speech emotion recognition using recurrent neural networks with local attention. In: 2017 IEEE International Conference on Acoustics, Speech and Signal Processing (ICASSP), pp. 2227–2231 (2017)
14. Lalitha, S., Geyasruti, D., Narayanan, R., Shravani, M.: Emotion detection using MFCC and cepstrum features. Procedia Comput. Sci. **70**, 29–35 (2015)
15. Araño, K.A., Gloor, P., Orsenigo, C., Vercellis, C.: When old meets new: emotion recognition from speech signals. Cogn. Comput. **13**(3), 771–783 (2021). https://doi.org/10.1007/s12559-021-09865-2
16. Schuller, B., Steidl, S., Batliner, A.: The interspeech 2009 emotion challenge. In: INTERSPEECH (2010)
17. Schuller, B., et al.: The INTERSPEECH 2010 paralinguistic challenge. In: INTERSPEECH (2010)
18. Koolagudi, S.G., Murthy, Y.V.S., Bhaskar, S.P.: Choice of a classifier, based on properties of a dataset: case study-speech emotion recognition. Int. J. Speech Technol. **21**(1), 167–183 (2018). https://doi.org/10.1007/s10772-018-9495-8
19. Bhavan, A., Chauhan, P., Shah, R.R.: Bagged support vector machines for emotion recognition from speech. Knowl.-Based Syst. **184**, 104886 (2019)
20. Yadav, S.P., Zaidi, S., Mishra, A., et al.: Survey on machine learning in speech emotion recognition and vision systems using a recurrent neural network (RNN). Arch. Comput. Methods Eng. **29**, 1753–1770 (2022)
21. Langari, S., Marvi, H., Zahedi, M.: Efficient speech emotion recognition using modified feature extraction. Inform. Med. Unlocked **20**, 100424 (2020)
22. https://librosa.org/doc/latest/index.html
23. About openSMILE—openSMILE Documentation. https://audeering.github.io/opensmile/about.html#capabilities. Accessed 18 May 2021
24. https://machinelearningmastery.com/information-gain-and-mutual-information/. Accessed 10 Dec 2020
25. Permutation feature importance with scikit-learn. https://scikit-learn.org/stable/modules/permutation_importance.html. Accessed 18 May 2021
26. Sefara, T.J.: The effects of normalisation methods on speech emotion recognition. In: Proceedings - 2019 International Multidisciplinary Information Technology and Engineering Conference, IMITEC 2019 (2019)
27. Zehra, W., Javed, A.R., Jalil, Z., Khan, H.U., Gadekallu, T.R.: Cross corpus multi-lingual speech emotion recognition using ensemble learning. Complex Intell. Syst. **7**(4), 1845–1854 (2021)
28. Koduru, A., Valiveti, H.B., Budati, A.K.: Feature extraction algorithms to improve the speech emotion recognition rate. Int. J. Speech Technol. **23**(1), 45–55 (2020). https://doi.org/10.1007/s10772-020-09672-4
29. Matsane, L., Jadhav, A., Ajoodha, R.: The use of automatic speech recognition in education for identifying attitudes of the speakers. In: IEEE Asia-Pacific (2020)
30. Bestelmeyer, P.E.G., Kotz, S.A., Belin, P.: Effects of emotional valence and arousal on the voice perception network. Soc. Cogn. Affect. Neurosci. **12**(8), 1351–1358 (2017). https://doi.org/10.1093/scan/nsx059. PMID: 28449127; PMCID: PMC5597854
31. Russell, J.A.: A circumplex model of affect. J. Personal Soc. Psychol. **39**(6), 1161–1178 (1980)

Modelling

Towards the Strengthening of Capella Modeling Semantics by Integrating Event-B: A Rigorous Model-Based Approach for Safety-Critical Systems

Khaoula Bouba[1(✉)], Abderrahim Ait Wakrime[1], Yassine Ouhammou[2], and Redouane Benaini[1]

[1] Computer Science Department, Faculty of Sciences, Mohammed V University in Rabat, Rabat, Morocco
khaoula.bouba@um5r.ac.ma,
{abderrahim.aitwakrime,redouane.benaini}@fsr.um5.ac.ma

[2] LIAS/ISAE - ENSMA, 86961 Futuroscope Chasseneuil Cedex, Chasseneuil-du-Poitou, France
yassine.ouhammou@ensma.fr

Abstract. Safety-critical systems are increasingly model-based, since model-based system engineering (MBSE) paradigm reduces the time-to-market and allows evolving systems at different abstraction levels. Different languages have been proposed recently enabling to facilitate the modeling process and shorten the development life-cycle. However, these languages may be used at one or many modeling steps regarding the semantics of their artefacts. Capella language is one of these languages that gained popularity recently. It is dedicated to system engineering and its use may very beneficial for safety-critical system. However, designing with Capella is considered as semi-formal. Thus, the approach presented in this paper stands for systematic formal verification of Capella's behavioral models using Event-B method in a transparent way. Our proposal translates Capella models into Event-B specifications using automatic model-to-model transformations dedicated to Capella designers. The verification of correctness of the transformed models is provided by the ProB model-checker. An automatic lighting system is treated as a case study to validate of our contribution.

Keywords: Model-based system engineering · Formal methods · Capella/arcadia · Event-b · Meta-model · Operational analysis

1 Introduction

With the rapid pace of change in our world, safety-critical systems became more and more complex, and the traditional engineering practices that are mostly document-driven are no longer adequate to address increasing complexity in

P. Fournier-Viger et al. (Eds.): MEDI 2022, LNAI 13761, pp. 119–132, 2023.
https://doi.org/10.1007/978-3-031-21595-7_9

systems architecture. Modeling is the first step in the software development process for understanding requirements relative to the system. As a result, partial models of the structure and behavior of the system are created, allowing the developers to work at various abstraction levels before beginning the programming process. The second step is the consistency verification, that must also be performed at an early design phase, meaning that each refined model must be consistent with itself, with the previous models, and with global constraints.

The core of the MBSE is to construct the appropriate models regarding given system specifications. For this reason, Capella [1] was a turning point for engineering environments such as energy, aerospace, and automotive industries, which in recent years has become increasingly recommended for modeling. Capella provides better architecture quality, expresses the systems in five different abstraction levels, each one is a refinement of the previous. These levels are Operational Analysis (OA), System Analysis (SA), Logical Architecture (LA), Physical Architecture (PA), and End Product Breakdown Structure (EPBS). In addition to that, Capella offers an automatic traceability between its various levels and supports an easier integration of structure and behaviour. In this paper, we focus on the first modeling level, which captures the relevant stakeholders of the operational context in which the system will be integrated. In this phase, we chose the Operational Architecture (OAB) diagram, that allocates the activities to the entities and actors in order to present a conceptual overview. The benefit of this diagram is that it groups both the functional and behavioral decompositions of the system. Hence, a rigorous model-based approach for safety-critical systems starting from OAB models is needful.

So, in our work, we address an innovative challenge which is the automatic transformation of Capella modeling to Event-B [2] models, so that Capella models can be verified using model checking, which is an important step towards establishing a reliable development process. Hence, the present solution starts with identifying the system requirements and presenting them as a Capella diagram. Then, a model transformation is applied to transform automatically the Capella models into Event-B specifications. Also, to the best of our knowledge, the proposed formalization, based on model transformation of a Capella OAB diagram to Event-B, presented in this paper has not been developed so far.

The reminder to the paper is organised as follows. The next section presents the state-of-the-art, followed by presenting the running example adopted in this work. Section 3 is devoted to explain the process and the methodology of our approach, in addition of the verification process of our case study. Section 4 is dedicated to the tooling of a proof of concept. Finally, Sect. 5 concludes the paper and presents the future work.

2 Related Works and Preliminaries

In this section we first start by discussing some existing related works, then we present briefly Capella and Event-B. We also present an example that will be treated in our contribution to facilitate the explanation.

2.1 Background

Software development methodologies, such as Model-Driven Engineering (MDE) [3], are considered an effective method to simplify the design process. Towards developing more abstract and more automated systems, MDE makes extensive and consistent use of models at different levels of abstraction while designing systems. By using models, it becomes possible to eliminate some useless details as well as to break down complicated systems into smaller, simpler and manageable units. Due to a separation between the business and technical components of the application, MDE automates the generation of applications following the modification of the target platforms. In this context, the [4] focuses on transforming an adaptive run-time system model interpreted with MARTE [5] elements to Event-B concepts using the Acceleo [6] transformation engine. The [7] suggests a Sewerage System which is represented by a UML activity diagram, and further converted into Nondeterministic Finite Automata (NFA) [8] to describe the system's behavior of water effectively. In the following step, a formal model is created from the automata model using TLA+ [9], which will be checked and validated by TLC, a model checking feature included in TLA+. The work proposed in [10] presents the formalization of a system called Railway Signaling System European Rail Traffic Management System/European Train Control System (ERTMS/ETCS). The functionalities and relationships of its several subsystems are modelled via UML, and then translated to Event-B language. The proof of correctness of the end code is provided by ProB [11].

Contrary to previous researches, the research proposed in [12] proposes Capella as a modeling tool for Distributed Integrated Modular Avionics (DIMA). The choice of Capella was due to the increasing maturity of the system, which makes the DIMA system architects confront several issues face several problems during the design process due to the high number of functions, such as functions allocation and device physical allocation. The authors in [13] offer a set of constructions and principles in order to abstract heterogeneous models with the intention of being able to synchronize and compare them; They trusted the System Structure Modeling Language [14] (S2ML) to ensure the consistency between system architecture models designed with Capella and safety models written in AltaRica 3.0 [15] of the power supply system. The [16] introduces a transformation approach from Capella physical architecture to software architecture in AADL [17] using the Acceleo plugin, that was applied and validated on a robotic demonstrator called TwIRTee, developed within the INGEQUIP project. Besides, each element of the source architecture is mapped to a new concept of the target architecture. The work in [18] shows a model-to-model transformation application, which aims to verify the dynamic behavior model of Capella systems using Simulink [19]. The use case chosen for this research is the "clock radio" system, which will be interpreted in the form of a Capella data-flow diagram (physical architecture data-flow), and subsequently transformed into an executable simulink model. Unlike the researches mentioned above, we propose through this paper, a new approach which consists of formalizing a Capella model, one of the most useful modeling solutions, to Event-B, one of the formal languages more used.

2.2 Overview of Capella and Event-B

Capella is model-based engineering tool originally implemented by Thales, that has been successfully implemented in many industrial contexts. It was inspired by UML/SysML and NAF standards, and provides rich methodological guidance to system, software and hardware architects using of ARChitecture Analysis and Design Integrated Approach (Arcadia), a model-based engineering method that relies on functional analysis and on the allocation of functions to architecture components. Considering that the Capella tool embeds the Arcadia Method, one of the biggest challenges of Capella Modeling is that in a sense it cannot be regarded as a general-purpose modelling language. As a result, it makes more difficult for users to transition from an existing SE process based on document-based approaches. Also, system engineers must maintain consistency with the Capella transition mechanism to meet the user's requirements.

Event-B is a state formal-method for modeling systems and analysis, based on set theory and predicate logic. It has two main features: The first one is the refinement, which describes the system at a high level of abstraction, and then making it more and more precise by adding new concepts and details in successively understandable steps. And the second one is the consistency checking, which ensures the validity of system properties via mathematical proof obligations. Formal models are particularly challenging because they require additional argumentation about their correctness and well-definedness.

2.3 Motivating Example: Adaptive Exterior Light System

We introduce an example that is inspired from a real-world system and available in many recent cars, called the adaptive exterior light system. We will discuss a case study that was proposed during the ABZ2020 conference [20]. The system outlines the many available lights and the conditions under which they are turned on/off in order to improve the driver's vision without blinding oncoming motorists. The system can be considered as a lights controller that reads various data from accessible sensors (key state, outside luminosity, etc.) and performs appropriate actions by acting on the light actuators to guarantee good visibility for the driver based on the data read. Also, we will introduce another aspect that is particularly relevant to the driver's modeling behavior [21], which was not included in the use case, but it will allow us to use cameras to examine drivers' conditions, such as their level of attention and interest, driving skills, aiming to improve transportation safety.

3 Proposed Approach

Our approach contains two steps: (i) Preparatory step and (ii) Transformation step. First, we start by defining the needs of the stakeholders and the system functionality requirements. Next, we switch these informal requirements to Capella model. During this stage, it is possible for users to inject rules into the model through constraints, which must be preserved by each state of the system.

Once the Capella model and the translation implementation are ready, we can proceed to the transformation step. As input for our generator, we pass a specific file containing the model without the graphical part. Some of the Capella components will not be formalized later, because so far they do not participate in the state/transition behavior of the system. As soon as our formal Event-B model is generated, it must be validated and verified. In summary, our approach is depicted in Fig. 1.

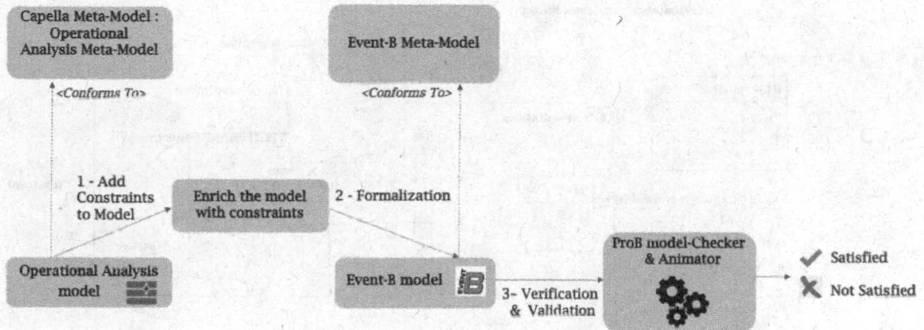

Fig. 1. Overview of the model transformation approach.

The manual construction of formal models is time-consuming, prone to error, and intractable for large systems. Therefore, automating model construction would make formal methods more effective for ensuring system correctness.

3.1 Preparatory Step

Operational Analysis Meta-model. The Capella meta-model is defined in ecore format, it is captured in an Ecore files (.ecore), which are basically a XML files that conforms to the XSD of Ecore. These files depend on each-other as elements frequently build on top of each-other via Generalization relationship. In our paper, we'll be focusing only on the Operational Analysis meta-model. It will contain only the classes that can be directly mapped to the target meta-model as showed in Fig. 2, which shows the main concepts of OA meta-model. We have the Entity class that represents an entity in the real world which is ready to interact with the system or its users. These entities could have a role in the system, for this reason, we could assign characters to them by associating them with roles, such as driver, pilot, etc. In addition to that, they can perform operational activities to achieve a specific objective, linked together by operational interactions that can present exchanges of information. Constraints are cross-perspective elements in the sense that they can be applied to components in multiple perspectives. These constraints help the users to inject the model with rules that should hold in each reachable state of the system. Finally, the operational processes are series of activities and interactions that are carried out successively to achieve a purpose.

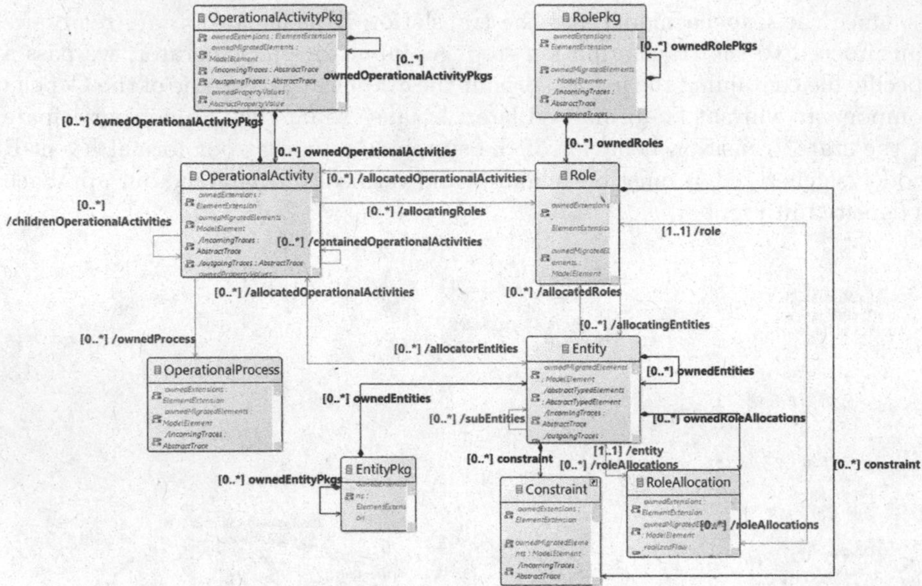

Fig. 2. Extract of Operational Analysis meta-model.

OAB Diagram: Adaptive Exterior Light System. We started by creating a model that describes the lighting system of cars, and we were keen that it contains all the components of the palette offered by the Capella workbench. The system's behaviour is described in a Capella operational architecture diagram by a set of actors/entities, interactions, constraints and operational activities. At this current stage, we have three entities and one actor to which we assign a role called Driver. We have also allocated them some activities, and these activities have different interactions with each other. Next, we assigned to them and to the activities different constraints. Each constraint have a specific interpretation, that we will discuss later. Also, we have an operational process that present a set of activities which are rolled out successively as shown in our diagram with the blue line; First and foremost, the car must be started so that the lights can turn on, or we can rely on the front camera detecting the condition of the driver as normal, so that he can adjust the parameters to set the state of lights. The Fig. 3 represents an operational architecture diagram of the adaptive exterior light.

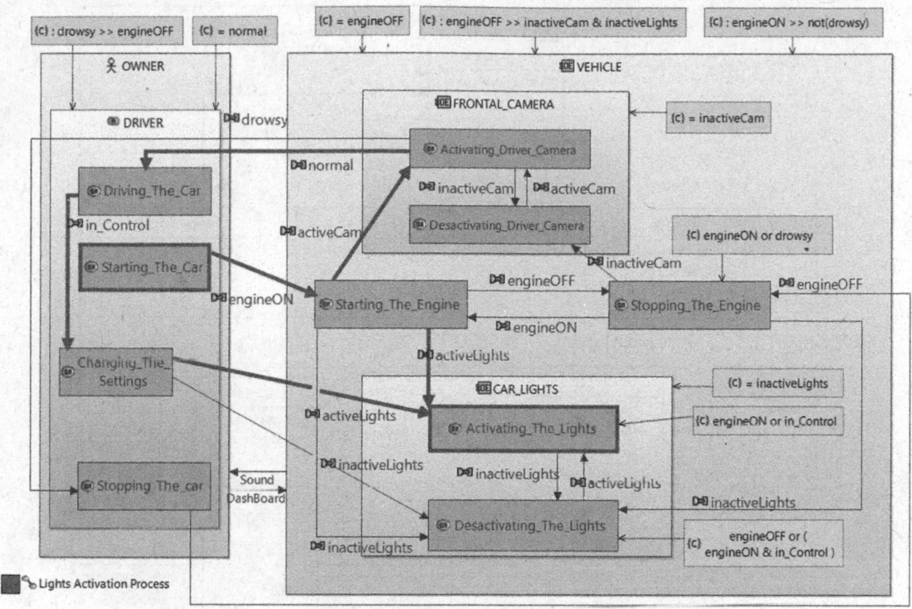

Fig. 3. Operational architecture diagram of the adaptive exterior light.

3.2 Transformation Step

Event-B Meta-model. An Event-B model [22] is composed of a machine, which models dynamic data and behavior, and zero or more contexts, which model static data structures or configurations. Events consist of guards and actions, provide the state transition mechanism for the machine's variables. The guard is a condition on the machine variables that determines whether or not an event is enabled; It is only enabled if all its guards are true. The action is to apply an update on a state variable. The invariants are state predicates that define the types of variables, specify properties of correctness that must always be true, and any violation of these invariants will cause the system to be inconsistent. The INITIALIZATION event is used to set the variables' initial values. The Fig. 4 represents the meta-model of Event-B that shows clearly Event-B meta-classes and structures.

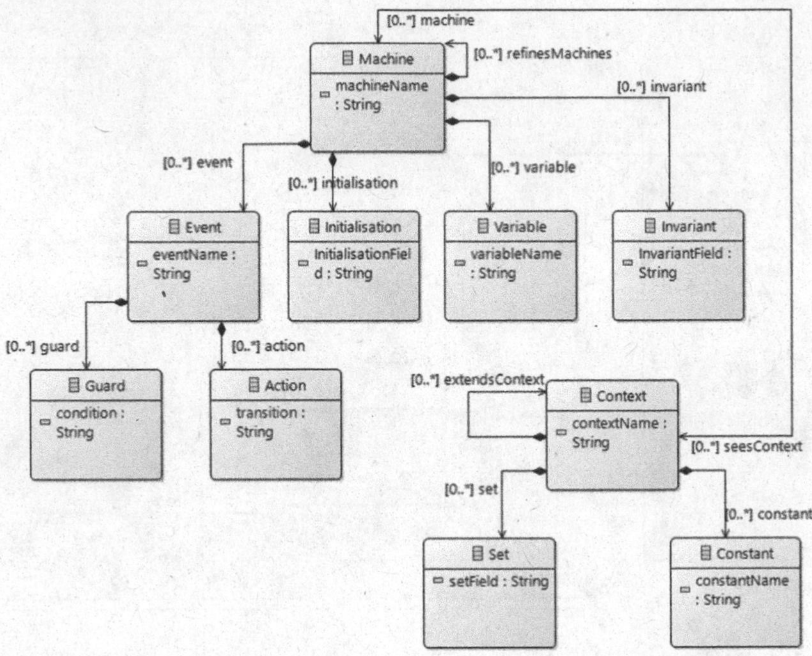

Fig. 4. Event-B meta-model.

Mapping of Capella/Event-B Concepts. Combining the two specification techniques Capella and Event-B is a well-studied topic. Our Capella to Event-B approach validates formally the modeled system's expected scenarios, and describes a simultaneous construction strategy that provides for more flexibility and the capacity to develop the specification while maintaining coherence and traceability between the two models. The Table 1 describes the mapping of the transformation between Capella and Event-B meta-models. The transformation process follows specific rules which are generally applied on the source model in order to generate the appropriate target model with respecting the mapping chosen above. For every Event-B component, we present its equivalent in the Capella side : **MACHINE:** The name of the Capella diagram is mapped as the machine name in Event-B. **SETS:** Each entity/actor or role element that contains at least one operational activity considered as the target of an operational interaction, is considered as a set (meaning that these activities have inputs). The elements of this set will be the values of the operational interactions coming to the activities of the entity/actor or role in question. **VARIABLES:** The concatenation of the constant "Var_" with the names of the selected entities/actors or roles (according to the aforementioned rule) forms the list of variables. **INVARIANTS:** constraints linked to entities/actors or roles, and beginning with the character ":" are automatically translated into invariants. **INITIALISATION:** constraints linked to entities/actors or roles, and beginning with the character "=", form the

Table 1. Concept mapping between Operational Analysis and Event-B meta-models.

Graphical Capella element	Capella concept	Event-B concept
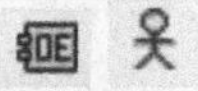	Operational Entity/Actor	Set, Variable
	Role	Set, Variable
	Operational Activity	Event
	Operational Interaction	Set's Element
	Constraint	Invariant, Initialisation

list of initializations. **EVENTS:** Operational activities that have at least one operational interaction as input are transformed into events. The inputs arriving at the activities form the post-conditions of its equivalent events. The precondition under which an event can occur is the conjunction of the postconditions of its previous activity (by default), but if the precondition have a specific logic (combination of the conjunctions and disjunctions operations), we must define a constraint associated with the activity, which carries the appropriate guard expression.

There is no mapping for the Operational Process component, because as we have already explained, it is a succession of activities carried out in a specific order, and we have also translated the activities into events (according to Table 1), therefore intuitively it is transformed into a series of events, which will be executed in a specific order (It will not be possible to execute the second event unless the first has already been executed). This order of execution is controlled by the the correctness of the precondition of the events (the precondition of the second event is the post-condition of the first).

Generated Results. The Fig. 5 shows an extract of the different parts of the generated Event-B model. Our case study is quite large, so we focused on a specific sub-systems for a clearer explanation.

In our case study, we have an entity called CAR_LIGHTS, which is responsible for turning on/off the vehicle lights. It includes a couple of activities, so for this reason a set denominated "CAR_LIGHTS" composed of two elements {activeLigts, inactiveLights}, a variable called "VAR_CAR_LIGHTS", an invariant indicating that the variable created belongs to the set (membership predicate), named "VAR_CAR_LIGHTS : CAR_LIGHTS", and initialisation event,

```
SETS
  FRONTAL_CAMERA = {activeCam,inactiveCam};
  CAR_LIGHTS = {activeLights,inactiveLights};
  VEHICLE = {engineOFF,engineON};
  DRIVER = {normal,drowsy,in_Control}

VARIABLES

  Var_VEHICLE,
  Var_CAR_LIGHTS,
  Var_DRIVER,
  Var_FRONTAL_CAMERA

INVARIANT

  Var_VEHICLE : VEHICLE &
  Var_CAR_LIGHTS : CAR_LIGHTS &
  Var_DRIVER : DRIVER &
  Var_FRONTAL_CAMERA : FRONTAL_CAMERA &
  (Var_VEHICLE = engineON => Var_DRIVER /= drowsy) &
  (Var_DRIVER = drowsy => Var_VEHICLE = engineOFF) &
  (Var_VEHICLE = engineOFF => Var_FRONTAL_CAMERA = inactiveCam
                            & Var_CAR_LIGHTS = inactiveLights)

INITIALISATION

    Var_VEHICLE := engineOFF
   || Var_CAR_LIGHTS := inactiveLights
   || Var_DRIVER := normal
   || Var_FRONTAL_CAMERA := inactiveCam
```

```
EVENTS

   Activating_Driver_Camera =
   ANY
     param
   WHERE
     param : FRONTAL_CAMERA & param /= Var_FRONTAL_CAMERA &
     Var_VEHICLE = engineON & Var_FRONTAL_CAMERA = inactiveCam
   THEN
     Var_FRONTAL_CAMERA := activeCam
   END;

   Desactivating_The_Lights =
   ANY
     param
   WHERE
     param : CAR_LIGHTS & param /= Var_CAR_LIGHTS &
     (Var_VEHICLE = engineOFF or (Var_VEHICLE = engineON &
                  Var_DRIVER = in_Control ))
   THEN
     Var_CAR_LIGHTS := inactiveLights
   END;

   Stopping_The_car =
   ANY
     param
   WHERE
     param : DRIVER & param /=  Var_DRIVER &
     Var_FRONTAL_CAMERA = activeCam
   THEN
     Var_DRIVER := drowsy;
     Var_VEHICLE := engineOFF;
     Var_FRONTAL_CAMERA := inactiveCam;
     Var_CAR_LIGHTS := inactiveLights
    END;
```

Fig. 5. An extract of the generated Event-B model.

denominated "Var_CAR_LIGHTS := inactiveLights", indicating the initial value of our variable are created in the MACHINE; Afterward, two events are created: the first one to active the lights of the engine, named "ActivateLights", with a disjunction expression "Var_VEHICLE = engineON or Var_DRIVER = INControl" as guard and "Var_CAR_LIGHTS := activeLights" as action, indicating that the lights can be activated only if the car is started (engineOn) or with the driver intervention (in_Control). Here the guard is defined in a constraint, because as we explained it before, the default operation between the conditions of the guard is the conjunction, but in this case we have some peculiarities related to the semantics of the lights activation process (see Fig. 7).

The invariant that expresses a predicate that must stay true during the whole execution must contain the character ">>". For the "VEHICLE" entity, we associated to it two invariants; The first one indicates that if the engine is off, the state of the lights and the driver frontal camera is also off, and the second expresses a safety property whose purpose is to protect the driver if its condition does not allow him to drive (see Fig. 6). It should be noted that the fifth and sixth invariants are not included in the approved case study (see Fig. 5). Nevertheless, we created them in order to have a sufficient number of invariants for the

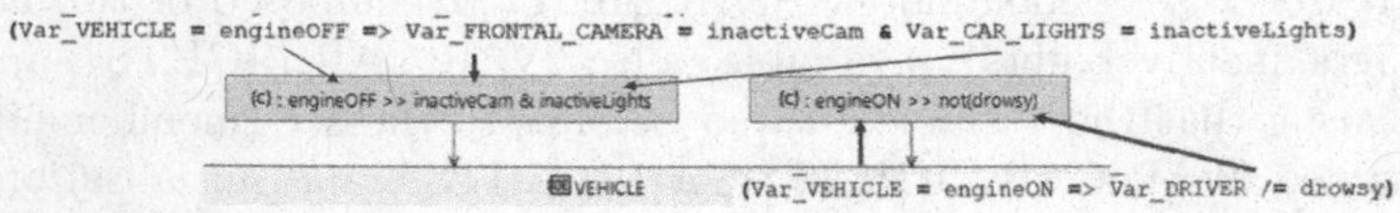

Fig. 6. The invariant concept of the VEHICLE sub-system.

verification simulation, and complete implementation grouping all the elements of the diagram (we added two concepts : the driver and the frontal camera that captures his actual state).

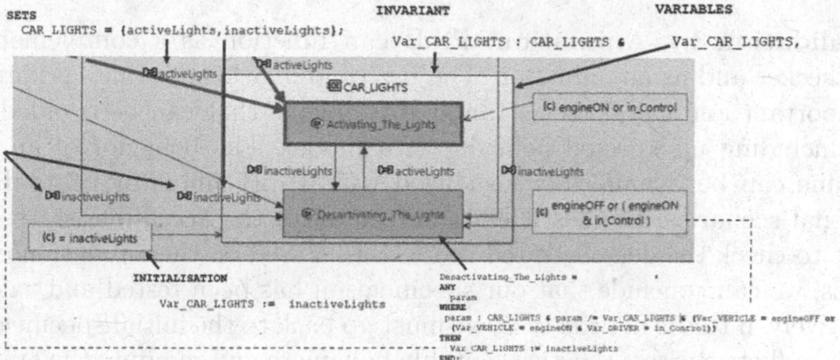

Fig. 7. The mapping of the Car_Lights sub-system elements into Event-B concepts.

Validation and Verification. Validating the Event-B model and ensuring that the invariants (typing and safety properties invariants) are preserved across all events is our intention in this section. It consists of checking whether a finite-state model of a system meets certain specifications. Also, there is no addition of new instances of the sets defined earlier in the model, so the objects set remains constant. To test whether our Event-B model is valid, we applied it to a simplified scenario derived from the use case study illustrated in Sect. 2.3. We launch our scenario by starting the engine, which activates the driver's front camera. This camera is used to predict the driver's state and fatigue levels. If the state of the driver is "normal", then he can change the settings of the car to turn on the lights, and as a consequence, his state is changed to "in_Control". On the other hand, if the detected state is "drowsy", which means that he's not capable of driving anymore, the vehicle, the lights and the frontal camera are turned off (to prevent accidents for example). For the purpose of demonstrating that the formal specifications of the adaptive exterior lighting model are correct, we'll use ProB in order to validate the Event-B model. Model checking is used here instead of theorem proving, since that requires more effort and training. Nevertheless, the model checking is sufficient to check system properties for a given initial state, since the system has a finite state space.

A - Verification Using Model Checking. Model-Checking [23] is a formal verification method that automatically and systematically checks whether a system description conforms specified properties. The behavior of the system is modeled formally, and the specifications expressing the expected properties (safety, security, etc.) of the system are also expressed formally using the first-order logic formulas. All experiments were conducted on a 64-bit PC, Windows 10 operating system, an Intel Core i7, 2.9 GHz Processor with 2 cores and 8 GB

RAM. Using the ProB model-checker and based on mixed breadth and depth search strategy, we have explored all states: 100% of checked states with 7 distinct states and 20 transitions. No invariant violation was found, and all the operations were covered. This verification ensures that invariants are preserved by each event. Otherwise, a counter-example would be generated.

B - Validation by Animation. ProB can function as a complement to a model-checker and as an animator. The use of animations during verification is very important and can detect a range of problems that can be avoided in the future, including unexpected behavior of a model. The behavior of an Event-B machine can be dynamically visualized with ProB animator using different operational scenarios; Besides it can analyze all of the accessible states of the machine to check the demonstrated properties. Based on the animation of these scenarios, we can conclude that our specification has been tested and validated. Alternatively, if this is not the case, we must go back to the initial specification to find the conflicts, correct the unacceptable behaviors and re-apply the animation to ensure the specification is aligned with the requirements.

4 Tooling

The Fig. 8 represents each step of our approach with its equivalent tool. The preparatory step is devoted to the construction of the operational analysis model using the Capella studio tool. The transformation step is dedicated for the transformation of Capella model to Event-B model using Acceleo. The last step is committed to the Event-B textual specification and to the verification of this using ProB.

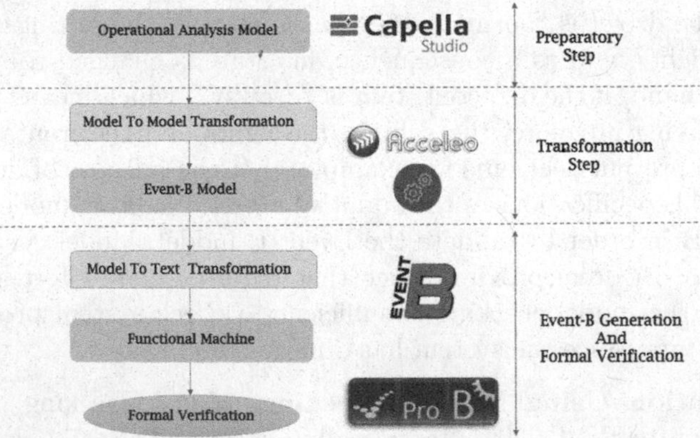

Fig. 8. The tools corresponding to each step of the proposed approach

With Capella Studio, extensions for Capella MBSE can be developed in an integrated development environment. It is based on Kitalpha, which is designed for creating model-based workbenches. Also, users can enhance and customize the Capella development artefacts (meta-models, diagrams) using the Capella development artefacts. There are many add-ons and viewpoints that are already integrated in the Capella studio. It includes EMF technology for the models management which are defined in the Ecore format. Ecore is a framework composed of a set of concepts, that can be manipulated by EMF to build a meta-model.

Ecore shares a lot of similarities with the class diagram of UML, that's why it can basically be seen as UML packages. Every package contains ontology elements (or UML Classes) and "local" element relationships (Associations). Moreover, it includes as well the Acceleo add-on, which is a language based on templates for creating code-generation templates. In addition to supporting OCL, this language provides a number of operations useful for working with text-based documents. There is a set of powerful tools bundled with Acceleo, including an editor, a code completion and refactoring tools, a debugger, error detection and a traceability API.

5 Conclusion and Future Works

This paper proposes the formalization and verification of a Capella Diagram named Operational Architecture Diagram using Event-B in order to develop a mechanism for automatic verification of these diagrams, with the potential to bring the benefits of formal methods to industrial practitioners. We used the Capella model to describe how components interact and the overall behavior of our system. Then, the Capella model is transformed into Event-B in order to ensure the validity of the functional properties of the system. The output of model transformation is verified using the ProB model-checker to monitor the invariants preservation for a given initial state. In our future work, we plan to make the construction of the invariants more automatic without the need of any constraint. Also, we are going to present a refinement methodology of Capella and Event-B models. Using this approach, the system is incrementally developed starting from a very abstract model that may be considered as a system specifications. The model of the system is gradually developed (using correct-by-construction process) by adding more details in a concrete model that must maintain the properties and functionality of the previous abstract models.

References

1. Roques P.: Modélisation architecturale des systèmes avec la méthode Arcadia: guide pratique de Capella, vol. 2, ISTE Group, 2018
2. Abrial, J.R.: Modeling in Event-B: system and software engineering. Cambridge University Press (2010)
3. Schmidt, C.: D.: Model-driven engineering. Computer-IEEE Computer Society-**39**(2), 25 (2006)
4. Fredj, N., Hadj Kacem, Y., Abid, M.: An event-based approach for formally verifying runtime adaptive real-time systems. The Journal of Supercomputing **77**(3), 3110–3143 (2021)

5. The ProMARTE consortium, UML profile for MARTE, beta 2, June 2008, OMG document number : ptc/08-06-08
6. Brambilla, M., Cabot, J., Wimmer, M.: Model driven software engineering in practice. SynthLect. Softw. Eng. **3**(1), 1–207 (2012)
7. Latif, S., Rehman, A., Zafar, N.A.: Modeling of sewerage system linking UML, automata and TLA+. In 2018 International Conference on Computing, Electronic and Electrical Engineering (ICE Cube), pp 1–6. IEEE (2018)
8. Hopcroft, J.E., Motwani, R., Ullman, J.D.: Introduction to Automata Theory, Language and Computation, Addison-Wesley, Reading (2001)
9. Cristiá, M.: A TLA+ encoding of DEVS models. In: Proceedings of the International Modeling and Simulation Multiconference, pp. 17–22 (2007)
10. Ait Wakrime, A., Ben Ayed, R., Collart-Dutilleul, S., Ledru, Y., Idani, A.: Formalizing railway signaling system ERTMS/ETCS using UML/Event-B. In: Abdelwahed, E.H., Bellatreche, L., Golfarelli, M., Méry, D., Ordonez, C. (eds.) MEDI 2018. LNCS, vol. 11163, pp. 321–330. Springer, Cham (2018). https://doi.org/10.1007/978-3-030-00856-7_21
11. Leuschel, M., Butler, M.: Prob: an automated analysis toolset for the b method. Int. J. Softw. Tools Technol. Transf. **10**(2), 185–203 (2008)
12. Batista, L., Hammami, O.: Capella based system engineering modelling and multi-objective optimization of avionics systems. In: IEEE International Symposium on Systems Engineering (ISSE), pp. 1–8. IEEE (2016)
13. Batteux, M., Prosvirnova, T., Rauzy, A.: Model synchronization: a formal framework for the management of heterogeneous models. In: Papadopoulos, Y., Aslansefat, K., Katsaros, P., Bozzano, M. (eds.) IMBSA 2019. LNCS, vol. 11842, pp. 157–172. Springer, Cham (2019). https://doi.org/10.1007/978-3-030-32872-6_11
14. Batteux, M., Prosvirnova, T., Rauzy, A.: System Structure Modeling Language (S2ML) (2015)
15. Batteux, M., Prosvirnova, T., Rauzy, A.: Altarica 3.0 in 10 modeling patterns. Int. J. Critic. Comput. Based Syst. (IJCCBS). **9**, 133 (2019). https://doi.org/10.1504/IJCCBS.2019.10020023
16. Ouni, B, Gaufillet, P., Jenn, E., Hugues, J.: Model driven engineering with Capella and aadl. In: ERTSS 2016 (2016)
17. Architecture Analysis and Design Language (AADL), SAE standards .http://standards.sae.org/as5506/
18. Duhil, C., Babau, J.P., Lépicier, E., Voirin, J.L., Navas, J.: Chaining model transformations for system model verification: application to verify Capella model with Simulink. In: 8th International Conference on Model-Driven Engineering and Software Development, pp. 279–286. SCITEPRESS-Science and Technology Publications (2020)
19. Klee, H., Allen, R.: Simulation of Dynamic Systems with MATLAB and Simulink. CRC Press, Boca Raton, February 2011
20. Houdek, F., Raschke, A.: Adaptive exterior light and speed control system. In: Raschke, A., Méry, D., Houdek, F. (eds.) ABZ 2020. LNCS, vol. 12071, pp. 281–301. Springer, Cham (2020). https://doi.org/10.1007/978-3-030-48077-6_24
21. AbuAli, N., Abou-zeid, H.: Driver behavior modeling: Developments and future directions. Int. J. Veh. Technol. **2016**, 1–12 (2016)
22. Weixuan, S., Hong, Z., Chao, F., Yangzhen, F.: A method based on meta-model for the translation from UML into Event-B. In: 2016 IEEE International Conference on Software Quality, Reliability and Security Companion, pp. 271–277 (2016)
23. M Clarke Jr., E., Grumberg, O., Kroening, D., Peled, D., Veith, H.: Model checking. Cyber Physical Systems Series (2018)

A Reverse Design Framework for Modifiable-off-the-Shelf Embedded Systems: Application to Open-Source Autopilots

Soulimane Kamni[1](✉), Yassine Ouhammou[1], Emmanuel Grolleau[1], Antoine Bertout[2], and Gautier Hattenberger[3]

[1] LIAS, ISAE-ENSMA, Futuroscope, France
{soulimane.kamni,yassine.ouhammou,grolleau}@ensma.fr

[2] Université de Poitiers, Futuroscope, France
antoine.bertout@univ-poitiers.fr

[3] Ecole Nationale de l'Aviation Civile, Université de Toulouse, Toulouse, France
gautier.hattenberger@enac.fr

Abstract. The development of real-time embedded systems is usually preceded by an important design phase to ensure that functional and behavioural constraints are met. However, the modification of some systems, especially Unmanned Air Vehicles that need to be frequently customised, is typically done in an ad-hoc way. Indeed, the design information may not be available, which may affect the proper functioning of the system. This paper aims to propose a framework helping reverse-engineering a Modifiable Off-The-Shelf (MOTS) embedded system in order to be able to ease its modification. In other words, our objective is to point out where modifications have to happen, and allow smooth use of third-party analysis and/or architecture exploration tools to re-analyse non-functional properties (safety, performances, etc.) regarding the customisation. This framework extracts functional-chains from the source code and represents them visually as a model-based design by using model-driven engineering settings.

Keywords: MOTS · Reverse engineering · Capella · Model-based design

1 Introduction

Nowadays, the re-usability is an aspect that becomes more and more requested when developing new systems. Indeed, many systems are being constructed by integrated existing independent systems, of different stakeholders, leading to new system of systems (SoS), like in UAV (Unmanned Air Vehicles) domain.

There is a growing interest in open and flexible architecture for UAV systems. A lot of small and medium stakeholders propose new drone-based innovative services by customising hardware and/or software parts. Moreover, from

P. Fournier-Viger et al. (Eds.): MEDI 2022, LNAI 13761, pp. 133–146, 2023.
https://doi.org/10.1007/978-3-031-21595-7_10

bare-metal autopilots executed on small microcontrollers, the hardware evolution of embedded systems on a chip allows multiprocessor chips to be embedded and used as hardware platforms for autopilots [1–3]. These platforms may also rely on companion boards to add extra and specialised computing performance, such as GPUs (Graphics Processing Unit) to execute machine learning methods. Sensors are also rapidly evolving, as well as many types of frames are available: fixed wings, helicopter, tri, quad, hexa, hepta or octo-copters, vertical take-off and landing vehicles, etc. As a result, the autopilot has to cope with fast-growing orthogonal dimensions: over a dozen of types of frames, dozens of hardware computing platforms, supporting different operating systems (or none at all), hundreds of different sensors and actuators, and an infinite number of customised functions. All these elements should be embedded as a payload or directly integrated in the heart of the autopilot. Starting the software development of an autopilot from scratch, unless supported by dozens of developers, is a lost cause against fast evolution in each of these dimensions [4]. It is therefore no surprise to observe that drone manufacturers which are mostly SME (small and medium-sized enterprises) are relying on existing open-source autopilots and adapt them to their needs.

One of the first open source autopilots were Paparazzi [3], Ardupilot (initially meant to target Arduino-based platforms) [1], and PX4 [2]. They all benefited from contributions from dozens to hundreds of users and developers. These open source autopilots have been designed in order to be portable as much as possible. For example, Paparazzi can be executed on bare metal platforms, but also on POSIX [5] compliant operating systems, while PX4 can be deployed on Nuttx Operating Systems [6] (OS) as well as on POSIX ones. Every specification depends on a specific frame, on a specific target hardware platform and on a specific operating system, on specific sensors and actuators, leads to a final customised autopilot, that can be considered as an instance of the original open source autopilot.

Since innovation relies on differentiation compared to the market, then custom behaviours, purely software or requiring specific hardware, have to be added to the autopilot instead of communicating with it as a black-box. This is the reason open-source autopilots should be considered as MOTS (Modifiable Off-The-Shelf).

In this paper, we propose a reverse design framework that allows engineers to understand and to modify the MOTS software easily. In other words, we propose an extraction of the open-source code and its visualisation as design models. These models enable to have an overview of the developed code and allows analysing the schedulability and timing performance of a modified design draft before being implemented and integrated. Our framework is model-based and implemented using model-driven engineering settings. It is equipped with a parsing engine, which extracts data and generates XML models. Those models are hence visualised using a specific Capella [7] view-point dedicated to embedded systems. In this paper, we consider mostly open-source autopilots as target MOTS, but we believe that the proposed methodology can be extended to other kinds of MOTS software.

The rest of this paper is organised as follows. Section 2 presents the background and motivations of this work. Section 3 presents our contribution. Section 4 and 5 present respectively the developed framework and its validation via Paparazzi autopilot. Finally, a conclusion and some perspectives are given.

2 Background and Work Positioning

This section aims to define the main concepts that are related to this work such as the UAVs, COTS and MOTS, and reverse engineering.

2.1 COTS and MOTS

Commercial off-the-shelf (COTS) software corresponds to products that are ready to use after configuration (without any code modification) by the user. Thus, they can be straightly integrated into a composite application, or used standalone. This facility has the disadvantage of being a source of potential security failure when it is used as a black box. Furthermore, its extension (e.g., adding functionalities) or customisation may be difficult, if not impossible. Even if the source code is available, an extension or modification may require important efforts to fully understand the design of the application. Raising the level of abstraction, by using reverse engineering, may be a key solution when design documentation is not provided. In contrast, a Modifiable off-the-shelf (MOTS) product is a software whose source code is modifiable and/or adapted. It requires design and documentation efforts from the vendor, but is adapted to the requirements or potential needs of the customer [8,9].

2.2 Autopilots of Unmanned Aerial Vehicles

These days, Unmanned Aerial Vehicles (UAV), or drones, are widespread and are used in both the civilian and military sectors. As explained in the previous section, open source autopilots are highly configurable, and are frameworks able to generate specific autopilots, which are meant to be extended with custom functions, software and/or hardware. In this regard, they are MOTS. Modifying such autopilots requires therefore a high expertise in computer science and knowledge of the source code of the autopilots. Integrating new sensors is made easy for the non-expert, with the usually well documented process of integrating modules conforming to an interface. Nevertheless, modifying the behaviour of the UAV, and especially modifying any functionality having to take place between the state estimation and the actuation is a very tedious process, with possible dangerous side effects.

Figure 1 represents the loop which is the core of a typical autopilot: at a frequency depending on the aircraft's inherent stability, a setpoint defines the wanted state of the drone. The setpoint can be given in some modes by the user, such as in assisted flight modes, or can be given in more autonomous modes by another loop, the flight guidance, which makes the trajectory of the UAV comply

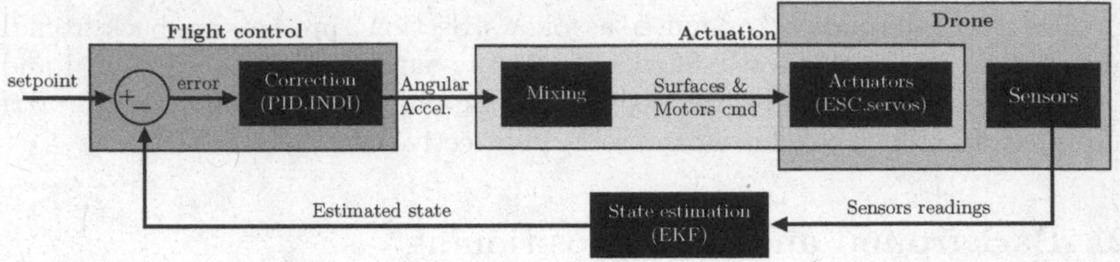

Fig. 1. Internal stability loop in an autopilot

to a plan. Giving the sensors readings, an estimation of the actual state is done. A state can be, for example, the angular speeds on each of the three-dimensional axis, as well as air and/or ground horizontal speeds, climb rate speed, as well as, for some specific cases, actual angle values on the three-dimensional axis. The difference between estimated state and setpoint is the error, which is usually corrected by several PID (Proportional, Integrator, Derivative) or PD, and more recently using INDI (Incremental Non-Linear Dynamic Inversion) controllers. These controllers compute torques to apply on each axis, converted to individual positions of surfaces or commands of rotors through a mixing process. This allows controllers to be relatively independent of the frame geometry.

2.3 UAV Autopilot Design

The design methods of a custom autopilot are thus different from what we observe for other embedded systems. Most other embedded systems use a top-down approach in their design life-cycle: from requirements, a functional decomposition can be derived independently of the hardware. Then the functions are mapped onto executable entities, at the low-level corresponding to processes and threads, themselves mapped to CPU, either with or without an operating system. This top-down approach has been used for decades now in several fields of embedded systems. Several methodologies are based on this top-down approach, starting with Structured Analysis for Real-Time (SA/RT) [10] in the 1980s, to Model-Driven Architecture (MDA) [11] launched by the Object Management Group (OMG) in the early 2000s, or the ARCADIA method tooled by Capella in the 2010s [12]. We also find this top-down approach in the automotive standard AUTOSAR [13], as well as in avionics with the DO-178C standards [14].

The development of UAV does not fit the top-down approach. It is usually requiring an instance of an autopilot, which implies a hardware platform supported by the chosen autopilot, an OS (or no OS) supported by the platform and the autopilot, compatible and supported sensors and actuators, a frame which can be completely custom-made or based on an existing Commercial off-the-shelf (COTS) frame. Then the added value can range from the specificity of the frame, to additional functions which are not implemented in the open source autopilot, to specific hardware. The development efforts consist in extending the instance of the autopilot to support the custom parts. In some cases, for some autopilots, this extension can be easy to integrate. The problem is that some custom parts are not only difficult to integrate, but may also compromise the smooth operation of the original autopilot or cause it to stop working.

2.4 Reverse Engineering

Chikofsky et al. [15] define reverse engineering as the process of analysing a system to identify its components and the relationships between them, and to create presentations at a higher level of abstraction. In computer science, especially for software, reverse engineering may be employed to retrieve source code from an executable, with the help of a decompiler. However, decompilers are generally not able to exactly reconstruct the original source code. This task is even more complex when the code has been obfuscated. Embedded systems software are mainly written in compiled languages (as C) which are then decompiled (disassembled) to low level and platform dependent assembly code.

When the source code is available, it can also be used to obtain a visualisation of the software design, for example via graphical modelling languages [16,17]. Several methods have been proposed in the 2000s s to represent abstract syntax trees of C/C++ code [18], and more recently, a lot of authors addressed the problem of getting this information from a binary. Nevertheless, we are interested in the multi-threaded program, and the ability to distinguish between native functions (e.g., operating system functions, drivers, low level input/output functions) and functions that embed functionality (e.g., navigation). For this, it is necessary to find the right level of granularity to obtain a model that is readable by a human.

2.5 Capella in a Nutshell

ARCADIA (Architecture Analysis & Design Integrated Approach) is a system and software architecture engineering method, based on the use of models with a focus on the collaborative definition, evaluation, and exploitation of its architecture. ARCADIA consists of four phases, as depicted in Fig. 2:

- **Customer operational requirements analysis**: defines the system users needs to be accomplished.
- **System requirements analysis**: defines what the system must accomplish for its users.
- **Logical architecture**: defines how the system will work to satisfy the system requirements.
- **Physical architecture**: defines how the system will be built.

The Capella workbench [7] is an open-source Eclipse application (Polarsys project). Capella implements the ARCADIA method, providing both a DSML (Domain Specific Modeling Language) and a dedicated toolset. It offers an entry point as a methodological guide of the ARCADIA method. This solution is used mainly for modeling complex and safety-critical systems development such as aerospace, avionics, transportation, space, communications, and security and automotive.

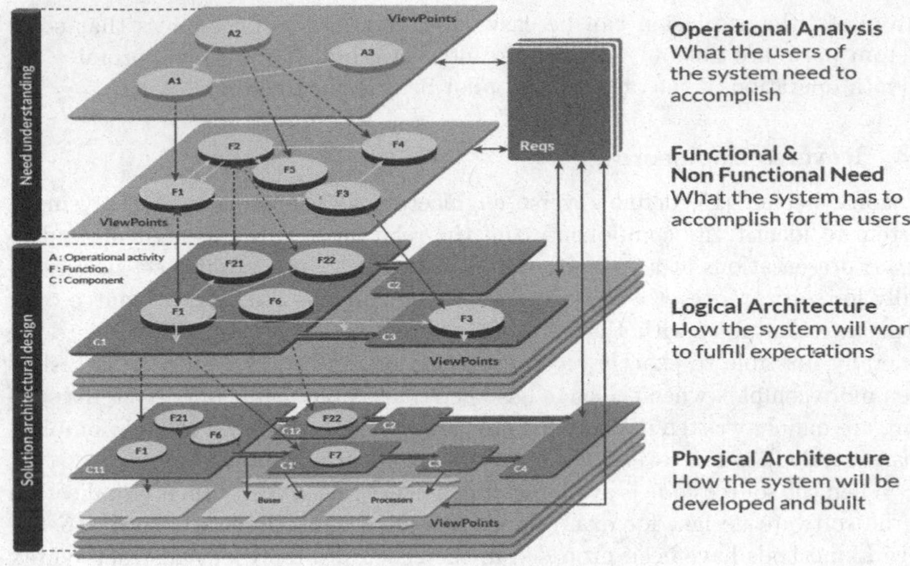

Fig. 2. ARCADIA engineering phases [12]

3 Model-Based Reverse-Engineering Framework

3.1 Overview

This section is dedicated to present a framework which is able, from a simply modified Makefile and C/C++ code, to extract from GNU SIMPLE (GIMPLE) [19] files generated during the compilation process, to represent in a Domain Specific Language (DSL) close to AADL [20], the set of threads, as well as their internal functions. In the process, middleware accesses are identified and represented. To echo the classic granularity problem in reverse engineering, some modules are considered as important modules, while others are considered as service modules. Service modules (e.g., module implementing functions to read and write on a serial bus) present functions seen as services, that should not be decomposed in sub-functions. On the contrary, important modules contain functions that are in general worth decomposing onto their sub-functions to allow the end-user to understand which functions are called. Choosing, for each module, if it is important or service, is done by hand once and for all for an autopilot framework by an expert.

By the following, we present our two contributions (see Fig. 3) based on the use of Model-Driven Engineering (MDE) tools, namely the customisation of the grammar and the extraction of the parse tree from the code behind the autopilots, and the transformation of the necessary elements of the code into components compatible with Capella and AADL-like extension in order to visualise them in a Capella diagram.

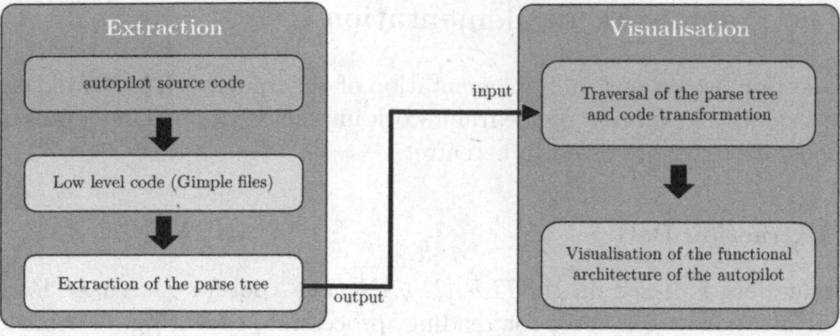

Fig. 3. Illustration of the framework's contributions

3.2 Extraction

After configuration of the target hardware and frame, COTS autopilots binaries are generated from C/C++ source code. In order to model an autopilot, it is thus necessary to extract a model from source code. The extraction step consists in retrieving the necessary information from the source code, in the form of a parse tree, for the purpose of visualisation. To retrieve the parse tree, it is necessary to parse the code with a parser such as Lex and Yacc, or more recent technologies such as Xtext, ANTLR, etc., that generate parsers directly by giving them as input a grammar expressed with a compatible DSL. Once the parser is set up, we input the code and generate the parse tree. The code behind the autopilots contains several C/C++ (.c/.cpp) and header (.h) files. This makes the task of extracting the information needed for visualisation difficult. To overcome this problem, we use the GCC compiler to generate GIMPLE code, the low-level three-address abstract code generated during the compilation process, that contains all the necessary information, including metadata about the functions at the time of their definition, inside files with the same extension which is ".lower".

3.3 Visualisation

This step consists in visualising the multithreaded and functional composition as well as the execution dataflow of the functions by showing the communications between them. It takes as input the parse tree generated at the end of the extraction step and gives as output a model that shows the set of threads, the set of functions and their sub-functions, the order of the function calls and the communications that happen between them.

The visualisation requires a text-to-model (T2M) transformation of the code, and this must be done all along the traversal of the parse tree. As the tree is being traversed, the elements of the tree are evaluated, and they are transformed into equivalent model elements compatible with the chosen visualisation tool. The objective of the visualisation is not for graphical aspects only but to also be able to analyse the modified code and check if it meets the non-functional requirements (like deadlines, end-to-end delays, etc.)

4 The Framework Implementation

In this section, we present an implementation of our framework presented earlier. Figure 4 shows an overview of the framework implementation. The details of the implementation are discussed in hereafter.

4.1 Extraction Part

Our framework is based on *ANTLR* [21](ANother Tool for Language Recognition). It is a parser generator for reading, processing, executing, or translating structured text or binary files. It's widely used to build languages, tools, and frameworks. From a grammar, *ANTLR* generates a parser that can build and walk parse trees.

The framework is composed of a parsing engine, which is the program that is responsible for the traversal and the transformation of the parse tree. The

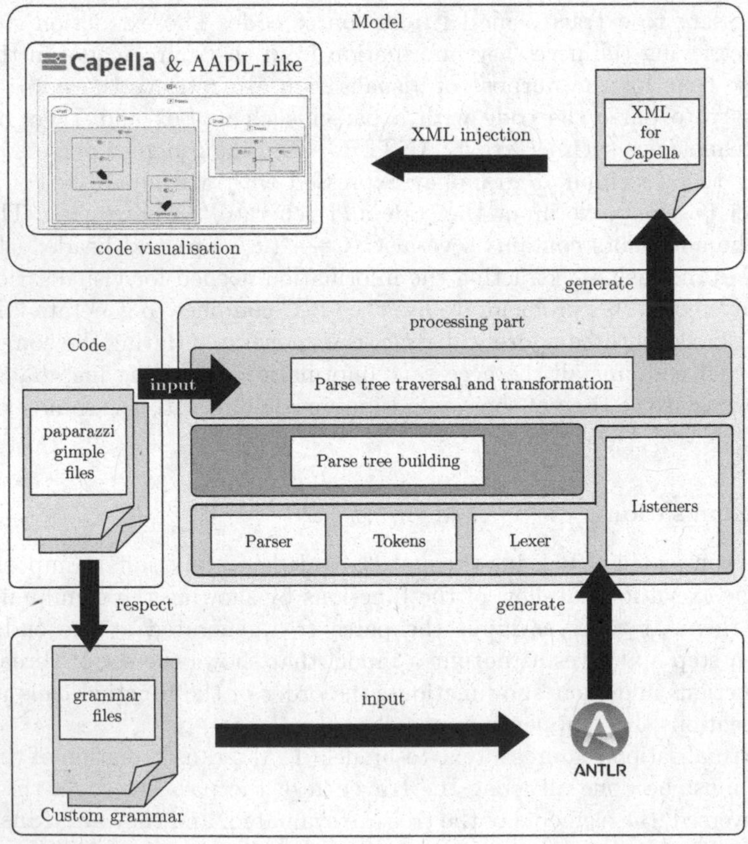

Fig. 4. Overview of the framework implementation

processing consists of three layers. From top to bottom, the program that performs the tree traversal and its text-to-text transformation layer. This program is built on top of the two other layers, which are provided by *ANTLR*, namely the built parse tree as well as the generated bricks (lexer, parser, tokens, and the listeners).

Building the parse tree consists in parsing the GIMPLE code (e.g., Paparazzi GIMPLE files) that is conforming to the GIMPLE grammar (see Listing 1.1) and requires the three given components of the first layer, namely the Parser, the Lexer, and the Tokens. Once the parse tree is built, it is then transformed into XML code. This process requires the generated listeners of the first layer.

```
      ...
functionDefinition
  : attributeSpecifierSeq? declSpecifierSeq? declarator
    virtualSpecifierSeq? functionBody
  | gimplePreamble? declarator virtualSpecifierSeq ?
    functionBody
  ;

functionBody:
  constructorInitializer? compoundStatement
  | functionTryBlock | Assign (Default | Delete) Semi;
      ...
}
```

Listing 1.1. Code snippet of the GIMPLE grammar.

4.2 Visualisation Part

As basis, we opt for Capella [7] as a tool to represent function decomposition, since it has a lot of facilities. Moreover, it is a tool implementing the ARCADIA [12] method, a well adopted top-down approach for designing embedded systems. It allows the creation of specific viewpoints. The choice of Capella also relies on the observation that in the embedded industry in France, this tool and supported methodology is increasingly used. We claim that one of its limitations is that the physical point of view (threads, processes, CPUs, networks) is not as readable as an AADL representation.

We therefore created an AADL viewpoint, since this DSL is an Architecture Description Language well suited to describe software and hardware architectures. Compared to the standard AADL, we added some specific modelling artefacts related to autopilot architectures, such as an abstract view of middlewares inputs and outputs.

The metamodel of the AADL-Like view point is shown in Fig. 5. This metamodel extends the metamodel of Capella by leveraging the existing concepts such as the `Physical component`, and it introduces the AADL elements, namely the `AADLProcess`, `AADLThread`, `AADLFunction`, `AADLThread ports`, `AADLFunction ports`, etc.

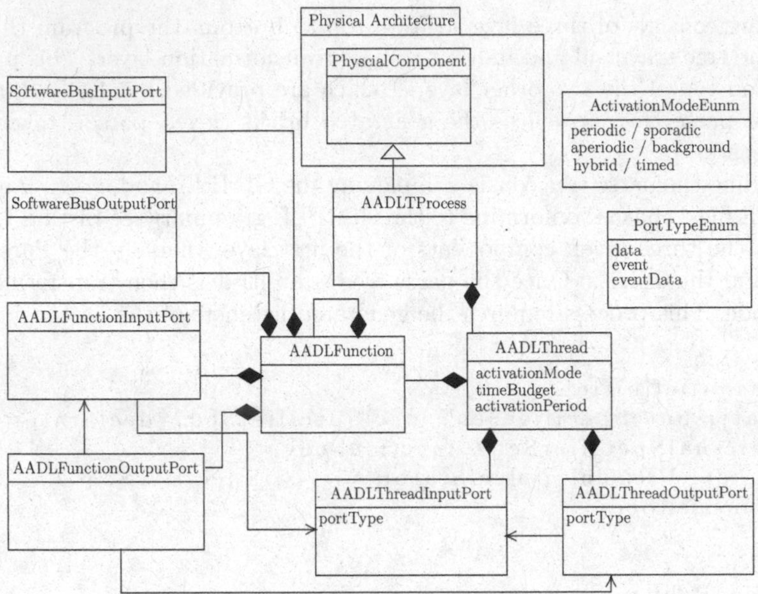

Fig. 5. AADL-Like metamodel

Table 1 summarises the mapping relationships between the GIMPLE statements and Capella concepts.

Table 1. Transformation mapping table

GIMPLE statement	Capella concept
pthread_create	AADLThread
pthread_create 3rd parameter	AADLFunction
Function call inside function definition	Sub-function (AADLFunction)
Global variable write access	SoftwareBusOutputPort
Global variable read access	SoftwareBusInputPort
Global variable read and write access in the same function	FunctionExchange

The tool palette shown in Fig. 6 is inspired from the graphical aspect of AADL [20]. It consists of some Capella existing artefacts such as the physical component that we consider as a processing unit, the physical actor or device which represents the sensors and actuators, and the physical link. It provides the AADL-like artefacts such as threads with different variants (periodic, sporadic, aperiodic, etc.), the different inter-thread communications (synchronous, asynchronous, and reset), the function component or sub-program as well as the functional exchange that connects two different functions.

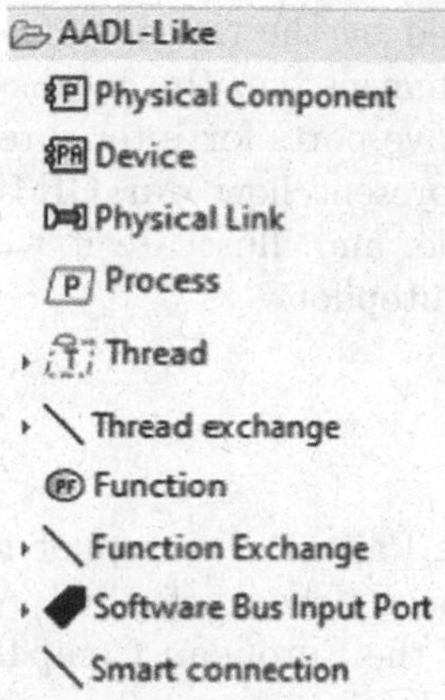

Fig. 6. AADL-like tool palette

The software bus is a newly introduced concept coming from the Comp4Drones European project. It does not exist in AADL, yet it is an important and widely used in the existing autopilots. This latter must exist in the diagram, since it represents a special kind of communication between functions. The last element is the smart connection, which creates the connection between the different components (thread-thread, function-thread, function-function, port-port, etc.) two functions that belong to two different threads can be connected simply by selecting the source and then destination functions, which will create different ports and different edges, thus the whole connection.

Other concepts missing from most existing ADLs, which have been identified by the members of the Comp4Drones European project, are mainly the lack of a description of a non-preemptive cyclic executive at the core of the autopilot, including "tasks" (function synchronous calls) which have a period of several cycles of the executive. Another important missing concept is the lack of an ontology used to classify functions in corresponding classic autopilots' blocks. We did not yet address this latter part in our framework.

The viewpoint allows a multi-level functional breakdown, i.e., functions can have a set of sub-functions, and every sub-function can be composed of another set of sub-functions and so on. Indeed, this functionality is not supported by *Capella* at the physical architecture level, it is necessary to show the different levels of function calls.

The visualisation is performed in the *physical architecture* step of the ARCADIA method inside a *Physical Architecture Blank* diagram (PAB). It requires the activation of the layer provided by the AADL-like viewpoint. The graphical representation consists of the physical component, which we consider as the processor. The process, the threads, the functions, software buses and different exchanges.

The processor is composed of the process that is executed on top of it. The process is composed of threads. The threads are composed of functions or subprograms that are executed by the threads. The functions have ports that constitute the functional exchanges, i.e. the means of communication between the functions. Threads also have ports for inter-thread communication.

In the next section, we present how our GIMPLE interpreter creates an AADL model from a Makefile, and illustrate it with examples obtained from retro-engineering Paparazzi autopilot.

5 Validation

This work has been tested on Paparazzi autopilot for validation purposes with the research group of National School of Civil Aviation[1] (ENAC) and the Paparazzi founders as part of the European Comp4Drones project[2]. The developers believe that the autopilot instances customisation using the framework presented in this work will be much easier than by direct accessing to the code. We present hereafter some excerpts of our framework utilisation to modify the Paparazzi code.

The diagram elements are extracted along the traversal of the parsing tree. This operation requires the recognition of the code statements corresponding to the diagram elements. For instance, the statement `pthread_create ()` (see Line 2, Listing 1.2) responsible for the creation of the POSIX thread will be translated to a thread in the diagram. The created thread takes the name of the function executed by the thread designated by the third argument of the `pthread_create ()` statement. Once the thread is created, the function is created inside it. To create the sub-functions, the definition of this later must be found first. Then, every function call is transformed into a sub-function (see Listing 1.2, Lines 5–9).

```
1 // i2c_thread thread creation
2 _1 = pthread_create (&tid, 0B, i2c_thread, p);
3 ...
4 // i2c_thread function exectued by the thread
5 i2c_thread (void * data) {
6 // sub-function
7     get_rt_prio();
8     ...
9 }
```

Listing 1.2. GIMPLE code to be transformed

Figure 7 shows the result of the reverse engineering process applied to the Paparazzi code. It consists of multiple threads, where each thread is composed of interconnected functions. Due to space limitations, the figure cannot be presented in its entirety in this article. However, we have zoomed in to show some details.

[1] http://optim.recherche.enac.fr/.
[2] https://www.comp4drones.eu/.

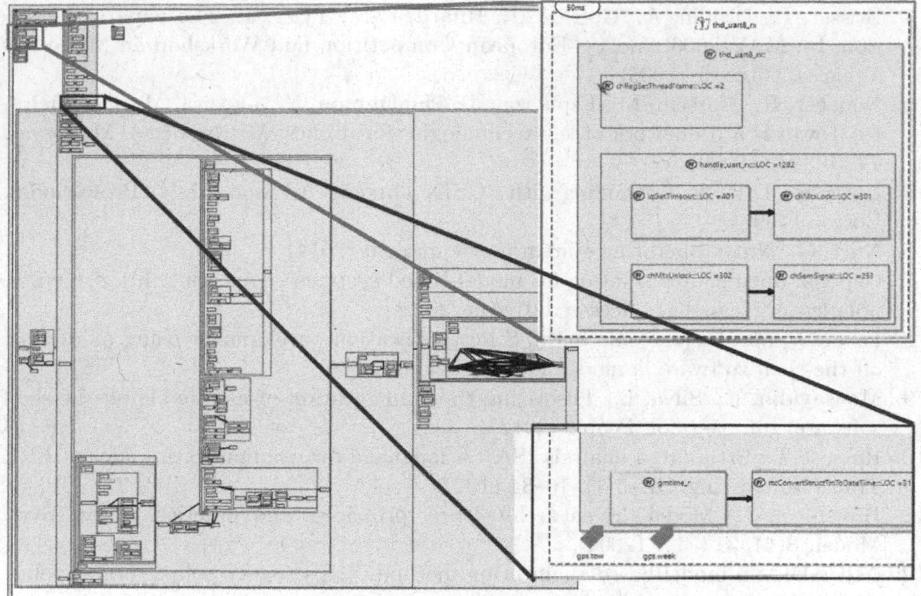

Fig. 7. Result of the reverse engineering process under Capella.

6 Conclusion

This paper presented a model-based framework for reverse engineering allowing the visualisation of the functional structure of a given input source code and more precisely the source code of autopilots. The objective behind this work is to make autopilots MOTS, i.e., software that can be customised according to the user's needs. Indeed, the source code of an autopilot can be visualised to be well understood and to analyse the performance and non-functional properties of the modification at an early design step. We believe that this framework can shorten sharply the design process of MOTS software. The framework has been demonstrated on a concrete example, namely the open source autopilot Paparazzi, in the context of the European project Comp4drones.

Acknowledgement. This work has received funding from the European Union's Horizon 2020 research and innovation program under grant agreement N. 826610.

References

1. Bin, H., Justice, A.: The design of an unmanned aerial vehicle based on the ardupilot. Indian J. Sci. Technol. **2**(4), 12–15 (2009)
2. Meier, L., Honegger, D., Pollefeys, M.: PX4: a node-based multithreaded open source robotics framework for deeply embedded platforms. In: 2015 IEEE International Conference on Robotics and Automation (ICRA), pp. 6235–6240. IEEE (2015)

3. Brisset, P., Drouin, A., Gorraz, M., Huard, P.-S., Tyler, J.: The paparazzi solution. In: MAV 2006, 2nd US-European Competition and Workshop on Micro Air Vehicles. Citeseer (2006)
4. Nouacer, R., Hussein, M., Espinoza, H., Ouhammou, Y., Ladeira, M., Castiñeira, R.: Towards a framework of key technologies for drones. Microprocess. Microsyst. **77**, 103142 (2020)
5. Butenhof, D.R.: Programming with POSIX Threads. Addison-Wesley Professional, Boston (1997)
6. Nutt, G.: Nuttx operating system user's manual (2014)
7. Capella: open source solution for model-based systems engineering. https://www.polarsys.org/capella/. Accessed 01 Aug 2022
8. Feng, Q., Mookerjee, V.S., Sethi, S.P.: Application development using modifiable off-the-shelf software: a model and extensions (2005)
9. Mousavidin, E., Silva, L.: Theorizing the configuration of modifiable off-the-shelf software. Inf. Technol. People (2017)
10. Ross, D.T.: Structured analysis (SA): a language for communicating ideas. IEEE Trans. Softw. Eng. **SE-3**(1), 16–34 (1977)
11. Brown, A.W.: Model driven architecture: principles and practice. Softw. Syst. Model. **3**(4), 314–327 (2004)
12. ARCADIA: a model-based engineering method. https://www.eclipse.org/capella/arcadia.html. Accessed 01 Aug 2022
13. AUTOSAR. The standardized software framework for intelligent mobility
14. Brosgol, B.: DO-178C: the next avionics safety standard. ACM SIGAda Ada Lett. **31**(3), 5–6 (2011)
15. Chikofsky, E.J., Cross, J.H.: Reverse engineering and design recovery: a taxonomy. IEEE Softw. **7**(1), 13–17 (1990)
16. Booch, G., Rumbaugh, J., Jackobson, I.: UML: unified modeling language. Versão (1997)
17. Wood, J., Silver, D.: Joint Application Development. Wiley, Hoboken (1995)
18. Ferenc, R., Sim, S.E., Holt, R.C., Koschke, R., Gyimóthy, T.: Towards a standard schema for C/C++. In: Proceedings Eighth Working Conference on Reverse Engineering, pp. 49–58. IEEE (2001)
19. Gimple (GNU compiler collection (GCC) internals). Accessed 01 Aug 2022
20. SAE. SAE. Architecture analysis and design language V2.0 (AS5506), September 2008. https://www.sei.cmu.edu/our-work/projects/display.cfm?customel_datapageid_4050=191439www.aadl.info
21. ANTLR (another tool for language recognition). https://www.antlr.org/. Accessed 01 Aug 2022

Efficient Checking of Timed Ordered Anti-patterns over Graph-Encoded Event Logs

Nesma M. Zaki[1], Iman M. A. Helal[1], Ehab E. Hassanein[1], and Ahmed Awad[1,2](✉)

[1] Cairo University, Giza, Egypt
{n.mostafa,i.helal,e.ezat,a.gaafar}@fci-cu.edu.eg
[2] University of Tartu, Tartu, Estonia

Abstract. Event logs are used for a plethora of process analytics and mining techniques. A class of these mining activities is conformance (compliance) checking. The goal is to identify the *violation* of such patterns, i.e., anti-patterns. Several approaches have been proposed to tackle this analysis task. These approaches have been based on different data models and storage technologies of the event log including relational databases, graph databases, and proprietary formats. Graph-based encoding of event logs is a promising direction that turns several process analytic tasks into queries on the underlying graph. Compliance checking is one class of such analysis tasks.

In this paper, we argue that encoding log data as graphs alone is not enough to guarantee efficient processing of queries on this data. Efficiency is important due to the interactive nature of compliance checking. Thus, anti-pattern detection would benefit from sub-linear scanning of the data. Moreover, as more data are added, e.g., new batches of logs arrive, the data size should grow sub-linearly to optimize both the space of storage and time for querying. We propose two encoding methods using graph representations, realized in Neo4J & SQL Graph Database, and show the benefits of these encoding on a special class of queries, namely timed ordered anti-patterns. Compared to several baseline encoding, our experiments show up to $5x$ speed up in the querying time as well as a $3x$ reduction in the graph size.

Keywords: Anti pattern detection · Process mining · Graph-encoded event logs

1 Introduction

Organizations strive to enhance their business processes to achieve several goals: increase customer satisfaction, gain more market share, reduce costs, and show adherence to regulations among other goals. Process mining techniques [1] collectively help organizations achieve these goals by analyzing execution logs of organizations' information systems. Execution logs, a.k.a. event logs, group events representing the execution of process steps into process instances (cases).

P. Fournier-Viger et al. (Eds.): MEDI 2022, LNAI 13761, pp. 147–161, 2023.
https://doi.org/10.1007/978-3-031-21595-7_11

Conformance checking [6], in specific, provides techniques to analyze the deviation of the recorded behavior against a predefined process model.

Compliance checking [22] is a specialization of conformance checking in which event logs are checked against compliance rules that might restrict process behavior w.r.t control flow, data, resources, and timing. Moreover, such rules are of a local nature. That is, they are not concerned with the end-to-end conformance of the process instance. Rather, they refer to the execution ordering of a subset of the activities and their timing constraints. For example, in a ticketing system, there might be a rule that the time taken by creating a ticket and the first contact with the client should not exceed three hours. Compliance checking is an interactive and repetitive task by nature due to changes in the compliance requirements. Compliance rules usually follow common patterns [22]. The objective of compliance checking is to identify process instances that violate the rules. As compliance checking is an interactive process, event logs should be stored following data models that allow efficient access. Moreover, user-friendly domain-specific languages, e.g., declarative query languages, allow non-technical users to access and analyze the data. Recently, the graph data model has been investigated to store and query event logs [5,9,14].

In this paper, we adopt the graph data model to represent event logs. Namely, we use the labeled property graph model [13]. We propose an encoding method of event logs as graphs that can efficiently check compliance by translating compliance rules into queries. We address a special type (pattern) of rules: *order* patterns. However, our graph representation can address the rest of the patterns. We leave this discussion out due to space limitations. Namely, we make the following contributions: – We propose a graph representation of event logs that help efficiently check for compliance with *order* rules, – We realize our encoding on top of two graph databases, Neo4J as a native graph database and graph extensions of Microsoft SQL Server, a layer on top of relational tables. – We empirically evaluate our method against a baseline graph representation and relational data models on a set of four real-life event logs, – We discuss the improvements in the stored graph sizes and the simplification of the queries to check compliance. Overall, the compliance checking, i.e., querying time is improved by $3x$ to $5x$ whereas the sizes of the graphs are $\sim 3x$ reduced compared to the baseline method.

The rest of this paper is organized as follows: Sect. 2 briefly discusses some of the background concepts and techniques that are used throughout the paper. Related work is discussed in Sect. 3. Section 4 presents our approach. In Sect. 5, we evaluate the proposed approach against the existing one.

2 Background

2.1 Events, Traces, Logs, and Graphs

We formalize the concepts of events, traces, logs and graphs to help in understanding the formalization introduced later in the paper.

Definition 1 (Event). *An event e is a tuple $(a_1, a_2, \ldots, a_n)$ where a_i is an attribute value drawn from a respective domain $a_i \in D_i$. At least three domains and their respective values must be defined for each event e: D_c, the set of case identifiers, D_a, the set of activity identifiers, and D_t, the set of timestamps. We denote these properties as e.c, e.a, and e.t respectively. Other properties and domains are optional such as D_r, the resources who perform the tasks, D_l, the lifecycle phase of the activity.*

We reserve the first three properties in the event tuple to reflect the case, the activity label, and the timestamp properties.

Definition 2 (Trace). *A trace is a finite sequence of events $\sigma = \langle e_1, e_2, \ldots, e_m \rangle$ where e_i is an event, $1 \leq i \leq m$ is a unique position for the event that identifies the event e_i in σ and explicitly positions it, and for any $e_i, e_j \in \sigma : e_i.c = e_j.c$*

Definition 3 (Event log). *An event log is a finite sequence of events $\mathcal{L} = \langle e_1, e_2, \ldots, e_m \rangle$ where events are ordered by their timestamps for any e_i and $e_{i+1} : e_i.t \leq e_{i+1}.t$.*

In general, graph data models can be classified into two major groups [13]: directed edge-labeled graphs, e.g., RDF, and labeled property graphs. In the context of this paper, we are interested in labeled property graphs as they provide a richer model that represents the same data in a smaller graph size.

Definition 4 (Labeled property graph). *Let L, K, and V be the sets of labels, keys, and values, respectively. A labeled property graph $\mathcal{G} = (N, E, label, prop)$ tuple, where N is a non-empty set of nodes, $E \subseteq N \times N$ is the set of edges. $label : (N \cup E) \rightarrow 2^L$ is a labeling function to nodes and edges. $prop : (N \cup E) \times K \rightarrow V$ is a function that assigns key-value pairs to either nodes or edges.*

When mapping from logs to graphs, we assume overloading of a function *node*() that identifies the corresponding node in the graph to the input parameter of the function. For instance for an event *e*, Definition 1, *node*(*e*) returns the corresponding node *n* in $\mathcal{G}.N$ that represents the encoding of *e*. Similarly, for the case identifier *e.c*, *node*(*e.c*) returns the node that corresponds to the case in the graph. Finally, for the activity label *e.a*, *node*(*e.a*) returns the node that corresponds to the respective activity label.

2.2 Activity Order Patterns

A trace is compared to a process model in traditional conformance checking [6] to quantify the deviation between the required behavior (the model) and the observed behavior (the trace). In many circumstances, checking deviations at a finer granularity, such as on the level of activities may be required, e.g., absence, existence, or pairs of activities, such as co-existence, mutual exclusion, along with time window constraints. Such finer granularity checks are referred to as compliance checking, and *compliance patterns* are used for categorizing the types of compliance requirements [22].

Occurrence patterns are concerned with activities having been executed (*Existence*) or not (*Absence*) within a process instance. *Order patterns* are concerned with the execution order between pairs of activities. The *Response* pattern (e.g., *Response(A, B)*) states that if the execution of activity A is observed at some point in a process instance, the execution of activity B must be observed at some future point of the same case before the process instance is terminated. A temporal window can further restrict these patterns. For instance, we need to observe B after A in no more than a certain amount of time. Alternatively, we need to observe B after observing A, where at least a certain amount of time has elapsed. Both patterns can be further restricted by so-called *exclude* constraint [4]. That is, between the observations of A and B, it is prohibited to observe any of the activities listed in the exclude constraint. Definition 5 formalizes the *Response* pattern.

Definition 5. ***Response*** – *Given two activities a and b and a trace* $\tau = \langle e_1, e_2, \ldots, e_n \rangle$, $\tau \in \mathcal{L}$, *we say that* $\tau \models Response(a, b, S, \Delta t, \theta)$ *if and only if* $\forall e_i \in \tau : e_i.a = a \; \exists \; e_j \in \tau : e_j.b = b \wedge e_i.t \leq e_j.t \wedge (e_j.t - e_i.t) \; \theta \; \Delta t \wedge \forall e_k$ *where* $i < k < j : e_k.a \notin S$, *where* Δt *represents the time window between occurrences of a and b,* $\theta \in \{<, =, >, \leq, \geq\}$ *represents when (e.g., after, before, or exactly at) we expect the observation of B after A with respect to* Δt, *and S is the set of excluded activities between a and b.*

Conversely, the *Precedes* pattern (e.g., *Precedes(A,B)*) states that if the execution of activity B is observed at some point in the trace, A must have been observed before (Definition 6).

Definition 6. ***Precedence*** – *Given two activities a and b and a trace* $\tau = \langle e_1, e_2, \ldots, e_n \rangle$, $\tau \in \mathcal{L}$, *we say that* $\tau \models Precedes(a, b, S, \Delta t, \theta)$ *if and only if* $\forall e_i \in \tau : e_i.b = b \; \exists \; e_j \in \tau : e_j.a = b \wedge e_j.t \leq e_i.t \wedge (e_j.t - e_i.t) \; \theta \; \Delta t \wedge \forall e_k$ *where* $i < k < j : e_k.a \notin S$. *where* Δt *represents the time window between occurrences of a and b,* $\theta \in \{<, =, >, \leq, \geq\}$ *represents when (e.g., after, before, or exactly at) we expect the observation of A before B with respect to* Δt, *and S is the set of excluded activities between a and b.*

Note that we get the unrestricted form of both patterns by setting Δt to a very large value and θ is set to $\leq$. That is, in $Response(A, B, \phi, \infty, \leq)$, B has to *eventually* be observed after A with no further restrictions on the time window nor restrictions on activities observed in between.

Response and precedence patterns cover the occurrence and absence patterns. For instance, if we reserve special activities, *start* and *end*, to indicate the beginning and the termination of a case respectively, we can model the occurrence pattern of some activity A as $Response(start, A, \phi, \infty, \leq)$. Similarly, we can specify it as $Precedence(A, end, \phi, \infty, \leq)$. We can specify the absence pattern as $Response(start, end, \{A\}, \infty, \leq)$. That is, we require not observing any event whose activity is A from the beginning to the end of the trace. In literature, the precedence and response patterns are indeed families of patterns [18,22].

However, in this paper, we focus on the core response and precedence patterns, due to space limitations.

When checking for compliance, analysts are interested in identifying process instances, i.e., cases that contain a violation, rather than those that are compliant. Therefore, it is common in the literature about compliance checking to use the term "anti-pattern" [16]. In the rest of this paper, we refer to anti-patterns rather than patterns when presenting our approach to detecting violations over graph-encoded event logs.

3 Related Work

There is vast literature about the business process compliance checking domain. For our purposes, we focus on compliance checking over event logs; we refer to this as auditing. For more details, the reader can check the survey in [12].

Auditing can be categorized in basic terms based on the perspective of the process, including control flow, data, resources, or time. We can also split these categories based on the formalism and technology that underpins them. Agrawal et al. [3] presented one of the first works on compliance auditing, in which process execution data is imported into relational databases and compliance is verified by recognizing anomalous behavior. Control-flow-related topics are covered by the technique.

Validating process logs against control-flow and resource-aware compliance requirements has been proposed while applying model checking techniques [2]. For control-flow and temporal rules, Ramezani et al. [19,20] suggest alignment-based detection of compliance violations.

De Murillas et al. [17] present a metamodel and toolset for extracting process-related data from operational systems logs, such as relational databases, and populating their metamodel. The authors show how different queries can be translated into SQL. However, such queries are complex (using nesting, joins, and unions). Relational databases have also been used for declarative process mining [23], which can be seen as an option for checking logs against compliance rules.

Compliance violations, i.e. anti-patterns can be checked by `Match_Recognize` (MR), the ANSI SQL operator. MR verifies patterns as regular expressions, where the tuples of a table are the symbols of the string to search for matches within. MR runs linearly through the number of tuples in the table. In our case, the tuples are the events in the log. In practice, the operational time can be enhanced by parallelizing the processing, e.g., partitioning the tuples by the case identifier. Still, this does not change the linearity of the match concerning the number of tuples in the table. A recent work speeds up MR by using indexes in relational databases [15] for strict contiguity patterns, i.e., patterns where events are in strict sequence. Order compliance patterns frequently refer to eventuality rather than strict order, limiting the use of indexes to accelerate the matching process.

Storing and querying event data into an integrated graph-based data structure has also been investigated. Esser et al. [9] provide a rich data model for multi-dimensional event data using labeled property graphs realized on Neo4j as a graph database engine. To check for compliance, the authors use path queries.

Such queries suffer from performance degradation when the distance between activities in the trace gets longer and when the whole graph size gets larger.

4 Graph-Encoded Event Logs for Efficient Compliance Checking

Graph representation of event logs is a promising approach for event logs analysis [5], especially for compliance checking [9]. This is due to the richness of this graph representation model, mature database engines supporting it, e.g., Neo4J[1], and the declarative style of the query languages embraced by such engines, e.g., Cypher[2]. In this sense, compliance checking can be mapped to queries against the encoded log to identify violations.

We show how encoding of the event log has a significant effect on the efficiency of answering compliance queries. We start from a baseline approach (Sect. 4.1) and propose a graph encoding method, Sect. 4.2, that leverages the finite nature of event logs to store the same event log in a smaller graph and answer compliance queries faster.

Table 1 shows an excerpt of a log that serves as the input to the different encoding methods. In the "Optional details" columns, the "StartTime" and "CompleteTime" columns are converted to Unix timestamp.

Table 1. Sample event log with additional attributes

C.ID	Activity	Resource	StartTime	CompleteTime	Position
1	A	Jack	1612172052	1612373652	1
1	B	John	1612360812	1612458012	2
2	A	Mark	1609491612	1609866012	1
1	E	Smith	1612602012	1612778412	3
3	A	George	1614589212	1614682812	1
2	C	Albert	1609678812	1609866012	2
1	D	Mark	1612954800	1613131200	4
2	E	Smith	1611838812	1612026012	3
3	E	Albert	1614934800	1615374000	2
3	C	Jack	1615107612	1615374012	3
2	D	John	1612256400	1612346400	4
3	E	Mark	1615539600	1615719600	4
3	D	George	1615546812	1615640412	5
⋮	⋮	⋮	⋮	⋮	⋮
Minimum details		Optional details			Added detail

[1] https://neo4j.com/.

[2] Cypher for Neo4J is like SQL for relational databases.

4.1 Baseline: Multi-dimensional Graph Modeling (BM)

Esser at al. [9] proposed a multi-dimensional graph data model to represent event logs. It uses labeled property graphs, cf. Definition 4, for the representation. *Multi-dimensionality* is proposed as a flexible definition of a *case* notion. However, for the scope of this paper, we will stick to the traditional definition of the case identifier. Yet, this simplification does not limit our contribution. Our proposed encoding methods can be generalized to any case notion embraced in the context of a specific compliance checking practice.

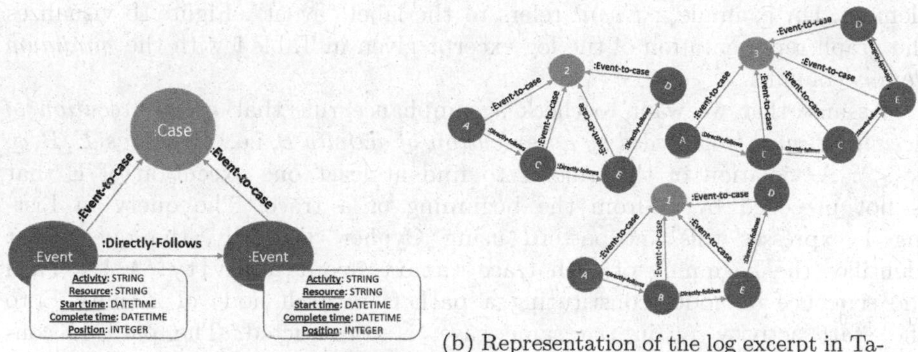

(a) Nodes, edges, labels, and properties

(b) Representation of the log excerpt in Table 1

Fig. 1. Baseline graph representation

Events and cases constitute the nodes of the graph. Node types, i.e., events, cases, etc., are distinguished through labels. Edges represent either structural or behavioral relations. Structural relations represent event-to-case relations. Behavioral relations represent the execution order among events in the same case, referred to as directly-follows relationships. Activity labels, resource names, activity lifecycle status, and timestamps are modeled as properties of the event nodes. Similarly, case-level attributes are modeled as case node properties. Figure 1a shows the representation of the Baseline graph.

Formally, for each log $\mathcal{L}$, cf. Definition 3, a labeled property graph $\mathcal{G}$, cf. Definition 4, is constructed by Esser et al. [9] approach as follows:

1. Labels for the graph elements are constituted of four literals. Formally, $L = \{event, case, event\_to\_case, directly\_follows\}$,
2. Keys for properties are the domain names from which values of the different event attributes are drawn. Formally, $K = \bigcup_{i=1}^{m}\{name(D_i)\} \cup \{ID\}$,
3. For each unique case in the log, there is a node in the graph that is labeled as "case" and has a property ID that takes the value of the case identifier. Formally, $\forall c \in D_c \; \exists \; n \in \mathcal{G}.N : \{case\} \in label(n) \wedge prop(n, ID) = c$,
4. For each event in the log, there is a node in the graph that is labeled as "event" and inherits this event's properties. Formally, $\forall e = (p_1, p_2, \ldots, p_m) \in \mathcal{L} \; \exists \; n \in \mathcal{G}.N : \{event\} \in label(n) \wedge \forall_{2 \leq i \leq m} D_i : prop(n, name(D_i)) = p_i$,

5. The structural relation between an event and its case is represented by a labeled relation. Formally, $\forall e \in \mathcal{L}, \exists\ r \in \mathcal{G}.E : r = (node(e), node(e.c)) \land \{event\_to\_case\} \in label(r)$,
6. The behavioral relationship between a pair of successive events in a trace is represented by a labeled relation between their respective nodes. Formally, for $e_i, e_{i+1} \in \tau\ \exists\ r \in \mathcal{G}.E : r = (node(e_i), node(e_{i+1})) \land \{directly\_follows\} \in label(r)$

In the following we adopt Cypher's notation to reflect on nodes, their labels, and their properties. We use the notation $: Label$ to refer to the "label" of a graph element. For example, $: Event$ refers to the label "event". Figure 1b visualizes the graph representation of the log excerpt given in Table 1 with the *minimum details* columns.

Assume that we want to check a compliance rule that *every execution of activity E must be preceded by an execution of activity B*, i.e., $Precedes(E, B, \phi, \infty, <)$. A violation of this rule is to find at least one execution of E that is not preceded by B from the beginning of a trace. The query in Listing 1 expresses this anti-pattern using Cypher. Basically, the query first identifies the beginning of each trace `(start:Event{activity:'A'})`. Then the sequence of nodes constituting a path from each node of activity E to the start activity A, in the same case, is constructed. The path is constructed by traversing the transitive closure of the *:Directly_follows* relation, `path=(e1:Event{activity:'E'}<-[:Directly_follows*]- (start))`. If the path does not include any node whose activity property refers to B, as in line 3, a violation exists, and this case is reported.

Listing 1. Precedes anti-pattern query using the baseline encoding

```
1| Match (c1:Case) <-[:Event_to_case]- (start:Event{activity:'A'})
2| Match path = (e1:Event{activity:'E'}<-[:Directly_follows*]-(start))
3| where none (n in nodes(path) where n.activity='B')
4| return c.ID
```

Although the query is expressive and captures the semantics of the violation, it is expensive to evaluate. To resolve those nodes, the processing engine must scan the *:Directly_follows* relation linearly. Another problem is the linear growth of the graph size w.r.t. the log size. In Fig. 1b, we observe that each time an activity occurs in the log, a distinct node is created in the graph.

In the following subsection, we propose a concise representation of the event log that improves both the space and time required to store and query the log.

4.2 Unique Activities (UA) Encoding

Many compliance patterns are concern is constructed in the same cases in process execution and their ordering. When checking such rules against event traces, we can exploit the finiteness of these traces and the positions of events within traces to simplify the queries and speed up their evaluation by utilizing indexes and skipping the linear scan of the *:Directly_follows* relation among events. So, we extend the baseline mapping by explicitly assigning a *position* property to each event node. Table 1 has a highlighted column, tagged as *added detail* column, where we assign each event to a position in the case (trace). For instance, the fourth row in Table 1 records that activity 'E' has been the third activity to be executed in case 1. Thus, the position property value is 3. We can observe that the check for ordering explicitly refers to the *position* property of the event nodes without the expensive transitive closure traversal. We follows the same formalism shown in Sect. 4.1, except for encoding the $directly\_follows$ relation as we add the explicit *position* property to event nodes. The dropping of such a relation positively affects the graph size.

Although the position property simplifies the processing of compliance queries, it inherits the linear growth of the graph size w.r.t the log size. To further limit the growth of the graph size, we modify the construction of the labeled property graph. This section's proposed encoding ensures a linear growth with the size of the set of activity labels, i.e., D_a. We generate a separate edge connecting a case node to the corresponding node representing the activity $a \in D_a$ and add properties to the edges that reflect the position, timestamp, resource, etc. These events' properties represent the activity's execution in the respective case. Formally, for each log $\mathcal{L}$, a labeled property graph $\mathcal{G}$ is constructed as follows:

1. Labels for the graph elements are constituted of case and activity labels. Formally, $\mathcal{L} = \{case, event\_to\_case\} \cup D_a$,
2. Keys for properties are the domain names from which values of the different event attributes are drawn. Formally, $K = \bigcup_{i=1}^{m} \{name(D_i)\} \cup \{ID\}$,
3. For each unique case in the log, there is a node in the graph labeled as "case" with a property ID that takes the value of the case identifier. Formally, $\forall c \in D_c \; \exists \; n \in \mathcal{G}.N : \{case\} \in label(n) \wedge prop(n, ID) = c$,
4. For each unique activity in the log, there is a node in the graph labeled as "activity". Formally, $\forall a \in D_a \; \exists \; n \in \mathcal{G}.N : \{activity\} \in label(n)$,
5. The structural relation between an event and its case is represented by a labeled relation between the activity node of the event's activity and the case node. Additionally, all event-level properties are mapped to properties on edge. Formally, $\forall e = (p_1, p_2, \ldots, p_m) \in \mathcal{L} \; \exists \; r \in \mathcal{G}.E : r = (node(e.a), node(e.c)) \wedge \{event\_to\_case\} \in label(r) \wedge \forall_{3 \leq i \leq m} D_i : prop(r, name(D_i)) = p_i$.

Figure 2 visualizes the graph resulting from encoding the log excerpt in Table 1 using the unique activities method. For example, for activity *E*, there is only one node and four different edges connecting to cases 1, 2, *and* 3. Two of these edges connect case 3, as activity *E* was executed twice in this case.

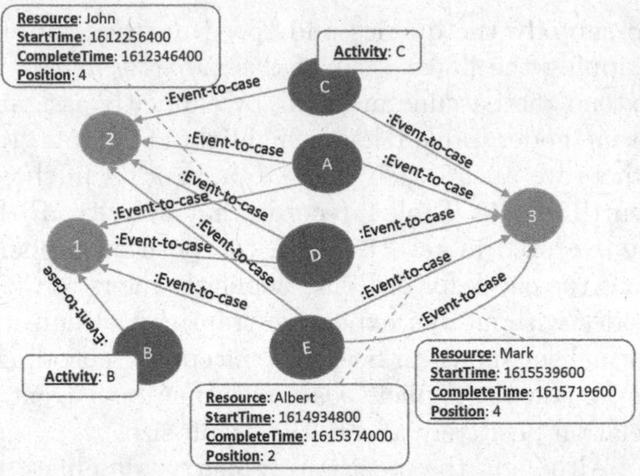

Fig. 2. UA Encoding of Table 1 log

Listing 2 shows the modification on the *Precedence* anti-pattern query. The query checks the ordering of the events using the *position* property, which is accessed in Line 2. With this encoding, the graph size grows sub-linearly w.r.t the log. In fact, the size, i.e., the number of nodes in the graph, grows linearly w.r.t $|D_c|$, the set of case identifiers, which is significantly smaller than the number of events recorded in the log. However, for compliance checking purposes, queries mostly refer to activity labels to resolve nodes.

Looking at the query in Listing 2, the database engine will handle this query by first binding variables $e3$ and $b3$ referring to activity labels 'E' and 'B', respectively. This binding would result in a single binding to explore for each variable. Next, relation variables $r1$ and $r2$ will be bound. The database engine can use the bindings of $e3$ and $b3$ and indexes on relation (edge) properties to prune the list of edges candidates for binding. Compared to the query in Listing 1, the node variable $e1$ will have as many bindings as there are events executed for activity 'E'. The activity label 'B' filter cannot be used to prune the nodes traversed and stored in the *path* variable.

Listing 2. Precedence anti-pattern query using unique activities encoding

```
1| Match (c3:Case)
2| where ((c3:Case)<-[r1:Event_to_Case]-(e3:Event{event:'E'}) and (c3:case)
   <-[r2:Event_to_Case]-(b3:Event{event:'B'}) and  r2.position >
   r1.position) or (not exists((c3)<-[:Event_to_Case]-(:Event{event:'B'}))
3| return c.ID
```

Theoretically, the UA method outperforms the BM method; the evaluation results empirically prove that. In such situation, if migrating the data to the UA encoding is prohibitive, event nodes in the graph can be updated with the position property so that all queries related to comparing positions of the nodes within the trace, e.g., compliance queries, can be processed efficiently.

5 Evaluation

This section reports the evaluation of the method we proposed to encode event logs as graphs. We compare our method, UA, against the baseline method BM. In addition, we compare the storage of event logs in relational databases. The relational table consists of four columns to store the case ID, the activity, the timestamp of the event and the position of each event within a case. To detect compliance violations, we evaluate two approaches. The first uses common SQL operators such as joins and nested queries (NQ). The second uses the analytical `Match_Recognize` (MR) operator.

We selected four real life logs: three logs from the BPI challenges to evaluate our experiments, namely: BPIC'12 [8], BPIC'14 [7], BPIC'19 [10] and the log namely: RTFMP [21]. We considered these logs as they expose different characteristics as summarized in Table 2.

Table 2. Logs characteristics

Logs	#Traces	#Events	#UA
BPIC'12	13087	262200	24
BPIC'14	41353	369485	9
BPIC'19	220810	979942	8
RTFMP	150370	561470	11

5.1 Implementation and Experimental Setup

We have implemented the UA encoding method using Neo4j version 4.3.1 as a graph database and Cypher to query the logs. Neo4J was instantiated with the default configuration of 2 GB of heap maximum size and a page cache size of 512 MB. Also, we use the SQL Server graph database extension (SGD) version 15.0.2000. To evaluate the MR implementation, we use Oracle 12c[3]. The instance was given 1 GB of RAM with default configurations of the storage engine. The experiments were run on a laptop running Windows 10 64-bit with an Intel Core i7 processor and 16 GB of RAM.

We prepared queries for *Response* and *Precedence* anti-patterns. For each pattern, we created two variants. The two variants enforce a time limit with an upper and lower bound on the time window. For instance, in case of a response query, we have $Response(A, B, \phi, \Delta t, <)$, and $Response(A, B, \phi, \Delta t, >)$[4]. Therefore, in total, we have four queries for each log. The actual values for A, B, and Δt vary depending on the log. The anti-pattern queries for each variant are translated to Cypher and SQL for the respective encoding method to test. All the details for the patterns (queries) variants and run details of experiments are available on Github[5].

5.2 Results and Discussion

In the first experiment, we report on the loading time of the logs following the respective encoding, i.e., loading into Neo4J, SQL graph database (SGD), and the relational database (RDB). For each log, we report the loading time and the number of nodes and edges created in the graph database (Table 3).

Loading large logs following the BM method, Neo4J crashed with an out-of-memory error due to the large amount of data. This is the case for the BPIC'14, BPIC'19, and the RTFMP logs. We have examined several subsets of these logs. The number of cases reported in Table 3 corresponds to the maximum size that could be loaded using the Neo4J configuration we mentioned earlier. For the UA, SGD, and RDB encoding, all the data are loaded into the database for the full log sizes. For the common log sizes, graph-based encoding using SGD is superior

[3] SQL Server does not support Match_Recognize. Thus, we used Oracle.
[4] We will report later on experiments with exclude property, i.e. $S \neq \phi$.
[5] https://github.com/nesmayoussef/Graph-Encoded-Methods.

Table 3. Loading time (seconds) for each encoding method. [LT: Loading Time, # N: number of nodes, # E: number of edges]

Methods		BM				UA				RDB
		Graph Details		Neo4j	SGD	Graph Details		Neo4j	SGD	
Logs	# Cases	#N	#E	LT	LT	#N	#E	LT	LT	LT
BPIC'12	13087	177597	315933	16	56.6	13111	164510	6.7	17.4	1641
BPIC'14	15000	148883	252766	13	3.7	15009	133883	9.9	1.2	890
	41353	410833	697607	—	11.7	41362	369480	9.4	2.9	2447
BPIC'19	25000	135933	196866	13	2.3	25008	110933	4.9	1.3	960
	220810	1197804	1733178	—	49	220818	976994	19.7	11.8	8153
RTFMP	50000	236633	323266	21	3.8	50011	186633	10	1.7	1250
	150370	711810	972510	—	55.9	150011	560046	19	6.7	3731

to the relational database and Neo4J. Additionally, UA is the fastest in graph encoding methods due to the smaller number of nodes and edges compared to the BM method.

Turning to graph sizes, we can observe the reduction of their sizes when encoding with the UA method. It reduces the number of edges, as it does not store edges for the *directly follows* relation and reduces the size of the nodes.

In the second experiment, we run the four compliance anti-pattern queries against the respective logs, two for response and two for precedes. We have run the queries five times and report the average execution time of the *Precedes* and *Response* anti-pattern queries for the different encoding methods in Table 4 and Table 5, respectively. Overall, the execution time is reduced using the proposed encoding method, especially Neo4J, compared to the baseline method BM, NQ, and MR methods. The magnitude of gain differs, though.

Table 4. Execution time (msec) for the variants of the *Precedes* queries [B: Before time window, W: Within time window]

Methods		SGD				Neo4J				NQ		MR	
		BM		UA		BM		UA					
Logs	# Cases	B	W	B	W	B	W	B	W	B	W	B	W
BPIC'12	**13087**	668	665	309	296	74	138	47	30	124	251	571	723
BPIC'14	**15000**	526	775	281	351	253	85	58	28	137	178	432	633
	41373	1354	1626	596	762	—	—	17	29	352	925	1202	1759
BPIC'19	**25000**	694	668	575	174	76	68	36	54	206	96	1006	634
	220810	6112	5885	3638	2519	—	—	12	78	2355	813	9154	57704
RTFMP	**50000**	1015	778	437	378	138	118	61	92	203	187	1106	477
	150370	3584	2551	1403	1087			71	219	649	484	3352	1447

For the precedence anti-patterns, in the case of the UA method, the reduction of execution time goes up to $14x$, as in the case of the BPIC'12 log for the Before

Table 5. Execution time (msec) for the variants of the *Response* queries [A: After time window, W: Within time window]

Methods		SGD				Neo4J				NQ		MR	
		BM		UA		BM		UA					
Logs	**# Cases**	**A**	**W**	**A**	**W**	**A**	**W**	**A**	**W**	**A**	**W**	**A**	**W**
BPIC'12	**13087**	374	381	274	484	112	39	59	12	1374	293	747	953
BPIC'14	**15000**	391	399	291	278	71	46	17	26	307	151	425	712
	41373	962	1122	651	824	—	—	28	32	838	366	1181	1979
BPIC'19	**25000**	554	412	315	275	29	20	12	16	6077	93	1267	931
	220810	4874	3629	3544	2814	—	—	54	123	824	1009	11523	8461
RTFMP	**50000**	1362	542	931	368	62	91	14	31	916	147	1433	942
	150370	4357	1669	2082	1062	—	—	51	121	1355	434	4345	2824

time limit, B, query in Table 4 compared to execution time of SGD, MR and NQ. In NQ, we use nested queries and self joins which leads the query engine to perform additional tasks to retrieve data. In SGD, querying tables works much the same way as querying relational tables in NQ. Using MR, the database has to scan all the records and match them to the non-deterministic finite automata (NFA) to check for matches. Comparing UA to the BM graph encoding, we still get an improvement in query time. In Neo4J, the gain goes up to $2x$ as in the case of BPIC'14 log for the W query.

For the response anti-patterns, in BPIC'12 log, for within time window, W, SGD, MR and NQ perform worse than the other methods, Table 5. The improvement in query time goes up to $79x$ comparing UA in Neo4J to MR. The lowest improvement is about $3x$, comparing UA in Neo4J to other graph encoding methods. Note that this gain is on a small subset of the log. It is not clear how fast the processing would be if the full log was loaded using the BM method. This is left as a future work when testing with higher hardware specification.

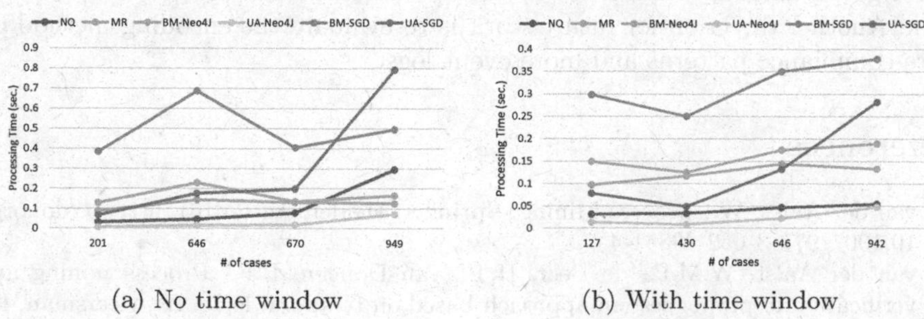

(a) No time window (b) With time window

Fig. 3. Comparing the results of Exclude queries

Comparing the Neo4J to SGD, Neo4J is faster. The best gain of UA in Neo4J compared to SGD is $303x$ in the case of the BPIC'19 log for the Before time window and $65x$ in the case of the BPIC'19 log for the After time window,

Table 4 and Table 5, respectively. This shows the superiority of the native graph databases compared to graph extension of relational databases.

In the third experiment, we run *Response with exclude*, i.e., *Response* $(A, B, \{C\}, \Delta t, <)$ anti-pattern queries against BPIC'15 [11] log. This log contains 1199 cases with 52217 events and 398 unique activities. We chose this log due to its large number of unique activities. We empirically validate that the proposed method still gives the best execution time. This experiment was run five times with different activities and time windows for the different encoding methods/storage engines.

Figures 3a and b report the execution time of the queries, with and without time window, respectively. We show on the x-axis the query results sorted by the matching number of cases. Obviously, the UA method shows the best scalability as the number of matching cases (process instances) is a function in both the input log size and the anti-pattern query.

Overall, for the different types of anti-pattern queries, the graph-based encoding of event logs outperforms the relational database encoding. This aligns with recent directions to employ graph databases for process analytics [9]. Additionally, the UA encoding method we propose improves query time and storage space against the baseline BM graph encoding method.

6 Conclusion and Future Work

We propose a graph-based encoding method for event logs to efficiently check their compliance with timed order patterns. The encoding enhances checking time and reduces the size of the stored graphs. Experimental evaluation empirically confirms the gain in both directions. In addition, the evaluation shows the superiority of graph encoding of event logs over relational encoding. Moreover, native graph databases, e.g. Neo4J, outperform graph extensions over relational databases, e.g. MS SQL Server.

A limitation of this work is that it has been evaluated using Neo4J only as a native graph database engine. We intend to address this limitation in our future work. Another direction for future work is to evaluate the encoding method on more compliance patterns and more event logs.

References

1. van der Aalst, W.: Process Mining. Springer, Heidelberg (2016). https://doi.org/10.1007/978-3-662-49851-4
2. van der Aalst, W.M.P., de Beer, H.T., van Dongen, B.F.: Process mining and verification of properties: an approach based on temporal logic. In: Meersman, R., Tari, Z. (eds.) OTM 2005. LNCS, vol. 3760, pp. 130–147. Springer, Heidelberg (2005). https://doi.org/10.1007/11575771_11
3. Agrawal, R., et al.: Taming compliance with sarbanes-oxley internal controls using database technology. In: ICDE, pp. 92–92. IEEE (2006)
4. Awad, A., Weidlich, M., Weske, M.: Visually specifying compliance rules and explaining their violations for business processes. J. Vis. Lang. Comput. **22**(1), 30–55 (2011)

5. Beheshti, A., Benatallah, B., Motahari-Nezhad, H.R.: Processatlas: a scalable and extensible platform for business process analytics. Softw. Pract. Exp. **48**(4), 842–866 (2018)
6. Carmona, J., van Dongen, B., Solti, A., Weidlich, M.: Conformance Checking - Relating Processes and Models. Springer, Cham (2018). https://doi.org/10.1007/978-3-319-99414-7
7. van Dongen, B.B.: BPI Challenge 2014 (2014). https://doi.org/10.4121/uuid:c3e5d162-0cfd-4bb0-bd82-af5268819c35
8. van Dongen, B.: BPI Challenge 2012 (2012). https://doi.org/10.4121/uuid:3926db30-f712-4394-aebc-75976070e91f
9. Esser, S., Fahland, D.: Multi-dimensional event data in graph databases. J. Data Semant. **10**(1), 109–141 (2021)
10. Fahland, D.: Event Graph of BPI Challenge 2019 (2021). https://doi.org/10.4121/14169614.v1
11. Fahland, D., Esser, S.: Event Graph of BPI Challenge 2015 (2021). https://doi.org/10.4121/14169569.v1
12. Hashmi, M., Governatori, G., Lam, H.-P., Wynn, M.T.: Are we done with business process compliance: state of the art and challenges ahead. Knowl. Inf. Syst. **57**(1), 79–133 (2018). https://doi.org/10.1007/s10115-017-1142-1
13. Hogan, A., et al.: Knowledge graphs. ACM Comput. Surv. **54**(4), 1–37 (2022)
14. Jalali, A.: Graph-based process mining. arXiv preprint arXiv:2007.09352 (2020)
15. Körber, M., Glombiewski, N., Seeger, B.: Index-accelerated pattern matching in event stores. In: SIGMOD, pp. 1023–1036. ACM (2021)
16. Koschmider, A., Laue, R., Fellmann, M.: Business process model anti-patterns: a bibliography and taxonomy of published work. In: ECIS (2019)
17. González López de Murillas, E., Reijers, H., van der Aalst, W.: Connecting databases with process mining: a meta model and toolset. Softw. Syst. Model. **18**(2), 1209–1247 (2019)
18. Pesic, M., Schonenberg, H., van der Aalst, W.M.P.: DECLARE: full support for loosely-structured processes. In: EDOC, pp. 287–300. IEEE (2007)
19. Ramezani, E., Fahland, D., van der Aalst, W.M.P.: Where did i misbehave? Diagnostic information in compliance checking. In: Barros, A., Gal, A., Kindler, E. (eds.) BPM 2012. LNCS, vol. 7481, pp. 262–278. Springer, Heidelberg (2012). https://doi.org/10.1007/978-3-642-32885-5_21
20. Ramezani Taghiabadi, E., Fahland, D., van Dongen, B.F., van der Aalst, W.M.P.: Diagnostic information for compliance checking of temporal compliance requirements. In: Salinesi, C., Norrie, M.C., Pastor, Ó. (eds.) CAiSE 2013. LNCS, vol. 7908, pp. 304–320. Springer, Heidelberg (2013). https://doi.org/10.1007/978-3-642-38709-8_20
21. Reissner, D.: Public benchmark dataset for Conformance Checking in Process Mining (2022). https://doi.org/10.26188/5cd91d0d3adaa
22. Saralaya, S., Saralaya, V., D'Souza, R.: Compliance management in business processes. In: Patnaik, S., Yang, X.-S., Tavana, M., Popentiu-Vlădicescu, F., Qiao, F. (eds.) Digital Business. LNDECT, vol. 21, pp. 53–91. Springer, Cham (2019). https://doi.org/10.1007/978-3-319-93940-7_3
23. Schönig, S., Rogge-Solti, A., Cabanillas, C., Jablonski, S., Mendling, J.: Efficient and customisable declarative process mining with SQL. In: Nurcan, S., Soffer, P., Bajec, M., Eder, J. (eds.) CAiSE 2016. LNCS, vol. 9694, pp. 290–305. Springer, Cham (2016). https://doi.org/10.1007/978-3-319-39696-5_18

Trans-Compiler-Based Database Code Conversion Model for Native Platforms and Languages

Rameez Barakat[1,2(✉)], Moataz-Bellah A. Radwan[1,2], Walaa M. Medhat[1,2,4], and Ahmed H. Yousef[1,3]

[1] School of Information Technology and Computer Science, Nile University, Giza, Egypt
{rbarakat,Mo.Radwan,wmedhat,ahassan}@nu.edu.eg

[2] Center for Informatics Science, Nile University, Giza, Egypt

[3] Department of Computer and Systems, Faculty of Engineering, Ain Shams University, Cairo, Egypt
ahassan@eng.asu.edu.eg

[4] Faculty of computers and artificial intelligence, Benha University, Banha, Egypt
walaa.medhat@fci.bu.edu.eg

Abstract. Cross-platform mobile application development frameworks are now widely used among software companies and developers. Despite their time and cost-effectiveness, they still lack the performance and experience of natively developed applications. Many research tools have been proposed to solve this problem by converting a natively developed application from one platform to another. The Trans-Compiler Based Android to iOS Converter (TCAIOSC) was proposed to convert the front-end and back-end code of Android Java applications to iOS applications. Since databases are essential for mobile applications, this paper proposes a new database code conversion model based on trans-compilation and pattern matching. It proposes a model that can be used to support database code conversion between native languages and platforms and applies the proposed model to support the conversion of Firebase Firestore database code from Android (Java) to iOS (Swift) using TCAIOSC. The enhanced tool's results show high accuracy for the converted database code and a noticeable improvement in the overall conversion results for TCAIOSC.

Keywords: Cross-platform mobile development · Trans-compilation approach · Databases · Firebase · Firestore · Android · iOS

1 Introduction

Mobile application development is one of the most highly required and demanded areas in information technology. This is a result of the huge number of smartphones and tablets that have created a huge and competitive market for mobile applications. Consequently, the demand for mobile application developers is increasing all over the world [1]. As a result of the variety of mobile operating systems and the need for the mobile application

P. Fournier-Viger et al. (Eds.): MEDI 2022, LNAI 13761, pp. 162–175, 2023.
https://doi.org/10.1007/978-3-031-21595-7_12

to operate similarly on all these operating systems, there exist different cross-platform solutions that enable developers to write their application once and run it on different platforms.

Although they are time and cost-effective, cross-platform solutions are known for their relatively low performance when compared to native applications. Several commercial/research tools have been introduced as a solution for cross-platform development [2–6]. In [5] Salama et al. proposed a trans-compiler-based approach for converting native Android applications' source code to native IOS applications' source code. However, they only succeeded in converting the backend Java source code to swift source code without including Android features, which resulted in a low conversion rate. In [6] the solution in [5] was enhanced to support some of the Android features like connecting UI views to the backend, responding to user events, and Android intents.

However, the enhanced solution did not support converting database connections, which is an essential module in most of our daily used mobile applications. According to [7], the number of available mobile applications until the first quarter of 2021 on the two main big app stores (Google play and Apple app store) are 3.48 million and 2.22 million, respectively. Out of these millions of apps in the app stores, it would be difficult to find one that does not require a database or some sort of storing or handling data in a particular way. Therefore, almost all our daily used mobile applications require storage and management of data including querying the stored data to retrieve certain information.

Hence, mobile databases are considered an essential part of most mobile application development. Therefore, a native mobile application converter tool that does not support database connections' conversion is missing an essential functionality for mobile applications.

Enhancing the existing solution to support database connections' conversion will noticeably enhance the performance of the converter and will help make the needed human modifications to the converted code minimal.

The contribution of this paper is to propose a new trans-compiler-based database code conversion model that aims to extend the solutions in [5] and [6] to support the conversion of Firebase Firestore [8] database connections, evaluating the conversion accuracy results and the improvement achieved from the proposed solution extension.

The outline structure for the rest of this paper is as follows: Sect. 2 represents the related work. Section 3 gives a background on mobile databases. Section 4 presents the methodology for developing the proposed model and applying it to enhance TCAIOSC. Section 5 presents the results for the database code conversion model and the effect of these results on the solution performance. Finally, Sect. 6 presents the conclusion and future work.

2 Literature Reviews

There are many papers that present and categorize either the mobile application development approaches [9–15] or the different mobile application types [15–18] or evaluate different approaches, tools, or solutions [12, 19–24]. His section presents common approaches and application types for mobile application development. Then it mentions the former attempts at supporting database-related code conversion for tools converting native-to-native applications.

2.1 Cross-platform Development Approaches

Much work has been done to identify and describe the different approaches in mobile application development. Many papers describe the main approaches and sub-approaches for mobile development. The following are the main and most common approaches among these papers.

Compilation Approach. In this approach, the application code is transformed from one high-level language to another (trans-compilation) or executable code is generated for a platform other than the platform on which the compiler is running (cross-compilation).

Modeling Driven Approach. In this approach, the model-driven development (MDD) methodology is adopted where abstract models are developed based on the system's requirements using a domain-specific modelling language. Then, a platform-specific mobile application is generated from these models.

Eric and Marco [25] surveyed model-driven based approaches in mobile development and defined and classified them based on a defined classification schema. This classification included research approaches in MDD like MD2 [26] where a textual model for the application is first defined using a textual modelling language; then, a platform-specific code generator transforms the model into the required platform.

SInterpretation Approach. In this approach, a common language is used to write the code, then an interpreter translates this code, and the native code for each intended platform is generated. An abstract layer is used to provide the native features for each platform that interprets the code during runtime and provides the platform-specific features.

2.2 Cross-platform Native-to-Native Tools Supporting Database Conversion

J2ObjC [4] which is an open-source command-line tool developed by Google using the trans-compilation approach, provides SQL code conversion by supporting the conversion of Java.sql. The source and destination for J2ObjC are different from TCAIOSC as it translates a source code written in Java to the equivalent source code in Objective-C, which is different than the scope of TCAIOSC. Also, J2ObjC doesn't provide any support for converting user interfaces.

MechDome [27] was a commercial tool that converted applications from native Android to native iOS. It operated at the level of the binary executables, not the source code itself, which is different from the way TCAIOSC operates. It supported some of the Google Firebase APIs, like the Realtime database and authentication. However, it has been folded since 2018, two years after it was founded and is no longer available.

ICPMD [13, 14], a proposed tool by El-Kassas et al., succeeded in converting shared preferences, which is a way of storing local data in Android. They also proposed a methodology that combines different development approaches and enhanced it by using a new conversion approach using XSLT and regular expressions. However, ICPMD converted applications from Windows Phone to Android, which is not in the same scope as TCAIOSC, in addition to the fact that Windows Phone is no longer used and no longer supported by Microsoft.

3 Background

There are two main types of databases that are used in mobile application development which are SQL and NoSQL databases. Each type has its own advantages and disadvantages. Therefore, selecting the right database for a mobile application depends on the type of developed application. Table 1 shows the advantages and disadvantages of each type.

Table 1. Advantages and disadvantages of databases used in mobile application development.

Comparison	SQL databases (Relational databases)	NoSQL databases
Schema	Fixed schema	Flexible schema
Data storage	Store data in tables with fixed columns and rows	Store data as documents
Data duplication	Less duplication of data	More duplication of data
Scaling	Vertical scaling	Horizontal scaling
Update queries	Faster for update -heavy applications	Slower for update-heavy applications
Read/write queries	Relatively slower for read/write heavy applications	Faster for read/write heavy applications
Examples	PostgreSQL, SQLite, MySQL	Firebase, MongoDB, Realm, Neo4j

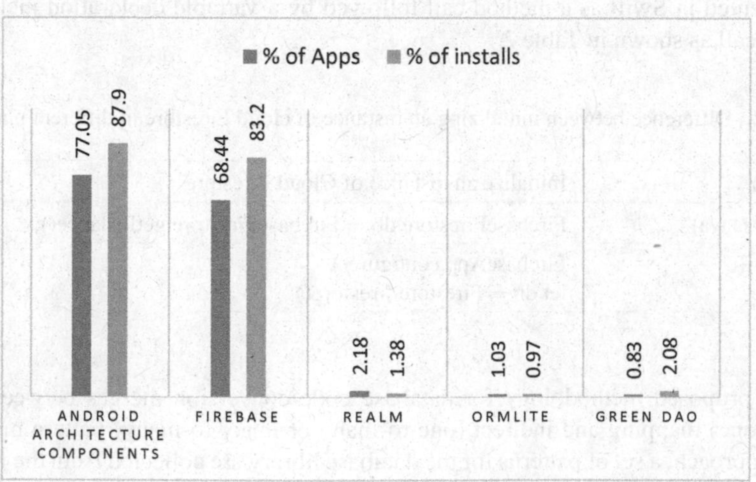

Fig. 1. Percentages of used Android database libraries in total apps and total number of installs on play store

In the world of Android and iOS development, there are many database frameworks under different database types like SQLite, Realm, Firebase, CoreData (iOS), and others. According to AppBrain [9], among database libraries used in Android applications on the play store, Firebase came in second place after the Android Architecture Components with a percentage of 68.44% of apps and 83.20% of installs on the play store. Therefore, converting Firebase Firestore database code to test the applicability of the solution was chosen over the other libraries. Figure 1 shows the AppBrain statistics for the most used database libraries in Android.

4 Methodology

TCAIOSC has successfully provided two code conversion units, one for backend code conversion and the other for UI code conversion. These code conversion units have proven to successfully convert backend and frontend code when tested to convert simple applications that didn't use database connections from Android to iOS. Although the architecture for TCAIOSC implies that these converters can be easily extended to support any library/API, this supposition is built on the assumption that the Android and iOS code will always have a one-to-one lexical mapping. In this section, the enhanced methodology to support database libraries and the proposed enhanced solution's architecture are presented.

4.1 Proposed Enhanced Methodology

In practice, mapping between the two platforms or languages is not always a direct one-to-one relationship. For example, an Android Java code to achieve a simple functionality such as initializing an instance of Cloud Firestore, from a compiler's perspective, can be implemented as a variable declaration including a method call in Java while it is implemented in Swift as a method call followed by a variable declaration including a method call as shown in Table 2.

Table 2. Difference between initializing an instance of cloud Firestore in different platforms

Platform	Initialize an instance of Cloud Firestore
Android (Java)	FirebaseFirestore db = FirebaseFirestore.getInstance();
iOS (Swift)	FirebaseApp.configure() let db = Firestore.firestore()

The proposed methodology for database code conversion merges between direct (one-to-one) mapping and indirect (one-to-many or many-to-many) pattern matching. In this approach, a set of patterns for the database library are collected from the Firebase Firestore official documentation [28] and predefined to match the input source code against.

Whenever a database-related statement is detected in the input source code, first a one-to-one mapping is attempted in TCAIOSC's backend code conversion unit. If there was no direct match, then a pattern matching is attempted against a predefined set of patterns. If there was no pattern matching, finally, the input code is commented as a non-converted code in the generated output source code.

Table 3 shows an example of each of the three mapping types. In the first example (one-to-one), the compiler translates the Java statement consisting of a Firestore object followed by two method calls to the corresponding Swift code consisting of the Firestore object followed by the corresponding two method calls in Swift. This type of mapping is handled completely by the trans-compiler without the need for pattern matching.

Table 3. Examples for direct and indirect mapping between Android and iOS

Mapping Type	Android (Java)	iOS (Swift)
Direct (one-to-one)	db.collection("users").add(user)	db.collection("users").addDocument(data: user)
Indirect (one-to-many)	FirebaseFirestore db = FirebaseFirestore.getInstance();	FirebaseApp.configure() let db = Firestore.firestore()

Indirect (many-to-many)	db.collection("users").add(user)
	.addOnSuccessListener(new OnSuccessListener<DocumentReference>()
	{
	@Override
	public void onSuccess(DocumentReference documentReference) {
	Log.d(TAG, "DocumentSnapshot added with ID: " + documentReference.getId());
	}
	})
	.addOnFailureListener(new OnFailureListener() {
	@Override
	public void onFailure(@NonNull Exception e) {
	Log.w(TAG, "Error adding document", e);
	}
	});
		ref =
		db.collection("users").addDocument(data: user
		) { err in
		if let err = err {
		print("Error adding document: \(err)")
		} else {
		print("Document added with ID: \(ref!.documentID)")
		}
		}

In the second example (one-to-many), the trans-compiler first tries to convert the statement as a normal variable declaration, including a method call. It will check the variable type and used method against the compiler's defined data types and methods to get the corresponding data type and method in Swift. Then, the pattern matcher will attempt to match the statement against the defined patterns and add the app configuration statement in Swift.

In the third example (many-to-many), the trans-compiler will try to convert the statement using normal trans-compilation. It will match the object type and methods against the compiler's defined data types and methods. Then, it will reach the listener statements with no equivalent direct mapping. Then, the pattern matcher module will read the pattern and match it to the defined corresponding pattern in Swift.

4.2 Proposed Enhanced Solution Architecture

The enhanced solution's architecture, illustrated in Fig. 2, consists of the following components:

Controller. The Controller is an interface unit which handles all communication between different components. It is responsible for receiving the source project files, categorizing the files into backend and UI files, and then passing the files to either the code conversion unit or UI conversion unit based on the file category. Finally, the controller collects the generated output files, packages them, and generates the output iOS Swift project.

Java Lexer and Parser. The Java lexer and parser components are generated using ANTLR (ANother Tool for Language Recognition) [29], which is a parser generating tool that is passed the grammar file for the Java language to produce the source language tokens. The generated tokens are passed to the parser, which constructs the parse tree according to the grammar file and generates interfaces that are used in the backend converter to traverse the parse tree during the backend code conversion process.

Backend Converter. The backend converter implements the interfaces generated by the parser to perform the backend code generation from the source language (Java) to the destination language (Swift). During the process of code generation, the backend converter interacts with the database to get the mapped data types, operators, built-in functions,…etc. from Java to Swift.

During the backend code conversion, the backend converter uses a firebase detector module to detect whether a certain statement is a Firestore statement. If a statement is detected as a Firestore, firstly, the code converter checks for an equivalent code using the available Firestore mappings for functions, data types, or classes in the database (one-to-one mapping); secondly, if the first approach didn't find a match, the backend converter interacts with the firebase pattern matching module which matches the statement against predefined patterns (one-to-many mapping); thirdly, if the second approach didn't find a match, the backend converter interacts with the firebase detector to check whether the current statement is a part of a larger set of consecutive Firestore statements, if found true, the backend converter interacts with the firebase pattern matching module to match a larger set of statements against predefined patterns (many-to-many mapping); lastly, if no match was found, the statement is marked as non-converted code and commented in the output code file.

The backend converter takes as an input all the backend code files of the project, then processes these files one by one to convert the code files from the source platform to the target platform, then passes the converted files to the controller unit. It also produces

UI-related information found in activity files that are used by the UI converter during UI code conversion.

Firebase Pattern Matching Module. This module is introduced as a solution for indirect Firestore statements' mapping. It uses a set of predefined Firestore patterns in Java and their equivalent Firestore code patterns in Swift. If a Firestore statement or statements were not converted using direct mapping, they are passed from the backend converter to the pattern matching module to be checked for one-to-many or many-to-many mapping.

Firebase Detector. The firebase detector has two main roles, one concerning the conversion of the code itself and the other concerning the testing and evaluation of the converted Firestore code and calculating the improvement rate for the tool after supporting the Firestore conversion.

This module is used by the backend converter to determine whether a certain statement is firestore-related or not. Whenever a statement is passed to the detector, it checks keywords in the statement against a set of firestore keywords that have been previously defined by a developer and stored in the database. This set of keywords includes one-level keywords like library names and data types and two-level keywords like methods that belong under a certain library, where the method name is considered the first level and the parent library is considered the second level.

The advantage of this module's design is that it is generic and extensible as it can be easily extended to be a general database statement detector that detects any database statement not only firestore by simply extending the set of keywords by adding keywords for other database frameworks that has keywords up to any number of levels.

XML Lexer and Parser. The XML lexer and parser components are also generated using ANTLR using the grammar file for XML language. Then it generates tokens that are passed to the parser, which constructs the XML parse tree and generates interfaces that are used in the UI converter to traverse the parse tree during the UI code conversion process.

UI Converter. The UI converter maps each XML file that represents an activity in the Android project into a scene in the iOS project. The different scenes resulting from different XML activity files are then grouped together into one Storyboard file by the UI converter and passed to the control unit.

For the UI converter to convert the UI code, it needs both the UI parse tree and the backend parse tree in order to handle the UI-related code that existed in the backend (.JAVA) code files and was previously identified by the backend converter.

The UI converter, much like the backend converter, is responsible for implementing the interfaces that are produced by the XML parser. This implementation is then used to build the output UI code for iOS, which is then passed to the controller unit.

Databases. The database contains all the necessary mapping data to complete the backend, UI, and Firestore conversions. It includes mappings for data types, methods and their signatures (parameters and their types), methods return types, libraries, operators, static built-in functions, UI data types, observers, and a defined set of Firestore keywords that the Firestore detector module uses to determine whether a given statement is a Firestore statement.

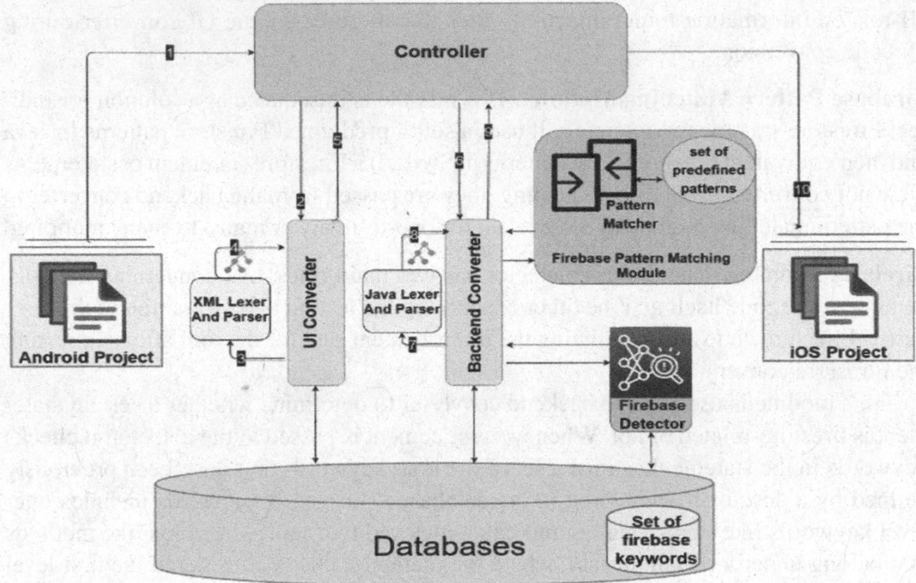

Fig. 2. Enhanced solution's architecture after supporting Firestore conversion

5 Results and Discussion

In this section, three evaluation requirements are presented. The first is to evaluate the success in converting the database code; the second is to evaluate different applications' code conversion rates before and after applying the proposed enhancement to measure the size of the enhancement; and the third is to compare the runtime of the solution on different applications before and after integrating the proposed enhancement.

A set of open-source native Android applications were selected from GitHub to test the solution. Table 4 lists the sample applications that are used to test the solution. The criteria for selecting the test applications set were:

Selecting Most Recent Open-Source Applications. To guarantee that the test samples include the most recent Firestore features and avoid deprecated code in old applications, all the selected applications were published on GitHub after 2019.

Selecting Applications from Different Categories. To test the tool's performance for different and broad types of applications and to test the generalization of the results.

Selecting Android Java Applications Only. Since TCAIOSC only supports the conversion of Android Java applications, not Kotlin.

Table 4. List of sample test applications

Application name	Application URL
Quiz App	https://github.com/tayyabmughal676/QuizAppJava
FirebaseFiresore Android	https://github.com/lspusta/FirebaseFirestoreAndroidJava
StPreacher-SignUp-With-PhoneNumber	https://github.com/StPreacher/StPreacher-SignUp-With-PhoneNumber-using-FirebaseCloudFirestore
NotesAppWithFireStore	https://github.com/mohamed00736/NotesAppWithFireStore
Trigger Push Notification	https://github.com/moataz-bellah/Firestore-Android-Trigger-Push-Notification
FitMe	https://github.com/brapana/FitMe
Firestore demo Android Java	https://github.com/mirodone/Firestore-demo-Android-Java
Shopping App	https://github.com/developersamuelakram/ShoppingApp-Firestore-MVVM-Navigation

5.1 The Percentage of the Successful Conversion of Firestore Code

The same metric used by TCAIOSC to calculate the percentage of converted code was adopted, that is, the percentage of successfully converted statements. This was done to establish consistency between TCAIOSC and the enhanced solution. Also, it was adopted to keep the integrity of the results when calculating the improvement in the second part of the evaluation. Table 5 presents the percentage of successfully converted statements for the set of test applications. The equation for calculating this percentage is as follows:

$$Firestore\ \ Conversion\,\% = \frac{Number\ \ of\ \ converted\ firestore\ \ statments}{Total\ \ number\ \ of\ \ firestorestataments} \times 100 \quad (1)$$

After analyzing the results for converting Firestore statements, the statements that were not converted were due to:

- Using Consecutive Multiple Listeners in Android: This pattern has no direct equivalent in iOS.
- Partially Converted Statements: Some statements were partially converted whereas the Firestore part of the statement was converted. However, the statement uses another unsupported feature/data type in TCAIOSC, which results in the whole statement being counted as not converted.
- Unmatched Statements: Some Firestore statements did not match any of the defined patterns (this can be improved by adding more patterns to the set of predefined patterns).

Table 5. Percentage of successfully converted firestore statements

Application name	Converted firestore statements (%)
Quiz App	100%
FirebaseFiresore Android	83%
StPreacher-SignUp-With-PhoneNumber	80%
NotesAppWithFireStore	75%
Trigger Push Notification	100%
FitMe	84.3%
Firestore demo Android Java	89%
Shopping App	71.75%

5.2 The Improvement of the Overall Conversion Rate for an Entire Application

The overall improvement in the conversion rate was calculated by converting the same set of test applications by TCAIOSC before and after the support of Firestore. Table 6 compares the backend conversion results for TCAIOSC before and after supporting the Firestore library.

Table 6. Comparison between TCAIOSC's results before and after supporting firestore

Application name	Conversion (%) before supporting Firestore	Conversion (%) After Supporting Firestore
Quiz App	23%	26.5%
FirebaseFiresore Android	59%	63%
StPreacher-SignUp-With-PhoneNumber	73%	89%
NotesAppWithFireStore	68%	70%
Trigger Push Notification	29%	44%
FitMe	47%	50%
Firestore demo Android Java	79%	84.5%
Shopping App	52%	63%

5.3 Comparison Between Conversion Runtime Before and After Supporting Firestore Code Conversion

A comparison for the runtime of TCAIOSC to generate the converted code for the list of test applications is presented in Table 7. The comparison shows that the conversion runtime after supporting Firestore is relatively higher with a minimum and maximum increases of approximately 1% and 30% respectively depending on the size of the application and the size of firestore-related statements per application.

Considering that the application conversion process is a one-time process that will be executed once, the observed increase could be accepted since the total conversion time is still found relatively small for producing a mobile application source code in seconds.

Table 7. Comparison between TCAIOSC's runtime before and after supporting Firestore

Application name	Runtime before supporting Firestore (milliseconds)	Runtime After supporting Firestore (milliseconds)
Quiz App	10,675.1	13,709.2
FirebaseFiresore Android	10,779.5	12,959.1
StPreacher-SignUp-With-PhoneNumber	11,947.2	12,151.7
NotesAppWithFireStore	11,215.1	11,782
Trigger Push Notification	13,143.1	14,688.2
FitMe	13,911	18,053.3
Firestore demo Android Java	13,182.1	16,233.3
Shopping App	11,260.9	14701.3

6 Conclusion and Future Work

Since mobile databases are an essential part of mobile applications, database-related code exists in most mobile application source codes. The proposed model to enhance the TCAIOSC solution after supporting Firebase Firestore code from native Android (Java) to native iOS (Swift) shows a reasonable improvement rate in the overall successful code conversion for the application.

SThe proposed methodology succeeded in converting the selected database framework (firebase Firestore) and resulted in a relatively high conversion rate for the database-related code. However, it didn't achieve full conversion for all applications due to the following limitations:

- Using Consecutive Multiple Listeners in Android: the impact of this limitation can be minimized by defining a standardized way for the tool to handle certain code in the source language or platform that has no equivalent in the target language or platform.
- Partially Converted Statements: this can be solved/minimized by supporting more backend features in the original tool of TCAIOSC.
- Unmatched Statements: can be improved by defining and adding more patterns to the set of predefined patterns.

The future work includes:

- Generalizing the proposed enhanced solution to be database framework type-independent and testing this generalization by applying it to other tools.
- Extending the pre-defined pattern matching set to include more patterns by analyzing more source code patterns in Android projects.
- Developing a handling method to handle the issue of code with no equivalent between different platforms.
- Publicating the solution as an open-source project on GitHub to encourage developers to contribute other code methods/patterns that may not be widely used or documented.

References

1. Montandon, J.E., Politowski, C., Silva, L.L., Valente, M.T., Petrillo, F., Guéhéneuc, Y.G.: What skills do IT companies look for in new developers? A study with stack overflow jobs. Inf. Softw. Technol. **129**(August), 2021 (2020). https://doi.org/10.1016/j.infsof.2020.106429
2. Cordova. https://cordova.apache.org/
3. xamarin. https://dotnet.microsoft.com/en-us/apps/xamarin
4. J2OBJC. https://developers.google.com/j2objc
5. Salama, D.I., Hamza, R.B., Kamel, M.I., Muhammad, A.A., Yousef, A.H.: TCAIOSC: Trans-compiler based android to iOS converter. In: Hassanien, A.E., Shaalan, K., Tolba, M.F. (eds.) AISI 2019. AISC, vol. 1058, pp. 842–851. Springer, Cham (2020). https://doi.org/10.1007/978-3-030-31129-2_77
6. Hamza, R.B., Salama, D.I., Kamel, M.I., Yousef, A.H.: CAIOSC: application code conversion. In: 2019 Novel Intelligent and Leading Emerging Sciences Conference (NILES), vol. 1, pp. 230–234 (2019). https://doi.org/10.1109/NILES.2019.8909207
7. Google Play Store: number of apps 2021 | Statista. https://www.statista.com/statistics/266210/number-of-available-applications-in-the-google-play-store/ (accessed Sep. 04, 2021)
8. Firebase. https://firebase.google.com/
9. Perchat, J., Desertot, M., Lecomte, S.: Component based framework to create mobile cross-platform applications. Procedia Comput. Sci. **19**, 1004–1011 (2013). https://doi.org/10.1016/j.procs.2013.06.140
10. Rahul Raj, C.P., Tolety, S.B.: A study on approaches to build cross-platform mobile applications and criteria to select appropriate approach. In: 2012 Annual IEEE India Conference INDICON 2012, pp. 625–629 (2012). https://doi.org/10.1109/INDCON.2012.6420693
11. Ribeiro, A., Da Silva, A.R.: Survey on cross-platforms and languages for mobile apps. In: Proceedings of 2012 8th International Conference on Quality Information Communication Technology QUATIC 2012, pp. 255–260 (2012). https://doi.org/10.1109/QUATIC.2012.56
12. Heitkötter, H., Hanschke, S., Majchrzak, T.A.: Evaluating cross-platform development approaches for mobile applications. In: Web Information Systems and Technologies, pp. 120–138 (2013)
13. El-Kassas, W.S., Abdullah, B.A., Yousef, A.H., Wahba, A.: ICPMD: integrated cross-platform mobile development solution. In: Proceedings of 2014 9th IEEE International Conference on Computer Engineering and Systems, ICCES 2014, pp. 307–317 (2014). https://doi.org/10.1109/ICCES.2014.7030977
14. El-Kassas, W.S., Abdullah, B.A., Yousef, A.H., Wahba, A.M.: Enhanced code conversion approach for the Integrated Cross-Platform Mobile Development (ICPMD). IEEE Trans. Softw. Eng. **42**(11), 1036–1053 (2016). https://doi.org/10.1109/TSE.2016.2543223

15. El-Kassas, W.S., Abdullah, B.A., Yousef, A.H., Wahba, A.M.: Taxonomy of cross-platform mobile applications development approaches. Ain Shams Eng. J. **8**(2), 163–190 (2017). https://doi.org/10.1016/j.asej.2015.08.004
16. Smutný, P.: Mobile development tools and cross-platform solutions. In: 2012 13th International Carpathian Control Conferenc, ICCC 2012, pp. 653–656 (2012). https://doi.org/10.1109/CarpathianCC.2012.6228727
17. Ohrt, J., Turau, V.: Cross-platform development tools for smartphone applications. Comput. (Long. Beach. Calif) **45**(9), 72–79 (2012). https://doi.org/10.1109/MC.2012.121
18. Xanthopoulos, S., Xinogalos, S.: A comparative analysis of cross-platform development approaches for mobile applications. In: BCI 2013: Proceedings of the 6th Balkan Conference in Informatics, September 2013. https://doi.org/10.1145/2490257.2490292
19. Rieger, C., Majchrzak, T.A.: Weighted evaluation framework for cross-platform app development approaches. In: Wrycza, S. (ed.) SIGSAND/PLAIS 2016. LNBIP, vol. 264, pp. 18–39. Springer, Cham (2016). https://doi.org/10.1007/978-3-319-46642-2_2
20. Rieger, C., Majchrzak, T.A.: Towards the definitive evaluation framework for cross-platform app development approaches. J. Syst. Softw. **153**, 175–199 (2019). https://doi.org/10.1016/j.jss.2019.04.001
21. Jobe, W.: Native apps Vs. mobile web apps. Int. J. Interact. Mob. Technol. **7**(4), 27 (2013). https://doi.org/10.3991/ijim.v7i4.3226
22. Mohammadi, F., Jahid, J.: Comparing native and hybrid applications with focus on features. p. 49 (2016)
23. Pulasthi, L.K., Gunawardhana, D.: Native or web or Hybridwhich is better for mobile application. Turkish J. Comput. Math. Educ. Res. Artic. **12**(6), 4643–4649 (2021)
24. Nawrocki, P., Wrona, K., Marczak, M., Sniezynski, B.: A comparison of native and cross-platform frameworks for mobile applications. Comput. (Long. Beach. Calif). **54**(3), 18–27 (2021). https://doi.org/10.1109/MC.2020.2983893
25. Umuhoza, E., Brambilla, M.: Model driven development approaches for mobile applications: a survey. In: Younas, M., Awan, I., Kryvinska, N., Strauss, C., Thanh, D.V. (eds.) MobiWIS 2016. LNCS, vol. 9847, pp. 93–107. Springer, Cham (2016). https://doi.org/10.1007/978-3-319-44215-0_8
26. Heitkötter, H., Majchrzak, T.A., Kuchen, H.: Cross-platform model-driven development of mobile applications with MD 2. In: Proceedings on ACM Symposium on Applied Computing. SAC, pp. 526–533 (2013). https://doi.org/10.1145/2480362.2480464
27. Mechdome. http://www.mechdome.com/
28. Firestore. https://firebase.google.com/docs/firestore
29. Parr, T.: The Definitive ANTLR 4 Reference. Pragmatic Bookshelf (2013)

MDMSD4IoT a Model Driven Microservice Development for IoT Systems

Meriem Belguidoum(✉), Aya Gourari, and Ines Sehili

LIRE Laboratory University of Constantine 2, Algiers, Algeria
{meriem.belguidoum,aya.gourari,ines.sehili}@univ-constantine2.dz

Abstract. Nowadays, IoT systems are widely used, they are embedded with sensors, software, and technologies enabling communication and automated control. The development of such applications is a complex task. Therefore, we have to use a simplified methodology and a flexible and scalable architecture to build and run such applications. On the one hand, Model-driven development (MDD) provides significant advantages over traditional development methods in terms of abstraction, automation, and ease of conception. On the other hand, microservice architecture (MSA) is one of the booming concepts for large-scale and complex IoT systems, it promises quick and flawless software management compared to monolithic architectures. In this paper, we present MDMSD4IoT, Model-driven Microservice architecture development for IoT based on the profile SysML4IoT and combined Model Driven Development and microservice architecture. We illustrate our contribution through a smart classroom case study.

Keywords: Internet of Things · Model driven development · Model driven architecture · SysML · Microservice architecture · M2T transformation · Smart classroom

1 Introduction

The field of IoT is growing exponentially and has sparked a revolution in the industrial world [16]. It refers to an emerging paradigm that allows the interconnection of physical devices equipped with sensing, networking, and processing capabilities to collect and exchange data. These things connected to each other form a much larger system and enable new ubiquitous and pervasive computing services [23]. Therefore, the development of IoT systems is very challenging due to their complexity [6] and the lack of IoT development methodologies and appropriate application architecture style.

Several modeling languages and tools based on the Model Development Engineering approach [26] have been proposed to design and develop complex software systems through meta-modeling, model transformation, code generation, and automatic execution. Thus, Model-Driven Development (MDD) [24] and

P. Fournier-Viger et al. (Eds.): MEDI 2022, LNAI 13761, pp. 176–189, 2023.
https://doi.org/10.1007/978-3-031-21595-7_13

Model Driven Architecture (MDA) [5] are classified as Model Driven Engineering (MDE) [25], MDD is a development paradigm that uses models as the primary artifact of the development process and is a subset of MDE, MDA is a subset of MDD and relies on the use of OMG standards [4]. This latter proposes several specification languages, the most relevant one for IoT-based applications is the OMG System Modeling Language SySML [12], it is a general purpose graphical modelling language for specifying, analyzing, designing, and verifying complex systems that may include hardware, software, information, etc. It has been conceived as a profile of UML [13].

Furthermore, IoT requires heavy integration between devices, data, and applications. This integration is becoming increasingly costly, complex, and time-consuming, especially with a monolithic architecture. These problems could be reduced considerably by using microservices architecture since it structures an application as a collection of services as small, modular, independently deployable, and loosely-coupled services [14,22].

In order to provide some solutions to these IoT systems engineering issues, we believe that we have to take advantage of the most convenient and relevant approaches, methodologies, languages, and tools for IoT application development and combine them in an original way. Therefore, our proposed approach is based on: 1) SysML profile for IoT [18] based on the architecture reference model IoT-ARM [3,8], 2) microservices architecture (MSA) [20], 3) Model Driven Architecture [5] and 4) a methodology for IoT development process adopted in recent IoT projects [7]. Our approach is called MDMSD4IoT for *Model Driven Microservices Architecture Development for IoT systems*. It aims to design and develop IoT systems in an efficient and flexible way. Indeed, firstly, Systems Modeling Language (SysML) [18] is the most popular tool for model-based development, it allows modeling physical aspects like hardware, sensors, etc. Secondly, the microservices architecture [20] defines an application as a set of small autonomous services, which is very suitable for IoT systems thanks to its characteristics such as weak coupling, modularity, flexibility, independent deployment, and resilience. Thirdly, model-driven architecture (MDA) [5] is a paradigm that promotes the use of models to solve software engineering problems through techniques such as abstraction and automation. Finally, IoT Design Methodology proposed in [7,18] is a useful methodology that consists of a set of development steps and is based on the Architecture Reference Model (ARM) [8]. In this paper, we extended this design methodology with two steps related to the definition and the specification of microservices.

We illustrate our approach through the development of "UC2Smart Classroom" using the fingerprint module for secure access and automatic resource management, eg. smart lighting, etc. in our university.

This paper is outlined as follows: Sect. 2 presents a summary of some related work. Section 3 describes the MDMSD4IoT approach. Section 5 illustrates the proposed approach through *UC2SmartClassroom* case study which is a smart classroom prototype for the University of Constantine 2. Finally, we conclude the paper and present some future work.

2 Related Work

In order to address the above-mentioned challenges and be able to develop efficiently IoT applications, we focus on the most relevant related work, namely based on: IoT Architecture Reference Model, SysML4IoT profile, Microservice Architecture (MSA), and Model Driven Development (MDD).

The goal of the IoT Architectural Reference Model project [3] is to provide developers with common technical basics and a set of guidelines for building interoperable IoT systems. The architectural reference model (ARM) [8] provides the highest level of abstraction for the definition of IoT systems. Besides models, the IoT domain model provides the concepts and definitions on which IoT architectures can be built.

In [11] IDeA (IoT DevProcess & IoT AppFramework) is proposed as a model-based systems engineering methodology for IoT application development. It focuses on modelling and considers the model as the primary artifact for systems development. The main objective of this methodology is to provide a high-level abstraction to deal with the heterogeneity of software and hardware components.

IDeA is composed of a method, called *IoT DevProcess*, and a support tool, called *IoT AppFramework*. The IoT DevProcess is used for the design of IoT applications, to support IoT DevProcess activities, the IoT AppFramework provides a SysML profile for IoT applications called SysML4IoT [11], which is strongly based on the IoT-ARM domain model presented previously.

In [19], a model-driven approach is proposed to ease the modeling and realization of adaptive IoT systems, it is based on SysML4IoT (an extension of the SysML) to specify the system functions and adaptations, the generated code is deployed later on to the hardware platform of the system, a smart lighting system is developed.

A model-driven environment called CHESSIoT is presented in [21] to design and analysis of industrial IoT systems. It follows a multi-view, component-based modeling approach with a comprehensive way to perform event-based modeling on system components for code generation purposes employing an intermediate ThingML model [17]. An Industrial real-time safety use case is designed and analysed.

The authors of [7] have proposed a generic design methodology for IoT systems independent from the specific product, service, or programming language and allow designers to compare various alternatives for loT system components. The presented methodology is generally based on the IoT-ARM reference model [8], it focuses on the domain model to describe the main concepts of the system to be designed and to help designers understand the IoT domain for which the system must be designed. The methodology consists of the specification of objectives and requirements, the process, the domain model, the information model service, IoT level, functional view, operational view, integration of devices and components, and finally, application development.

Authors in [9] propose the FloWare approach and its toolchain, it combines Software Product Line and Flow-Based Programming paradigms to manage the complexity in the various stages of the IoT application development process. The final IoT application and the executable Node-RED flow are generated using an automatic transformation procedure starting from a configuration of the designed Feature Models.

In [10] a model-driven integrated approach is provided to exploit traceability relationships between the monitored data of a microservice-based running system and its architectural model to derive recommended refactoring actions that lead to performance improvement. The approach has been applied and validated on e-commerce and ticket reservation, and the architectural models have been designed in UML profiled with MARTE.

In [15], the authors explained how typical Microservice Service Architecture problems can be addressed using Model Driven Development such as abstraction, model transformation, and modelling viewpoints. Indeed, MDD offers several advantages in terms of service development, integration and migration. However, MDD is rarely applied in SOA engineering. Nonetheless, the authors claim that the use of MSA greatly facilitates and fosters the usage of MDD in microservice development processes. They list these characteristics according to Newman [22] and correlates them with the means of MDD, for example, *Service Identification* is supported by model transformation, *Technology Heterogeneity* is supported by abstraction and code generation techniques, while *Organizational Alignment* is supported by modelling viewpoints.

Table 1 presents a comparison with some related work. We have classified them according to the following four criteria: IoT context, Model driven development and MDA approach, microservice architecture style, and modelling with SysML language. We noticed that none of the existing approaches takes into account the four criteria at the same time, namely the architecture of microservices with the SysML modelling (for the IoT hardware parts, etc.) and the use of the MDA approach.

Table 1. Comparison with some related work

Work	IoT	MDD	Microservices	SysML
[8]	✓	✓		
[11]	✓	✓		✓
[7]	✓			
[15]		✓	✓	
[19]	✓	✓		✓
[21]	✓	✓		
[9]	✓	✓		
[10]		✓	✓	
Our approach	✓	✓	✓	✓

3 The MDMSD4IoT Approach

MDMSD4IoT is a model-driven approach based on microservice architecture for the development of IoT systems. First, depending on the system requirements, a design model is specified, this model captures the functionalities of the system that are modelled with an extended SysML profile based on microservices and IoT concepts. Then, the design model is transformed into an IoT platform-specific model to generate the system implementation stubs. The code generation (M2T transformation) can interpret the models and generate the source code. At last, the generated code is deployed on a hardware platform to have a functional software system. MDMSD4IoT is composed of: a SysML profile called *SysML4IoTMSA*, a development method called *IoTMSADev* and a code generator called *SysML2IoTMSA* (see Fig. 1).

The SysML4IoTMSA profile (SysML for IoT systems based on microservice architecture) is a SysML extension inspired by the SysML4IoT profile [11], which is based on the IoT-ARM domain model. SysML4IoTMSA encompasses both the concepts of the IoT domain and those of the microservice architecture to provide high-level abstractions. To create the SysML4IoTMSA, we used Papyrus [2] which is an open-source plugin for the Eclipse platform, which provides SysML compliant meta-modelling functionalities. While the IoTMSADev method includes a set of steps, that represent an MDA approach based on the microservice architecture for IoT system development. This latter is inspired by the methodology proposed in [7], we proposed an extended method to be able to develop microservices and to use SysML diagrams. This method is supported by the SysML4IoTMSA profile to elaborate the models, and by the code generator to perform M2T transformations. The SysML2IoTMSA code generator is an Eclipse plugin based on the Acceleo tool, which is an open-source code generator. It uses elaborate models as input and generates text files based on a programming language as output.

3.1 SysML4IoTMSA Profile

MDMSD4IoT facilitates the specification and the design of complex IoT systems through the SysML4IoTMSA profile. This profile provides stereotypes to represent the concepts of IoT and microservices and their associations. According to SysML4IoTMSA, a microservices-based IoT system is made up of four parts: user, microservices, environment, and hardware. Figure 2 details the stereotypes defined in the profile. SysML Block is the principal extended element since SysML conceptualizes it as a modular unit of the system that is used to define a type of software, hardware, or data elements of the system or its composition of them [11]. *The user part* (with the yellow color) represents the users of the system (client-side application). *The environment and materials part* (with the green color) represents the physical and hardware aspects. It is related to the application domain concepts. For example, the building automation domain is expressed in terms of floors and rooms. The main concepts are as follows:

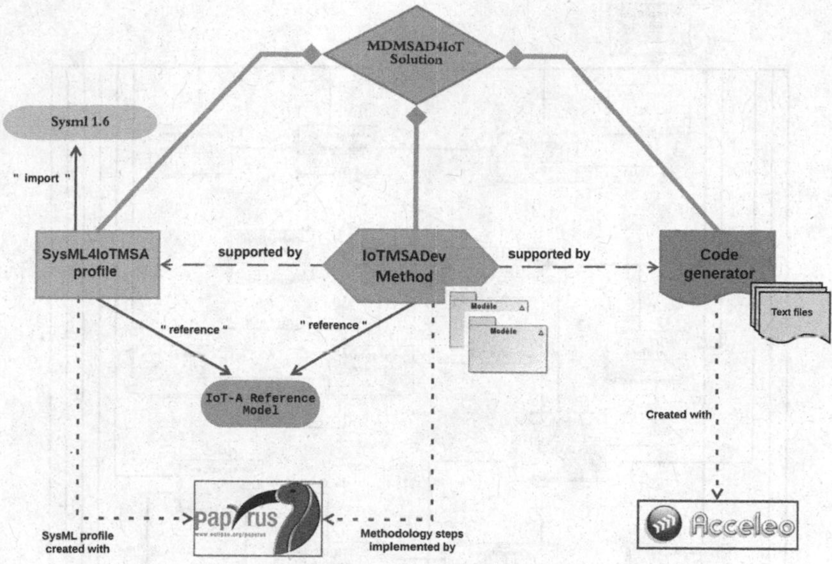

Fig. 1. The MDMSD4IoT approach

- *PhysicalEntity_zone*: is a region, which has an observable and controllable property called *phenomenon*. A typical example is a room with the value of its temperature.
- *PhysicalEntity_Device*: is a device in *PhysicalEntity_zone*, which has an attribute that describes its state. For example an air conditioner with its state (ON or OFF).
- *Sensor*: is used to detect changes in the environment. The sensor observes a phenomenon of a *PhysicalEntity_zone*. For instance, a temperature sensor observes the room temperature phenomenon.
- *Actuator*: is used to make changes in the environment through action. Heating or cooling elements are examples of actuators. The actuator affects the phenomenon of a *PhysicalEntity_zone* by performing actions. For instance, a heater is set to control a room temperature level.
- *ComputingDevice* is a hardware component that hosts resources, sends the collected data, and receives commands. An Arduino or raspberry card are examples of *ComputingDevice*.

Microservice part (with the pink color) represents the logical aspect of the application. A system consists of multiple *Microservices* which are either Functional (*FunctionalMS*), which realize business capabilities, or Infrastructure (*InfrastructureMS*), which provide the system architecture with infrastructure properties such as configuration and registration.

- A *FunctionalMS* consist of: *Entity*, *Interface* and *Service*.
- *VirtualEntity_zone* and *VirtualEntity_Device* represent respectively *PhysicalEntity_zone* with its phenomenon and *PhysicalEntity_Device* with its state.

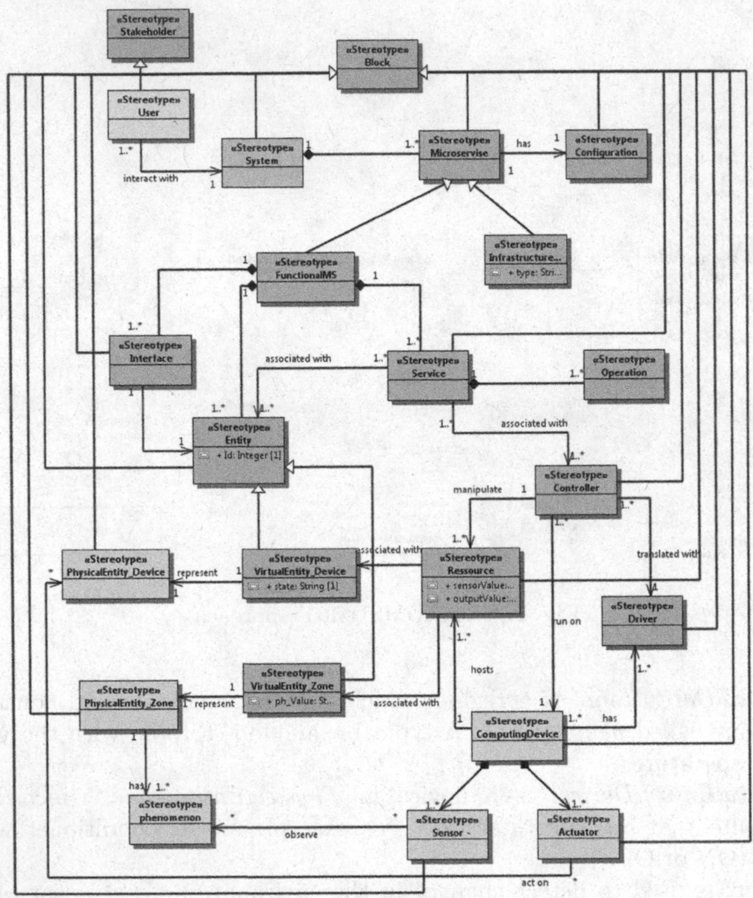

Fig. 2. The SysML4IoTMSA profile

- *Resource* represents the sensor and the actuator.
- A *Service* encapsulates a set of *Operations* and represents the web service provided by the microservice. For example, *SwitchOnLED* is an operation of *LuminosityService*.
- A *Controller* is a processing element that encapsulates a set of functionalities. It consumes one or more units of information as inputs, processes them, and generates an output. It runs on *ComputingDevice* and interacts with web services, the *controller* sends the collected data to the web service and receives commands from the application (through web services) to control hardware or trigger actions.
- A *Driver* is a computer program that acts as a translator between a *ComputingDevice* and the *Controller*. For example, if a java code does not work directly on an Arduino board, it must use the JArduino (a Java API and an Arduino firmware) which is in our case the *Driver*.

3.2 IoTMSADev Method

In our approach, we have extended the methodology for the development of IoT proposed in [7] by adding new steps related to microservices (step *D*: Definition of the microservice architecture and step *E*: Microservice specifications) using SysML diagrams. In the following, we briefly explain the development steps:

a) *Objective and requirements analysis*: the first step is to define the objective and the main requirements of the system. In this step, we describe our system, why it is designed, and its expected functionalities.
b) *Requirement specification*: in this step, the use cases of the IoT system are described and derived from the first step. It consists of extracting the functional requirements of the system and modelling them with the SysML requirement diagram. In this step, the system functionalities are grouped into domains.
c) *Process specification*: describes the behavioral aspect of the system, and how it works through SysML activity diagram.
d) *Definition of the microservice architecture*: the fourth step is to define the logical architecture of the system. Designers must break down the system into fundamental architecture building blocks that represent microservices, each one is highly cohesive and encapsulates a unique business capability. A microservice can be a functional type or an infrastructure type. This step represents the interactions between the different microservices according to the topology of the architecture.
e) *Microservice Specification*: consists of defining the microservice specifications. It models each functional microservice within and architecture (defined in the previous step) using a definition block diagram and describes the main concepts (interface, entities, service, operations, and the relationships between them), in order to understand each microservice domain.
f) *Functional View Specification*: defines the Functional View (FV). It defines the different functionalities of the IoT system grouped into functional groups (FG). In this step the system is represented in layers, each layer will be mapped to one or more groups (FG) according to its functionalities.
g) *Operational view specification* : consists of defining the operational view. The various options related to the deployment and the system operation are defined, such as the options of hosting services, storage, and devices.
h) *Integration of devices and components*: it installs and integrates the various devices.
i) *Application Development*: this step involves developing the entire IoT application (backend and frontend), testing it and deploying it.

3.3 SysML2IoTMSA Code Generator

The proposed SysML2IoTMSA code generator is presented in Fig. 3, it is responsible for transforming the models developed in step "E. Microservice specification" into source code. The generator is an Eclipse plugin based on the Papyrus and Acceleo Framework for an M2C (Model to Code) transformation. Acceleo [1] is an implementation of the MTL (Model-to-Text Language) standard, defined by the OMG. It uses a transformation by the *template approach.* A template is a text containing reserved spaces in which information is taken from the input model. These placeholders are generally expressions specified on the entities that will be used in the model from which information is extracted. Templates are written in *modules*, a module corresponds to a *.mtl file* and it contains a set of templates and queries. The *.mtl file* specifies the name of the module and the type of the meta-model used.

The generated code related to the microservices is based on Spring Boot and Spring Cloud frameworks. SysML2oTMSA is based on the SysML metamodel, which is available in the Papyrus plugin. Finally, SysML2IoTMSA aims to strengthen reuse to facilitate the development of the MSA in order to avoid heavy and redundant coding.

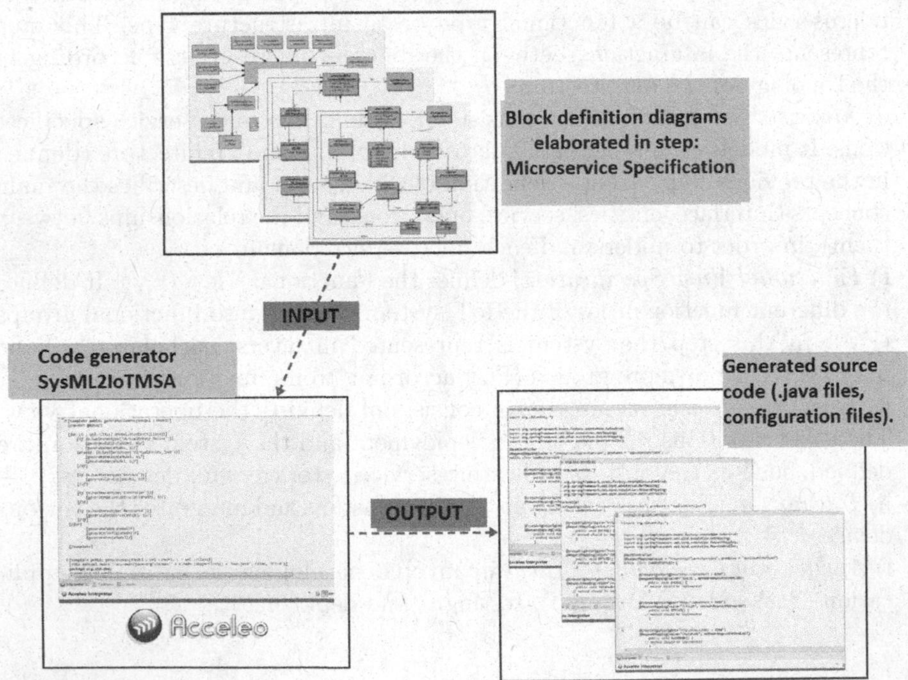

Fig. 3. The SysML2IoTMSA code generator

To generate the code related to the microservice architecture, we created an Acceleo project with two modules *(two .mtl files)*: *FunctionalMSgenerator* for the generation of the code related to the functional microservices type and *InfrastructureMSgenerator* for the one related to the infrastructure microservices type. Each module:

- Specifies the metamodel types using (SysML).
- Consists of a set of templates (generateEntity, generateRestService, generateDiscovery, etc.) which is called by the main template at runtime.
- Queries written in AQL (Acceleo Query Language), a language used to navigate and query an EMF model. Their syntax is very similar to OCL syntax. For example, the query *hasStereotype (value: String)* tests whether the element has the stereotype passed as a parameter.
- The input of the *FunctionalMSgenerator* is the SysML block definition diagram of the microservice.
- The input of the *InfrastructureMSgenerator* is the SysML block definition diagram of the overall system structure.
- The output is *.java files and configuration files.*

4 A Case Study: UC2SmartClassroom

To illustrate our proposed approach, we have used it to develop a prototype of a smart classroom system called *UC2SmartClassroom* for the University of Constantine 2 as a part of our master's project.

This system aims to ensure a certain degree of comfort, security and energy saving within the university, through:

- Access control using fingerprints;
- Resource management;
- Reducing energy consumption through the concept of smart lighting.

For the efficient management of classroom resources (lighting, heating, air conditioning, and data show), each teacher can access the classroom that it is assigned to him in his schedule. A sensor fingerprint allows him to open the door of the classroom and turn on its materials and devices, once he finished, the system will automatically turn off these resources. Each classroom is equipped with a touch screen tablet attached near the teacher's desk, allowing the control of the main devices.

The system automatically configures the lighting and temperature levels inside the classrooms according to the light and temperature values captured through different sensors and actuators.

The design and the development of *UC2SmartClassroom* system rely on the MDMSD4IoT approach. Thus, the SysML4IoTMSA profile is used to specify and model the system functionalities according to the IoTMSADev steps (from step a) to i)) and the SysML2IoTMSA code generator to get the system implementation stubs.

4.1 Modelling Microservices-based IoT Systems

Following the SysML4IoTMSA profile, we modeled the *UC2SmartClassrooms* system as a composed structure. The structure of the system through a SysML block definition diagram specifies the following entities:

- a set of functional microservices (LuminosityMS, TemperatureMS, AccessMS and ResourceMS)
- a set of infrastructure microservices (ConfigurationMS, DiscoveryMS and Getway MS)
- a set of devices (LuminosityDevice, TemperatureDevice, AccessDevice and RessourceDevice) which are *microcontrollers*. Each microcontroller is attached to its corresponding *sensor* (LumSensor, TemSensor, FingerprintSensor, MotionSensor) and its *actuator* (Relay, InfraRed LED, ServoMotor).

The system interacts with the classroom environment and with the system administrator or the teacher.

Microservices allow making the correspondence between each service with its corresponding resources and devices (sensors and actuators) in order to carry out this service. *For each functional microservice*, we develop a block definition diagram, to extract the offered services and to define the responsibilities, i.e., associate services with their required resources. The microservices specification also shows the devices that host the resources.

4.2 Microservice Architecture Implementation

In this part, we present the different technical steps to implement microservices architecture based on *Spring Boot, Spring Cloud and Netflix OSS (Eureka, zuul)* for our UC2SmartClassrooms system. Figure 4 shows the technical architecture. As mentionned before, we have four microservices which represent the functional aspect of the system through *REST APIs*. To ensure **horizontal scalability** (adding additional nodes) and **fault tolerance**, multiple instances of the same microservice can be started simultaneously. Each functional microservice:

- has its own database;
- is an application in the SpringBoot framework;
- has its own Maven module containing some Java classes and configuration files.

To operate in a distributed environment, microservices rely on a set of tools offered by Spring Cloud: centralized configuration management, automated discovery of other microservices, load balancing, and API routing. The front-end part is developed with Angular. The *Gateway API* is a microservice responsible for routing a request from one microservice instance to another one. This latter centralizes browser invocations. Although it can play an aggregator role, most calls go directly to functional microservices.

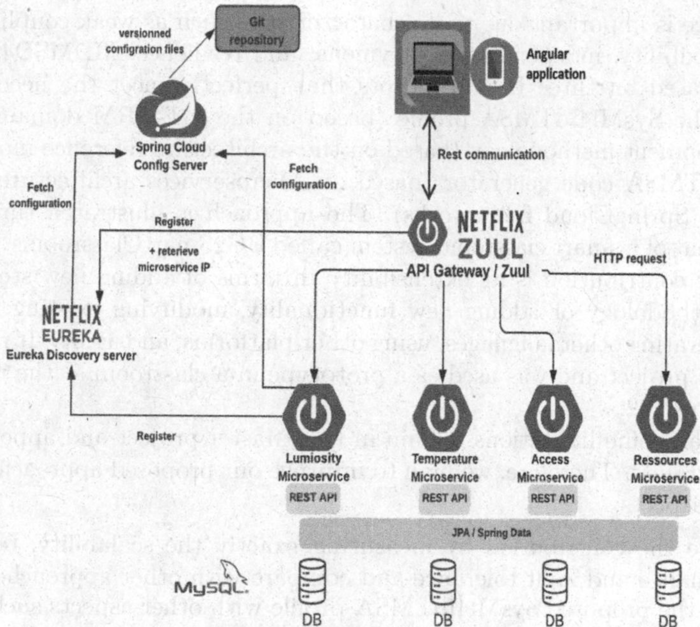

Fig. 4. The technical architecture of microservices

Spring Cloud Netflix offers to use *Zuul proxy* to forward requests received by the Gateway API to functional microservices. In a microservices architecture hosted in the Cloud, it is difficult to anticipate the number of instances of the same microservice (depending on the load) or even where they will be deployed (and therefore on which IP and which port they will be accessible). The role of *Eureka server*, is to connect microservices. Each microservice will register and retrieve the address of its adhesion. The microservices (as well as the other servers) will load their application configuration from the *Config Server*, whose role is to centralize the configuration files of microservices in a single place.

The configuration files are versioned on a *git directory* containing a common configuration for all microservices and a specific configuration for each microservice (in this case the *Git Bash* tool is used as a version manager). Changing the configuration no longer requires rebuilding applications or redeploying them, a simple restart is sufficient.

5 Conclusion and Future Work

In this paper, we have proposed a Model Driven MicroServices Architecture Development for IoT systems (MDMSD4IoT) based on microservices architecture (MSA) and driven by models. We opted for the MDA approach for its efficiency in modelling complex software systems through its sophisticated techniques including abstraction and automation. In addition, the microservices architecture is important due to its characteristics such as weak coupling, modularity, flexibility, independent deployment, and resilience. MDMSD4IoT approach is based on three basic concepts that perfectly meet the needs of IoT systems: the SysMl4IoTMSA profile (based on the IoT-ARM domain model), IoT development methodology (based on the architecture reference model), and SysML2IoTMSA code generator (based on Microservices architecture, Spring-Boot, and SpringCloud frameworks). The approach is illustrated through the development of a smart classroom system called UC2SmartClassrooms. The novelty of our contribution is its extensibility in terms of adding new steps in the design methodology or adding new functionality, modifying existing microservices, generating other languages, using other platforms, and agility. It represents a Master's project and was used as a prototype in a classroom at the university of Constantine 2.

However, some limitations remain in this Master project and appear due to time constraints. Therefore, we plan to improve our proposed approach with the following aspects:

- evaluate the contribution by measuring exactly the scalability, reusability, performance and fault tolerance and compare with other approaches
- extend the proposed SysMl4IoTMSA profile with other aspects such as security;
- analyse data and create a predictive model for forecasting future events in smart buildings
- validate and verify our system by transforming our meta-model to another formal language
- develop a DSL that integrates SysMl4IoTMSA profile as an input model with other IoT aspects and other Code generators.

References

1. Acceleo. https://www.eclipse.org/acceleo/
2. Eclipse papyrus. https://www.eclipse.org/papyrus/
3. Iot-a: internet of things architecture. https://portal.effra.eu/project/1470
4. Object management group (omg). https://www.omg.org/
5. OMG: object management group MDA (Model Driven Architecture) Guide Version 1.0.1. http://www.omg.org/mda/ (2001)
6. Aguilar-Calderón, J.A., Tripp-Barba, C., Zaldívar-Colado, A., Aguilar-Calderón, P.A.: Requirements engineering for internet of things (loT) software systems development: a systematic mapping study. Appl. Sci. **12**(15), 7582 (2022)
7. Bahga, A., Madisetti, V.: Internet of things: a hands-on approach, chap. 5, pp. 99–115. Bahga and Madisetti (2014)
8. Bassi, A., et al.: Enabling Things to Talk: Designing IoT Solutions with the IoT Architectural Reference Model. 1st edn. Springer, Berlin (2013). https://doi.org/10.1007/978-3-642-40403-0
9. Corradini, F., Fedeli, A., Fornari, F., Polini, A., Re, B.: FloWare: an approach for IoT support and application development. In: Augusto, A., Gill, A., Nurcan, S., Reinhartz-Berger, I., Schmidt, R., Zdravkovic, J. (eds.) BPMDS/EMMSAD -2021. LNBIP, vol. 421, pp. 350–365. Springer, Cham (2021). https://doi.org/10.1007/978-3-030-79186-5_23

10. Cortellessa, V., Pompeo, D.D., Eramo, R., Tucci, M.: A model-driven approach for continuous performance engineering in microservice-based systems. J. Syst. Softw. **183**, 111084 (2022). https://doi.org/10.1016/j.jss.2021.111084
11. Costa, B., Pires, P., Delicato, F.: Modeling IoT Applications with SysML4IoT. In: 42th Euromicro Conference on Software Engineering and Advanced Applications (SEAA), pp. 157–164 (2016)
12. Debbabi, M., Hassaïne, F., Jarraya, Y., Soeanu, A., Alawneh, L.: Verification and Validation in Systems Engineering. Springer, Berlin (2010). https://doi.org/10.1007/978-3-642-15228-3
13. Delsing, J., Kulcsár, G., Haugen, Ø.: SysML modeling of service-oriented system-of-systems. Innov. Syst. Softw. Eng. (2022). https://doi.org/10.1007/s11334-022-00455-5
14. Dragoni, N., et al.: Microservices: yesterday, today, and tomorrow. In: Present and Ulterior Software Engineering, pp. 195–216. Springer, Cham (2017). https://doi.org/10.1007/978-3-319-67425-4_12
15. F. Rademacher, J. Sorgalla, P.W.S.S., Zundorf, A.: Microservice architecture and model-driven development: yet singles, soon married (?). In: Proceedings of the 19th International Conference on Agile Software Development: Companion, p. 5, No. 23, ACM, New York, USA (2018)
16. Giannelli, C., Picone, M.: Editorial industrial IoT as it and OT convergence: challenges and opportunities. IoT **3**(1), 259–261 (2022)
17. Harrand, N., Fleurey, F., Morin, B., Husa, K.E.: ThingML: a language and code generation framework for heterogeneous targets. In: Proceedings of the 19th International Conference on Model Driven Engineering Languages and Systems, pp. 125–135. ACM (2016)
18. Holt, J., Perry, S.: SysML for Systems Engineering. 2nd edn. The Institution of Engineering and Technology, London (2013)
19. Hussein, M., Li, S., Radermacher, A.: Model-driven development of adaptive IoT systems. In: MoDELS (2017)
20. Nadareishvili, I.R., Mitra, M.M., Amundsen, M.: Microservice Architecture. 1st edn. O'Reilly Media, Sebastopol (2016)
21. Ihirwe, F., Ruscio, D.D., Mazzini, S., Pierantonio, A.: Towards a modeling and analysis environment for industrial IoT systems. In: Iovino, L., Kristensen, L.M. (eds.) STAF 2021 Software Technologies: Applications and Foundations. CEUR Workshop Proceedings, vol. 2999, pp. 90–104. CEUR-WS.org (2021). http://ceur-ws.org/Vol-2999/messpaper1.pdf
22. Newman, S.: Building Microservices. O'Reilly Media, Sebastopol (2015)
23. Sethi, P., Sarangi, S.: Internet of things: architectures, protocols, and applications. J. Electr. Comput. Eng. **1**, 1–25 (2017)
24. Picek, R., Strahonja, V.: Model driven development - future or failure of software development? (2007)
25. da Silva, A.R.: Model-driven engineering: a survey supported by the unified conceptual model. Comput. Lang. Syst. Struct. **43**, 139–155 (2015)
26. Stahl, T., Voelter, M., Czarnecki, K.: Model-Driven Software Development: Technology, Engineering. Management. Wiley, Hoboken (2006)

Database Systems

Parallel Skyline Query Processing of Massive Incomplete Activity-Trajectories Data

Amina Belhassena[1(✉)] and Wang Hongzhi[2]

[1] Audensiel Technologies, Boulogne-Billancourt, France
a.belhassena@audensiel.fr

[2] Harbin Institute of Technology, Heilongjiang, China
wangzh@hit.edu.cn

Abstract. The big spatial temporal data captured from technology tools produce massive amount of trajectories data collected from GPS devices. The top-k query was proposed by many researchers, on which they used distance and text parameters for processing. However, the information related to text parameter like activity is always not presented due to some reason like lack internet connection. Furthermore, with massive amount of keyword semantic activity-trajectories, user may enter the wrong activity to find its activity-trajectory. Therefore, it's hard to return the desirable results based on the exact keyword activity. Our previous work proposed an efficient algorithm to handle the trajectory fuzzy problem based on edit distance and activity weight. However, the algorithm proposed does not work with incomplete Trajectory DataBases (TDBs). Therefore, the present investigation focuses on handling the trajectory skyline problem based on distance and frequent activities in incomplete TDB. To accelerate the query processing, the massive trajectory objects is managed through Distributed Mining Trajectory R-Tree (DMTR-Tree index) based on R-tree indexes and inverted lists. Afterward, an efficient algorithm is developed to handle the query. For a rapid computation, a cluster-computing framework of Apache Spark with MapReduce is used. Theoretical analysis and the experimental results show a well agreement and both attest on the higher efficiency of the proposed algorithm.

Keywords: Distributed processing · Incomplete data · Skyline-trajectory · Fuzzy · Top-k spatial keyword queries

1 Introduction

Nowadays, more and more spatial temporal data are produced by means of new sensors and smartphones, etc., which embedded with GPS (Global Positioning System) produce huge volumes of trajectories data. To discover the knowledge and support the decision-making, these moving objects are usually stored and archived in TraJectory DataBases (TDBs) for in-depth analysis and processing [1]. Several application domains such as animal breeding, traffic jam, etc.

P. Fournier-Viger et al. (Eds.): MEDI 2022, LNAI 13761, pp. 193–206, 2023.
https://doi.org/10.1007/978-3-031-21595-7_14

exploit these massive TDBs. In particular, TDBs have been successfully applied to the tourism and marketing application domains, since end-users are usually equipped with smart phones, or GPS embedded vehicles, which are able to track the movements of these people at detailed spatial and temporal scales. The analysis of these human activities providing semantic trajectories data could be then used to improve the quality of offered services. Indeed, end-users can manually add to the trajectory data some description about their activities like shopping, working, etc. when they arrive at Point of Interest (POI) locations. This crowdsourcing approach allowing people to comment, edit POIs is very common, and adopted by commercial services enterprises (such as Google) or free platforms like Wikipedia. Although, these GPS datasets become more and more massive, data concerning the end-users activities is not always present, and does not associated to the GPS points, for different reasons: lack of internet connection, the end-user does not want publish the data for privacy issues, lack of time, etc. Such kinds of applications using POIs can also be translated in other contexts where the POI is a location with a particular relevance such as drinking points in the bravery applications, energy recharge points for agricultural and cars vehicles, etc. This makes the proposal of this paper open to different application domains different from the tourism one described in the next of the paper.

A real common and used query type on these trajectories datasets annotated with activities consists in finding out the frequent shortest activity-trajectory, near to user location including a set of preferable activities. For example, "What is the shortest activity-trajectory to reach Eiffel tower from La Bastille (in Paris) and do some shopping at the same time?" For such a challenging query, frequent mining activity-trajectory algorithms used for top-k queries are not trivial. Indeed, efficient similarity measures among activity-trajectories are necessary. Moreover, the huge volume of this data needs efficient computation and storage methods such as distributed indexes. Finally, the lack of some activity data must also be taken into account. So, due to some reasons like: lack adaptation for tactile tablets, users may make mistakes while inputting their activity queries. Thus, it's hard to return the desirable activity-trajectory based on the exact keyword activity. Indeed, approximate keyword search has to be considered.

Some efforts have been done about the top-k trajectory similarity with activities like [2], others have used some indexing methods like hybrid grid index [3]. However, to the best of our knowledge, processing the top-k frequent activity-trajectories query with missing data has not been researched so far. Therefore, the proposals of our work aim to process the top-k query based on the similarity of the spatial distance and the full activity information by taking into account the problems of missing data and approximate activity keywords search. In particular, to solve the incomplete information problem in top-k queries, we will use an optimized method of skyline queries. Skyline query is a database query where the skyline operator answers to an optimization problem, used to filter results and keep only those objects that are not worse than any other. In our context, a skyline query is based on both spatial and textual (i.e. activities) dimensions. Furthermore, our solution is promoted also to process the fuzzy problem for approximate activities in POI using edit distance and activity weights [4]. To answer the skyline query mentioned above, which returns the activity-trajectory that is not dominated by others based on the massive historical trajectory data, it's hard to process this query without passing by an efficient storage method as well as a distributed parallel computing approach. Therefore, we use our previous method to organize the massive trajectory data into Distributed Mining Trajectory R-Tree index (DMTR-TREE) [5,6], which is based on distributed R-tree indexes and inverted lists. Using an aggregate of data through distributed parallel operations, we have developed a new algorithm for skyline query processing. This algorithm is performed using a distributed cluster based on Spark and MapReduce model to accelerate the process of large data trajectories. The contributions of our paper are:

- The proposal of some new functions to evaluate the multi-frequent activity and distance between similar trajectories in TDB and the query.
- Based on our previous distributed index, an effective new algorithm in order to solve the trajectory skyline and incomplete data problems efficiently.

The remaining of our paper is organized as follows: Sect. 2 presents the related work. Section 3 presents the overview of our index. Section 4 introduce a set of functions that used for query processing. Section 5 explains the query algorithm proposed. Section 6 describes our experimental studies.

2 Related Work

Recently, many researchers have widely studied the top-k trajectory query processing on trajectory data [3,7–9]. In [10] they integrated social activity data into the spatial keyword query of semantic trajectory. Furthermore, to process the top-k trajectory queries, an efficient indexing method can be necessary. Recently, many studied have proposed to index massive data through spatial access methods like R-Tree index [11], which is more suitable to index the spatial movement objects. However, as data is rapidly increased, the indexing methods based on a single node are insufficient. Therefore, some researchers have proposed distributed indexes to handle the limitation of centralized methods based on clusters [12,13], where authors have used h-base. To perform MapReduce to index the massive spatial data, in [14], they have developed an efficient framework called SpatialHadoop. In order to improve the MapReduce limitation, GeoSpark [12] have been proposed to support spatial access methods.

Furthermore, the distance of semantic trajectories have to be considered. Thus, it is hard to process top-k query based on both approximate semantic keywords and distance. Such a problem could be solved using skyline query. Authors in [16] have firstly proposed the skyline query by introducing the skyline operator for a relational database system based on the B-Tree index. In trajectory data processing, authors in [17] addressed the problem of trajectory planning by exploring the skyline concept. [18] have proposed an efficient algorithm to retrieve stochastic skyline routes for a given source-destination pair and a start time. Recently,

[19] applied the skyline queries method for a personalized travel route recommendation scheme using the mining of the collected check-in data. [20] combined the skyline query and top-k method to recommend travel routes that cover different landscape categories and meet user preferences. With the important contributions made by the aforementioned work, none of them take into consideration the missing semantic trajectory data for skyline query processing.

3 DMTR-Tree Index Overview

To support the process of both spatial objects and textual activity information. The Distributed Mining Trajectory DMTR-Tree constructed distributed R-Tree indexes with mining inverted lists based on MapReduce model, in which the process is carried out by the Spark cluster. The master node interconnects with each slave in the cluster, on which each slave has an R-tree index [5,21]. We build the indexes in three phases:

a. Partition method. Consider TD trajectory data sets. The total number of partitions in the cluster is $n = \sum_1^m n$, where m is the total number of slave machines in the cluster and n is the maximum partition number in each slave. The trajectory $POI_i.T = \{poi.key.T, poi.s.T, poi.a.T\}$ are composed of spatial and non-spatial data. Each point POI_i is defined by a unique key $poi.key.T$, representing its spatial location $poi.s.T$ (the longitude and the latitude), and activity keywords $poi.a.T$ that are holding in POI_i. To efficiently store the points POI_i with their activities in the indexes, the purpose of this phase is to assign each point of TD to one partition according to $poi.key.T$ based on the range partition. We solved the skewing data problem, by re-configuring the ranges of all the partitions in the same slave. Denoting n as the maximum number of partitions and e as the maximum number of trajectory points in n. ideally, each partition *should have* e/n *elements*.

b. Local construction. The RDD mapped in the first phase is used to split the data set TD into N partitions. Since, N small R-tree indexes are then built separately at the same time by N reducers in each slave. Thus, each worker's machine constructs an index for each partition, on which, the index uses the Minimum Bounding Rectangles MBRs for POI_i storage.

c. Inverted list phase. To extract the frequent activities practiced by previous users, we have implemented the frequent mining algorithm "Apriori Algorithm" on each point. The results of the algorithm were stored in the mining inverted lists, where each point has its own inverted list. To obtain the whole inverted list, all the object' lists were combined together. Finally, to optimize the query when traversing the whole list, we have reduced the lists' number by collecting the objects that are belonged to the same trajectory in the same inverted list.

4 Problem Statement of Skyline Trajectory Query

The trajectory object o_i has two dimensions. The first dimension is the distance dimension, which represents the geographic similarity measured between the query q and an activity trajectory T or q and an activity sub-trajectory $subT$. The second dimension is the activity dimension, which represents the activity keywords similarity measured between q and T or q and $subT$. During query processing, in the case when T candidate has incomplete data, which is related to frequent activity set, the distance dimension will be calculated between T and other closest trajectory object candidates. Further, the activity dimension will also be measured between q and other closest trajectory objects candidates. Let us consider an example of TDB, depicted in Fig. 1, which contains three activity-trajectories: T_1, T_2, T_3. Each activity-trajectory has a set of points of interest. Each POI has a set of associated activities. For example, the activity-trajectory T_1 is composed of two different shopping centers denoted by P_2 and P_3, and a sports club denoted by P_4, and $\{Shop, Hair - Cut, Restaurant\}$ are activities involved in the POI P_2. A trajectory skyline query is also depicted in Fig. 1, that is characterized by a user location point, user destination point, a distance threshold, and a set of activities. Here, a tourist in a new city wants to know the frequent shortest trajectories, which are close to its location. She/he aims not only to visit the POIs of the specific trajectory but, she/he wants to practice some activities (i.e. $\{Hair - Cut, Restaurant, Fitness\}$) that are evaluated by previous users, included in this trajectory and doesn't exceed 3 km. To answer this query, the system evaluates the trajectories $\{T_1, T_2, T_3\}$ and returns the best one. T_3 has a shortest distance to the user position, however it does not include all the required activities. T_1 has a short distance than T_2, and it presents all the asked activities. The solution to this problem has already been presented in our previous work [6]. However, since the activity set is also important for the user, some users prefer to choose the trajectory with an ordered sequence of activities, even if the trajectory chosen is not the shortest one (e.g. the user could prefer T_1 or T_2 to T_3). Besides, the incomplete information of some activities in the trajectory POIs may occur while archiving data into TDB, as discussed in the previous section. In this case, the system cannot process the query without full information. Under this circumstance, our previous work [6] does not apply. For example, some activities are missing in the POI P_3 of T_1. Therefore, the system must sort another trajectory similar to T_1. As T_2 is good in the distance dimension compared to T_3. Further, T_2 is better than T_1 and T_3 in the textual dimension. Thus, T_2 is not dominated by either T_1 neither T_3. Therefore, T_2 is the best choice that could be returned to the user.

To handle the trajectory skyline problem efficiently, we have developed two functions based on distance and activity dimensions

a. The distance function: $\mathcal{F}$ is a function, which aims to estimate the similarity distance between q and T. Assuming $q(L, D)$ is the query determining by a user location point L, and a user destination point D. To measure the shortest distance, we have used the Euclidean distance due to its lightweight computation. The distance function $\mathcal{F} \in [0, 1]$ is represented by the formula 1:

$$\mathcal{F}(q, T) = 1 - N(L.q.o_i.T) + \frac{\sum_{i=1}^{n}[d(o_i.T, o_{i+1}, T)] + M(T_i, T_j) + d(o_n.T, D.q)}{\hat{d}} \quad (1)$$

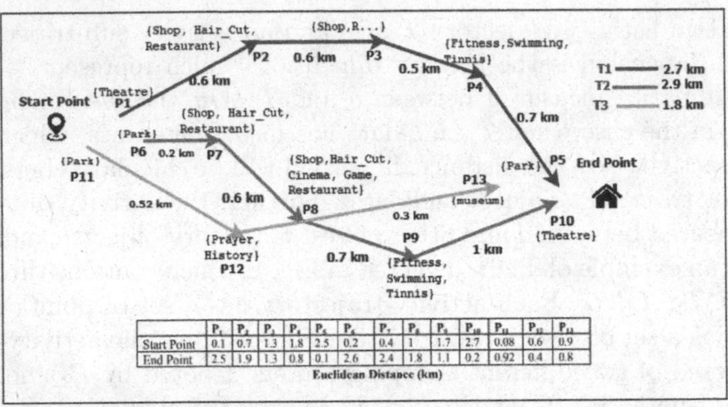

	P_1	P_2	P_3	P_4	P_5	P_6	P_7	P_8	P_9	P_{10}	P_{11}	P_{12}	P_{13}
Start Point	0.1	0.7	1.3	1.8	2.5	0.2	0.4	1	1.7	2.7	0.08	0.6	0.9
End Point	2.5	1.9	1.3	0.8	0.1	2.6	2.4	1.8	1.1	0.2	0.92	0.4	0.8

Euclidean Distance (km)

Fig. 1. Trajectory query motivating example

In the formula 1, $\widehat{d}$ is a distance threshold, the function $N(L.q.o_i.T)$ is used to extract the nearest Activity-Trajectory object T between user location L in q and the first object o_1 in each T. This function could be calculated using the following formula: $N\left(L.q.o_i.T\right) = 1 - \frac{d(L.q.o_i.T)}{\hat{d}}$. N could be calculated using the Euclidean distance table (Fig. 1). Assuming, the activity-trajectory query q is presented by the start point and the end point. We start by evaluating the trajectory T_1. The function $N(q, T_1)$=1- 0.1/3=0.96. Concerning the distance between the o_i in T, $\sum_{i=1}^{n}\left[d\left(o_i.T, o_{i+1}.T\right)\right]$ we could calculate it using the Euclidean distance table as: $\sum_{i=1}^{n}\left[d\left(o_i.T_1, o_{i+1}.T_1\right)\right] = d(p_1,p_2) + d(p_2,p_3) + d(p_3,p_4) + d(p_4,p_5) \quad = 0.6 + 0.6 + 0.5 + 0.7 = 2.4$ To handle the missing activity problem in the Activity-Trajectory candidate T_i, the function $M\left(T_i,\, T_j\right)$ is used to estimate the similar closest trajectory T_j, $M\left(T_i,\, T_j\right) = minDis\left(o.T_i,\, o.T_j\right)$. We specify, $M\left(T_i, T_j\right) > 0$. we have specified, $M\left(T_i, T_j\right) > 0$ to prune out the trajectories those are in the opposite direction. Finally, $d\left(o_n.T,\, D.q\right)$ is the distance computed between the last trajectory object o_n in T and D.

b. The activity function represented by G, which it aims to estimate the similarity of activity keywords. Initially, using edit distance [22], G allows the multi-keywords similarity query as is adopted in [15]. It aims to measure the similarity between the activities holding in the trajectory object of T and the activity keywords of q. This solution allows us to process the fuzzy query.

Given an example of an aged user who is not familiar with the system. Assuming, this user inputs in the system some desired activities like $\{shopp, Hair - Cut, Restaurant\}$, such activities are presented in the P_1 of T_1 (Fig. 1). As noticed, the user inputs the activity $\{shopp\}$ instead of $\{shop\}$. Here, we set the threshold of the distance to 3. The threshold of the distance on which the distance between two keywords is less or equal to it is set in advance. Here, the edit distance of two keywords: $\{shop\}$ and $\{shopp\}$ is 1, we should delete $\{p\}$ in $\{shopp\}$ to transform the string $\{shopp\}$ into $\{shop\}$. Thus, these two keywords are considered to be approximately similar. However, the threshold should change as the length of keywords varies. [4] proposed a method to handle the problem of edit distance threshold. They presented a function to quantify the text relevance between geo-textual objects and the query based on the edit distance and the keyword weight instead of setting the threshold in advance. This function is adopted in our paper.

Assuming the activity-query q lacks some activity that is involved in such a POI like is presented in P_3 of T_1 (Fig. 1). To efficiently handle the problem, the activities holding in the other trajectory candidates should be similar to q. Considering we have two activity-trajectories T_j and T_k with POI_x, POI_y., respectively. Each POI has its own activities φ_i. The activity-trajectory q has multiple activity keywords on which $\varphi_i.q = \{\varphi_1, \varphi_2, ...\varphi_i\}$.

$ed\,(\varphi_i.POI_x.T_j,\ \varphi_i.q)$ represents the edit distance between activities holding in POI_x T_j candidate and the query q. The weights of φ_i is presented by $w\,(\bar{\varphi}_i.POI_x.T_j)$. $ed\,(\varphi_i.POI_x.T_j, \varphi_i.POI_y.T_k)$ represents the edit distance between activities holding in POI_x, POI_y of T_j and T_k., respectively. We used formula 2 to estimate the activity function between activities $\varphi_i.T$ of T candidates and the query activity $\varphi_i.q$

$$\mathrm{G}\,(\varphi_\mathrm{i}.\mathrm{T}, \varphi_\mathrm{i.q}) = \sum_{\mathrm{i}=1}^{\mathrm{n}} \mathrm{w}(\bar{\varphi}_\mathrm{i}.\mathrm{T}){*}(1 - \frac{\mathrm{ed}(\varphi_\mathrm{i}.\mathrm{q}, \bar{\varphi}_\mathrm{i}.\mathrm{T})}{\mathrm{l_i}} {*} \frac{1}{\mathrm{n}} + \mathrm{ed}\,(\varphi_\mathrm{i}.\mathrm{T_j}, \varphi_\mathrm{i}.\mathrm{T_k}) \quad (2)$$

In the formula 2, $\bar{\varphi}_i.T \in \ \varphi_i.t$ are activities in T, which have a minimum edit distance with the query activity keyword $\varphi_i.q$, $w\,(\bar{\varphi}_i)$ represents its weight. $ed\,(\varphi_i, \bar{\varphi}_i)$ is the edit distance between φ_i and $\bar{\varphi}_i$, l_i is the length of φ_i. The number of activities is varied from 1 to n.

This formula is adopted from [4], but we have performed it adding the edit distance $ed(\varphi_i.T_j, \varphi_i.T_k) = ed(\varphi_i.POI_x.T_j, \varphi_i.POI_y.T_k)$ to handle the missing activity data problem between activity-trajectory candidates.

5 An Optimized Trajectory Skyline Algorithm

To process the skyline activity-trajectory query and solve the incomplete information activity problem, we present a parallel skyline algorithm for activity-trajectory data (Algorithm 1). The query q to be processed aims to find the frequent shortest activity-trajectory including the whole frequent activities φ_i located on POIs with distance no more than the distance threshold $\hat{d}$.

Initially, q is sent to the master node in the cluster with S and E as the start and the end points of q. In Algorithm 1, to obtain the activity-trajectory involved in q, we start by traversing the DMTR-tree index to find the Minimum Bounding Rectangles MBRs overlapping q. Each index is stored in HDFS files based on the partition method as is explained previously. To find which index partition should be visited and which node should be selected, this method may visit all the nodes of all the trees in the cluster. Thus, we have noticed that this process occurs at a high cost while computing the distance between each node and q. To accelerate the computation, we have efficiently pruned the

Algorithm 1. Trajectory skyline query processing

Input:

- $q(S, E, \varphi_i, \widehat{d})$, List $L < id_i, ind_i >$

Output:

- List D

```
1: L' = ⟨id_j, ind_i⟩
2: for each ind_i in L' do
3:     overlap(S, ind_i)
4: for each ind_j in L' do
5:     overlap(E, ind_j)
6: if dis[(S, ind_i)+ dis(ind_j, E)]<= d̂ then
7:     Return ind_i, ind_j
8:     if i=j then
9:         Algorithm 2 //short trajectory case
10:    else
11:        Algorithm 7 // long trajectory case
```

search space by using a list to store all keys of partitions with their MBRs (line 1). This list is in the master node in the cluster. Then, to select the covered MBRs (lines 2–5), the master node computes the distance between the query points and the MBRs before starting any traversing of the distributed indexes. The search space is pruned based on this distance formula: $dis(MBR, q) = dis(S, MBR) + dis(MBR, E)$ (line 6). Moreover, based on trajectory lengths and partition method, we have classified the activity-trajectories into two classes. The first class comprises the short trajectories (lines 8–9) while the second class comprises the long trajectories (line 10–11). In the following, we will explain the process of both classes.

5.1 Short Trajectories Class

In this class, the activity-trajectory matching is short, i.e., this trajectory belongs to one partition and is organized through one R-tree index. We have developed algorithm 2 to process the short activity-trajectory query. In algorithm 2, initially, we simultaneously start traversing the index partition. Using an RDD spark, we can read the activity-trajectory data from this tree (line 2). Then, based on this RDD, we process another RDD filters using a function FILTER to prune the search space and return the activity-trajectory candidates (line 3).

For the pseudo-code of the function FILTER presented in Algorithm 3, the distance σ is computed between R-tree and the query q (line 3). It aims to select the node that should be visited by pruning the search space while traversing along the tree (lines 5–9). In the end, we return a list of the nodes, which include the activity-trajectory objects (line 9). In the case when the activity-set information is missing in the selected node (lines 11), using the distance function, another node with a minimal distance has to be found (lines 12–15). This new node returned must contain similar activity-set keywords of q (lines 16–17).

Algorithm 2. Short Trajectory processing

Input:
$q(S, E, \varphi_i, \widehat{d})$, List L
Output:
– List D
$D <>= \emptyset$
$RDDtree = sc.parallelize(R[]).mapValue(R.Entry)$
$RDDfilters = Tree.Filter(tree).collect()$
$RDDresults[] = filters.Primary - Trajectory()$

Using the activity function, the SIMILAR function [6] is invoked in line 16. It allows multi-activity similarity based on the edit distance [4]. As the leaf node in the tree may store multiple activities of the same POI, this function aims to return the activity which has the minimum edit distance with the keywords of q. Afterward, the list K is updated (line 17). In the opposite case, i.e., when we have full activities, we just compare the similarity between activity-set holding in the visited node and q (line 19). Then, we return the new list K (line 20).

Algorithm 3. Function: FILTER

1: **function** FILTER(Tree tree)
2:     e: entry in tree,
3:     $\sigma = dis(e, q)$
4:     **if** $\sigma <= \widehat{d}$ **then**
5:         **if** e is a non-leaf **then**
6:             **for each** child e' of e **do**
7:                 $e'.FILTER()$ ▷ recursively
8:         **else**
9:             Update $K < e_i, \varphi_i >$ ▷ Updating the list of POI stored in e
10:         **for each** e_i in K **do**
11:             **if** φ_i of e_i is null **then** ▷ missing activity-set information
12:                 minD=dis(e_i, e_{i+1}) ▷ distance dimension
13:                 **if** $0 < minD < \sigma$ **then**
14:                     $\sigma = minD$
15:                     return e_{i+1}
16:                     **if** SIMILAR$(\varphi_i.e_{i+1}, \varphi_j.q)$ =true **then**
17:                         Update $K < e_{i+1}, \varphi_i >$
18:
19:             **if** SIMILAR$(\varphi_i.e_i, \varphi_j.q)$ =true **then**
20:                 Update $K < e_i, \varphi_i >$
21:     **else**
22:         T does not exist ▷ the activity-trajectory matching does not exist

To return the final result to user, we processed the data mining algorithm to choose the best top-k activity trajectories. We used Apriori algorithm to calculate the support Sup and the confidence $Conf$ of activities and stored them in the inverted lists. This method helps us to traverse the inverted list of each activity-trajectory candidate T and extract the Sup and $Conf$ of its activity-set. To collect T objects (the POI with activities), we use the Primary-Trajectory function [6] (line 4 in Algorithm 2) that is presented in Algorithm 4. It aims to return a list D (lines 2–8) containing collected activity-trajectories. The frequent trajectory is a trajectory on which its activity set has a higher Sup and a higher $Conf$. To extract the frequent activity-trajectory from the activity-trajectory candidates obtained in the previous step, which are stored in the list D, we need just to return the trajectory with a best Sup and $Conf$.

Algorithm 4. Function: Primary-Trajectory

1: **function** PRIMARY-TRAJECTORY(RDD filters: t_n)
2:     **for each** $t_n \in T$ **do**
3:         **while** $t_n.POI_j \in POI$ **do** ▷ Browsing the POI of each t_n candidate.
4:             s= $\sum[Sup(\varphi_i)]$
5:             r= extract the rules of φ_i
6:             c= $\sum[Conf(r)]$
7:         put s, c, t_n in $D <>$
8:         **return** $D <>$

5.2 Long Trajectories Class

In this class, the activity-trajectory has to be visited is long, i.e., it could be divided into sub-trajectories belonging to several index partitions. While query processing, if there is missing activity information, another similar long activity-trajectory should be extracted. Therefore, using Algorithm 5, we efficiently handle the long activity-trajectory problem. Initially, we extract the sub-activity-trajectories using FILTER, and Primary-Trajectory functions (lines 3–5) explained previously in Algorithms 3, 4, respectively. Then, we collect the sub-activity-trajectories to obtain the whole activity-trajectory candidates (lines 6–7). Afterward, we return the final activity-trajectory matching to the user.

6 Experimental Evaluation

We organized a series of experiments to evaluate the performance of the algorithm presented in the previous section. The experiments aim at:

- Evaluating the number of trajectory visited.
- Evaluating the effectiveness of the activity keywords.
- Evaluating the effectiveness of the distance threshold.

Algorithm 5. Long Trajectory processing

Input:

– $q(S, E, a_i, \widehat{d})$, List R_i

Output:

– List D

for each i **do**
  $RDDR_1 = sc.parallelize(R_i[]).mapValue(R_i.Entry)$
  $RDDfilters = Tree.FILTER(R_1).collect()$
for each $subT$ of T **do**
  $RDDPR[] = filters.Primary - Trajectory()$
for each element e of PR **do**
  $D = find - duplicate(element[e])$
  Return D

These experiments were implemented on Spark-1.6, using HDFS (version 2.6.0) and Hadoop (version 2.6.0). All experiments are conducted on a cluster of ten machines. Each machine has 32 GB (4*8 GB) RAM, 64-bit quad-core i7 processor, and four 7200 rpm SATA Disks (4*1TB). The computing cores are all running on UBUNTU (version 14.02) and Java 1.8 with Maven (3.0.4).

For these experiments, we used two historical Microsoft datasets with different characteristics: a short trajectory dataset called as GEO LIFE, and a long trajectory dataset of T-DRIVE. In the other hand, in database management system, adding, deleting or modifying tables or rows can be performed in a simple manner using some open source software like the couple of MySQL and PhpMyAdmin. These tools helped us to modify the above datasets adding new row representing the activity-set by the following way. Assuming that each file of the dataset is a table containing id, latitude, longitude as columns. As there is no open data in the net including the frequent activities, we used another dataset to fill the activity data rows. This dataset represents the frequent item-sets. [4 5 6 7] represents an example data row of this data. Using SQL statements, we added the activity row as another column in the table. Thus, each row represents a point in a tuple (id, latitude, longitude, activities).

6.1 The Performance of the Run-Time and Trajectory Visited Based on the Activity Keywords Number

To evaluate the effectiveness of the activity keywords in trajectories, we varied the activity keywords from 2 to 5 and fixed the distance threshold to 10. Figure 2 illustrates the runtime results. As noted, the runtime of the query becomes longer when the number of activity keywords increases. Since, initially, the trajectory query algorithm processing requires traversing more indexes, then, many inverted lists should be visited to select frequent activities. Furthermore, to handle the missing activity problem, other similar trajectories and sub-trajectories have to be selected. Afterward, to sort the best trajectory matching, the results should be collected and compared; thus, the runtime will become longer if activity keywords increase. Figure 3 illustrates the number of trajectories visited. The number of trajectories visited becomes smaller when the keywords number increases, since few trajectories are more likely to be candidates without incomplete activities.

6.2 The Performance of the Run-Time and Trajectory Visited Based on the Distance Threshold

To evaluate the effectiveness of the distance threshold $\hat{d}$, we varied $\hat{d}$ from 1 to 5 km. Figure 4 illustrates the runtime results. The runtime of the algorithm becomes longer when $\hat{d}$ increases, since it requires traversing more indexes and more inverted lists. In addition, if there are incomplete history activities, other similar trajectories and sub-trajectories want to be candidates with a large $\hat{d}$. Then, the results should be collected and compared to sort the shortest trajectory. Thus, the runtime will become longer when $\hat{d}$ increase. Figure 5 illustrates the number of trajectories visited. As noted, this number becomes larger when $\hat{d}$ increases, since many similar trajectories, including the short and the long ones, are more likely to be candidates with a significant distance.

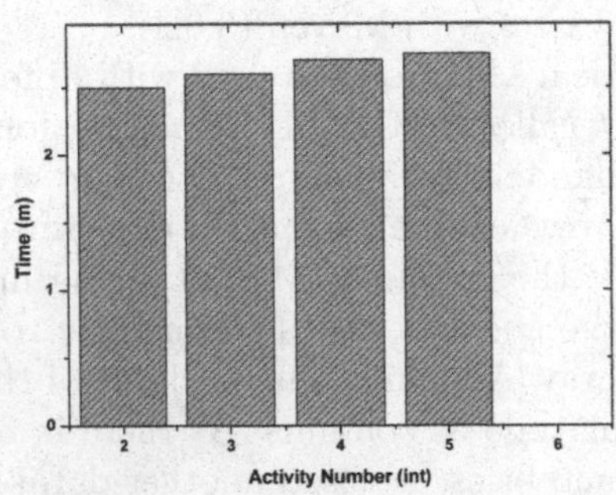

Fig. 2. Query performance (a)

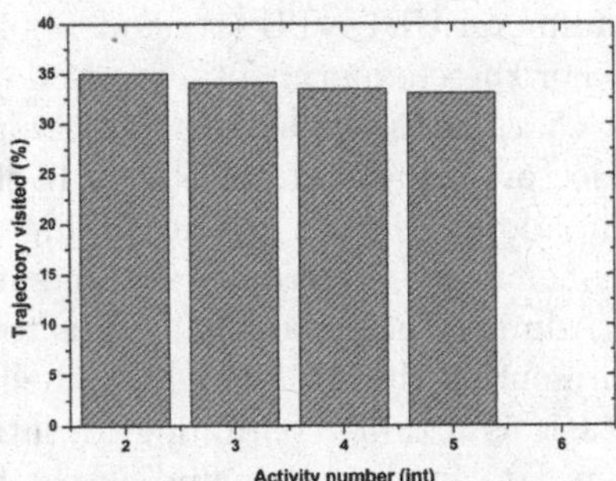

Fig. 3. Trajectory visited (a)

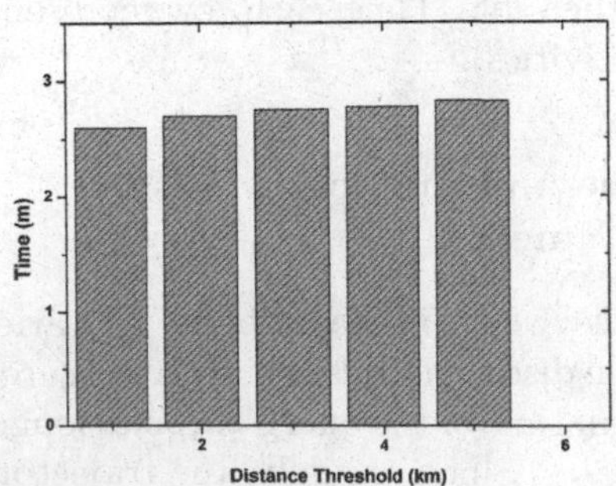

Fig. 4. Query performance (b)

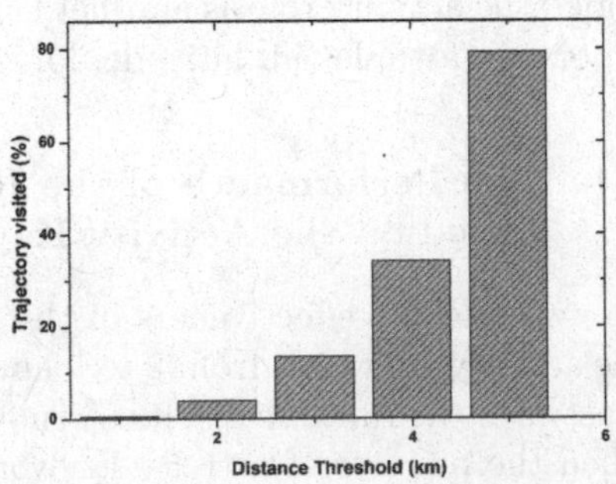

Fig. 5. Trajectory visited (b)

7 Conclusion

In this paper, we investigated a novel problem of skyline query in massive Trajectory DataBase (TDBs) with incomplete semantic trajectory. We studied the skyline query based on both spatial and textual (i.e. activities) dimensions. In other terms, our skyline approach aims to find out the best results based on the spatial distance and the number of activities that compose activity-trajectories. Further, with a massive amount of activities, such data is always not presented due to some reason like lack of internet connection. Further, users may make mistakes while typing activity text in the system keyboard. Thus, such problems make it hard to return the desirable results based on the exact keywords.

To handle the problem efficiently, we firstly re-used distributed indexes to organize the massive activity-trajectory data based on the R-tree index. Then, we developed a parallel activity-trajectory query algorithm based on approximate activity keywords and distance functions. These functions aim to evaluate three points. The first point is the multi-frequent activity, where we used the data-mining algorithm to find the frequent POI based on its activities. Further, we were promoted also to process the fuzzy query for approximate activities in POI using edit distance and activity weights. The second point is the distance measured between activity-trajectories in TDB and the query. The third point combines point 1 and point 2 to handle the missing activities problem by finding other good similar activity-trajectory to the query. To achieve scalability and fault tolerance, we used Apache Spark cluster to implement both distributed indexes and the query algorithm. The algorithm proposed solved the problem efficiently. Extensive experimental results show that our algorithm offers efficiency. As future studies, we plan to use a large dataset with tera-byte size, add more machines in our cluster, compare our work with existing methods and handle the temporal dimension problem in semantic trajectory skyline query for incomplete TDBs.

References

1. Htet, A.H., Long, G., Kian-Lee, T.: Mining sub-trajectory cliques to find frequent routes. In: Proceedings of the 13th of ISASTD, Munich, vol. 8098 (2013)
2. Kong, K., et al.: Trajectory query based on trajectory segments with activities. In: Proceedings of the 3rd ACM SIGSPATIAL ACM, pp. 1–8 (2017)
3. Zheng, K., Shang, S., Yuan, N.J., Yang, Y.: Towards efficient search for activity trajectories, pp. 230–241 (2013)
4. Li, J., Wang, H., Li, J., Gao, H.: Skyline for geo-textual data. GeoInformatica **20**(3), 453–469 (2016). https://doi.org/10.1007/s10707-015-0243-9
5. Belhassena, A., HongZhi, W.: Trajectory big data processing based on frequent activity. Tsinghua Sci. Technol. **24**, 317–332 (2019)
6. Belhassena, A., Wang, H.: Distributed skyline trajectory query processing. In: Proceedings of the ACM Turing 50th Celebration Conference, Shanghai (2017)
7. Chen, M., Wang, N., Lin, G., Shang, J.S.: Network-based trajectory search over time intervals. Big Data Res. 100221 (2021)

8. Rocha Junior, J.B., Nørvåg, K.: Top-k spatial keyword queries on road networks, USA, pp. 168–179 (2012)
9. Han, Y., Wang, L., Zhang, Y., Zhang, W., Lin, X.: Spatial keyword range search on trajectories. In: Renz, M., Shahabi, C., Zhou, X., Cheema, M.A. (eds.) DASFAA 2015. LNCS, vol. 9050, pp. 223–240. Springer, Cham (2015). https://doi.org/10.1007/978-3-319-18123-3_14
10. Cao, K., Sun, Q., Liu, H., Liu, Y., Meng, G., Guo, J.: Social space keyword query based on semantic trajectory. Neurocomputing **428**, 340–351 (2021)
11. Guttman, A.: R-trees: a dynamic index structure for spatial searching. In: Proceedings of ACM SIGMOD, vol. 14, pp. 47–57 (1984)
12. Yu, J., Wu, J., Sarwat, M.: GeoSpark: a cluster computing framework for processing large-scale spatial data. In: Proceedings of the ACM SIGSPATIAL GIS, USA (2015)
13. Wang, L., Chen, B., Liu, Y.: Distributed storage and index of vector spatial data based on h-base. In: Proceedings of Geoinformatics, pp. 1–5 (2013)
14. Eldawy, A., Mokbel, M.F.: A demonstration of spatialhadoop: an efficient mapreduce framework for spatial data. In: Proceedings of the VLDB, vol. 6, pp. 1230–1233 (2013)
15. Li, G., Deng, D., Feng, J.: A partition-based method for string similarity joins with edit-distance constraints. ACM Trans. Database Syst. **38**, 1–33 (2013)
16. Borzsony, S., Kossmann, D., Stocker, K.: The skyline operator. In: Proceedings of the 17th ICDE, pp. 421–430. IEEE (2001)
17. Hsu, W.T., Wen, Y.T., Wei, L.Y., Peng, W.C.: Skyline travel routes: exploring skyline for trip planning. In: Proceedings of the 15th ICMDM, vol. 2, pp. 31–36. IEEE (2014)
18. Yang, B., Guo, C., Jensen, C.S., Kaul, M., Shang, S.: Stochastic skyline route planning under time-varying uncertainty. In: Proceedings of the 30th ICDE, pp. 136–147 (2014)
19. Jiang, B., Du, X.: Personalized travel route recommendation with skyline query. In: Proceedings of the 9th DESSERT, pp. 549–554. IEEE (2018)
20. Ke, C.-K., Lai, S.-C., Chen, C.-Y., Huang, L.-T.: Travel route recommendation via location-based social network and skyline query. In: Hung, J.C., Yen, N.Y., Chang, J.-W. (eds.) FC 2019. LNEE, vol. 551, pp. 113–121. Springer, Singapore (2020). https://doi.org/10.1007/978-981-15-3250-4_14
21. HongZhi, W., Belhassena, A.: Parallel trajectory search based on distributed index. Inf. Sci. **388**, 62–83 (2017)
22. Ju, H., Ju, F., Guoliang, L., Shanshan, C.: Top-k fuzzy spatial keyword search. Chin. J. Comput. **35**(11), 2237–2246 (2012). (in Chinese)

Compact Data Structures for Efficient Processing of Distance-Based Join Queries

Guillermo de Bernardo[1], Miguel R. Penabad[1(✉)], Antonio Corral[2], and Nieves R. Brisaboa[1]

[1] Universidade da Coruña, Centro de investigación CITIC, 15071 A Coruña, Spain
{guillermo.debernardo,miguel.penabad,nieves.brisaboa}@udc.es
[2] Department of Informatics, University of Almeria, 04120 Almeria, Spain
acorral@ual.es

Abstract. Compact data structures can represent data with usually a much smaller memory footprint than its plain representation. In addition to maintaining the data in a form that uses less space, they allow us to efficiently access and query the data in its compact form. The k^2-tree is a self-indexed, compact data structure used to represent binary matrices, that can also be used to represent points in a spatial dataset. Efficient processing of the Distance-based Join Queries (DJQs) is of great importance in spatial databases due to its wide area of application. Two of the most representative and known DJQs are the K Closest Pairs Query (KCPQ) and the ε Distance Join Query (εDJQ). These types of join queries are executed over two spatial datasets and can be solved by plane-sweep algorithms, which are efficient but with great requirements of RAM, to be able to fit the whole datasets into main memory. In this work, we present new and efficient algorithms to implement DJQs over the k^2-tree representation of the spatial datasets, experimentally showing that these algorithms are competitive in query times, with much lower memory requirements.

Keywords: k^2-tree · K closest pairs · ε distance join · Spatial query evaluation

1 Introduction

The efficient storage and management of large datasets has been a research topic for decades. Spatial databases are an example of such datasets. Some of the methods used to efficiently manage and query them include distributed algorithms, streaming algorithms, or efficient secondary storage management [9], frequently accompanied by the use of indexes such as R^*-trees to speed up queries.

P. Fournier-Viger et al. (Eds.): MEDI 2022, LNAI 13761, pp. 207–221, 2023.
https://doi.org/10.1007/978-3-031-21595-7_15

Compression techniques, on the other hand, are focused on minimizing the storage needs of such datasets, but classical techniques (for example, any Huffman-based compressor) had a strong drawback: the datasets must usually be decompressed from the beginning in order to access any specific item of data. Therefore, compression has been mainly used for archival purposes, or when the whole dataset must be processed sequentially from beginning to end (for example, combining decompression with streaming algorithms to process the data).

Compact data structures [9] try to combine low space usage and processing efficiency. They store the information in a *compact* (compressed) form, thus requiring less space, and manage (query) it also in its compact form, without having to first decompress it. In this way, it is possible to store and process (query, navigate, and optionally modify) much larger datasets in main memory. This has the additional benefit of the higher speeds of higher levels of the memory hierarchy (more data in processor caches, for example). An example of such a compact data structure is the k^2-tree [1], which will be further discussed in Sect. 2.2. Initially designed to represent and query web graphs, the k^2-tree has proved to be a powerful tool to represent other kinds of graphs [4], being especially efficient when the graph is clustered. Spatial point datasets, as well as other raster spatial data, can also be managed by a k^2-tree [2].

Distance join queries (DJQs) have received considerable attention from the database community, due to their importance in numerous applications, such as spatial databases and GIS, location-based systems, continuous monitoring, etc. [8]. DJQs are costly queries because they combine two datasets taking into account a distance metric. Two of the most used DJQs are the K Closest Pair Query (KCPQ) and the ε Distance Join Query (εDJQ) [3]. Given two point datasets $\mathcal{P}$ and $\mathcal{Q}$, the KCPQ finds the K closest pairs of points from $\mathcal{P} \times \mathcal{Q}$ according to a certain distance function (e.g., Manhattan, Euclidean, Chebyshev, etc.). The εDJQ finds all the possible pairs of points from $\mathcal{P} \times \mathcal{Q}$ that are within a distance threshold ε of each other. DJQs are very useful in many applications that use spatial data for decision making and other demanding data handling operations. For example, we can use two spatial datasets that represent the hotels and the monuments in a touristic city. A KCPQ ($K = 10$) can discover the *10* closest pairs of hotels and monuments, sorted in increasing order by distance. On the other hand, an εDJQ ($\varepsilon = 200$) could return all possible pairs (hotel, monument) that are within *200* meters of each other.

DJQs have been extensively studied, and algorithms exist to answer KCPQ, εDJQ, and other similar queries over plain data [10], as well as taking advantage of indexes such as R-trees or Quadtrees [7]. In this paper, we explore the advantages of representing spatial data using a compact data structure, the k^2-tree, to implement DJQs, and test it with two of the most used DJQs: KCPQ and εDJQ. Thus, the most important contributions of this paper are the following:

– A detailed description of the algorithms to answer KCPQ and εDJQ over large datasets of points, using k^2-trees as the underlying representation for both datasets.

– The execution of a set of experiments using large real-world datasets for examining the efficiency and the scalability of the proposed strategy, considering performance parameters and measures.

This paper is organized as follows. In Sect. 2 we present preliminary concepts related to DJQs and k^2-trees, as well as previous contributions in these areas. In Sect. 3, the new algorithms for KCPQ and εDJQ using k^2-trees are proposed. In Sect. 4, we present the main results of our experiments, using large real-world datasets. Finally, in Sect. 5, we provide the conclusions arising from our work and discuss related future work directions.

2 Background and Related Work

In this section, we review some basic concepts about DJQs and the k^2-tree compact data structure, as well as a brief survey of the most representative contributions in both fields in the context of spatial query processing.

2.1 Distance-Based Join Queries - KCPQ and εDJQ

Distance-based Join Queries are special joins where two datasets are combined, taking into account a distance metric (*dist*). DJQs can be very costly when the size of the joined datasets is large, and for this reason, they have lately been thoroughly investigated. Two of the most representative and known DJQs are the K Closest Pairs Query (KCPQ) and the ε Distance Join Query (εDJQ)

The KCPQ discovers the K pairs of data formed from the elements of two datasets having the K smallest distances between them (i.e., it reports only the top K pairs from the combination of two datasets). Formally:

Definition 1. (K **Closest Pairs Query,** K**CPQ**)
Let $\mathcal{P} = \{p_1, p_2, \cdots, p_n\}$ and $\mathcal{Q} = \{q_1, q_2, \cdots, q_m\}$ be two set of points, and a number $K \in \mathbb{N}^+$. Then, the result of the K Closest Pairs Query is an ordered collection, $KCPQ(\mathcal{P}, \mathcal{Q}, K)$, containing K different pairs of points from $\mathcal{P} \times \mathcal{Q}$, ordered by distance, with the K smallest distances between all possible pairs: $KCPQ(\mathcal{P}, \mathcal{Q}, K) = ((p_1, q_1), (p_2, q_2), \cdots, (p_K, q_K))$, $(p_i, q_i) \in \mathcal{P} \times \mathcal{Q}$, $1 \leq i \leq K$, such that for any $(p, q) \in \mathcal{P} \times \mathcal{Q} \setminus KCPQ(\mathcal{P}, \mathcal{Q}, K)$ we have $dist(p_1, q_1) \leq dist(p_2, q_2) \leq \cdots \leq dist(p_K, q_K) \leq dist(p, q)$.

Three properties of KCPQ are: (i) it is symmetric (i.e., $KCPQ(\mathcal{P}, \mathcal{Q}, K) = KCPQ(\mathcal{Q}, \mathcal{P}, K)$); (ii) the cardinality of the query result is known beforehand $|KCPQ(\mathcal{P}, \mathcal{Q}, K)| = K$; and (iii) the distance of the K closest pairs of points is unknown a priori.

On the other hand, the εDJQ reports all the possible pairs of spatial objects from two different spatial objects datasets, $\mathcal{P}$ and $\mathcal{Q}$, having a distance not greater than a threshold ε of each other. Formally:

Definition 2. (ε **Distance Join Query,** ε**DJQ**)
Let $\mathcal{P} = \{p_1, p_2, \cdots, p_n\}$ *and* $\mathcal{Q} = \{q_1, q_2, \cdots, q_m\}$ *be two set of points, and a distance threshold* $\varepsilon \in \mathbb{R}_{\geq 0}$*. Then, the result of the* ε*DJQ is the set,* $\varepsilon DJQ(\mathcal{P}, \mathcal{Q}, \varepsilon) \subseteq \mathcal{P} \times \mathcal{Q}$*, containing all the possible different pairs of points from* $\mathcal{P} \times \mathcal{Q}$ *that have a distance of each other smaller than, or equal to* ε*:*
$\varepsilon DJQ(\mathcal{P}, \mathcal{Q}, \varepsilon) = \{(p_i, q_j) \in \mathcal{P} \times \mathcal{Q} : \ dist(p_i, q_j) \ \leq \ \varepsilon\}$

The εDJQ can be considered as an extension of the KCPQ, where the distance threshold of the pairs (ε) is known beforehand and the processing strategy (e.g., plane-sweep technique) can be the same as in the KCPQ for generating the candidate pairs of the final result.

If both $\mathcal{P}$ and $\mathcal{Q}$ are non-indexed, the KCPQ between two point sets that reside in main-memory can be solved using *plane-sweep-based* algorithms [10]. The *Classic plane-sweep* algorithm for KCPQ consists of two steps: (1) sorting the entries of the two points sets, based on the coordinates of one of the axes (e.g. X) and (2) combining the reference point of one set with all the comparison points of the other set satisfying that their distance on the X-axis is less than δ (distance of the Kth closest pair found so far), and choosing those pairs whose point distance is smaller than δ. A faster variant called *Reverse-Run plane-sweep* algorithm is based on the concept of *run* (a continuous sequence of points of the same set that does not contain any point from the other set) and the *reverse* order of processing of the comparison points with respect to the reference point. To reduce the search space and considering the reference point, three methods are applied in these two plane-sweep algorithms: Sliding Strip (δ on X-axis), Sliding Window and Sliding Semi-Circle. These DJQs have been recently designed and implemented in SpatialHadoop and LocationSpark, that are Hadoop-based and Spark-based distributed spatial data management systems (Big Spatial Data context) [5], respectively.

The problem of DJQs has also received research attention by the spatial database community in scenarios where at least one of the datasets is indexed. If both $\mathcal{P}$ and $\mathcal{Q}$ are indexed using R-Trees, the concept of synchronous tree traversal and Depth-First (DF) or Best-First (BF) traversal order can be combined for the query processing [3]. In [7], an extensive experimental study comparing the R*-tree and Quadtree-like index structures for DJQs together with index construction methods was presented. In the case that only one dataset is indexed, in [6] an algorithm is proposed for KCPQ, whose main idea is to partition the space occupied by the dataset without an index into several cells or subspaces (according to a grid-based data structure) and to make use of the properties of a set of distance functions defined between two MBRs [3].

2.2 k^2-tree

A k^2-tree [1] is a compact data structure used to store and query a binary matrix that can represent a graph or a set of points in discretized space. Figure 1 shows in (a) a set of points in a 2D discrete space, and its direct translation into a binary matrix in (b). For the k^2-tree representation, choosing $k = 2$, (c) is the conceptual tree and (d) the actual bitmaps that are stored. T represents the "tree" part (non leaf nodes), and L the leaves of the conceptual tree.

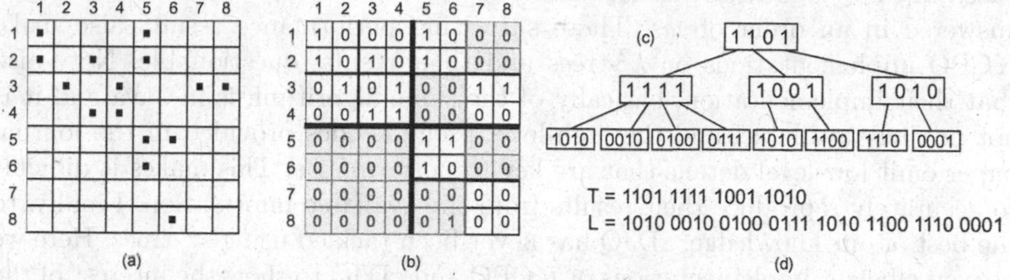

Fig. 1. A 2D-space model with its binary matrix and k^2-tree representations

Conceptually, the k^2-tree can be seen as an unbalanced tree, where each node has a bitmap of k^2 bits and up to k^2 children. This conceptual tree is built as follows: its root node corresponds to the full matrix, which is divided into $k \times k$ equal-sized submatrices (for $k = 2$, the matrix is decomposed in $k^2 = 4$ quadrants). For each submatrix, if there is at least one 1 in its cells, the corresponding bit in the conceptual tree node is set to 1, and the submatrix is included as a child of the node. If the submatrix is empty (there are no 1's) then the corresponding bit is set to 0 and the submatrix is discarded. See, for example, that the root node in Fig. 1 is 1101 because quadrants 1, 2 and 4 have 1's, but the third quadrant is full of 0's. The process continues recursively for all non empty submatrices until the individual cells (that correspond to leaves in the k^2-tree) are reached. The actual k^2-tree is just the bitmap that corresponds to the breadth-first traversal of the conceptual tree (usually split in two fragments T and L, as shown in Fig. 1(d)).

Points and regions of the original matrix can be easily found traversing the branches of the conceptual k^2-tree from the root node. This is achieved in practice using just the bitmaps by computing $getChild(T, i) = rank(T, i) \times k^2$, that returns, for the node at position i in T, the position in T where its children are located. The $rank(T, i)$ operation (which counts the number of 1's up to position i in T) can be computed in constant time by enhancing T with a small structure of counters.

Many variants and improvements have been proposed over the basic k^2-tree representation introduced here [1,2]. In this paper we focus on the basic k^2-tree representation described in this section, in order to provide a clear description of the algorithms.

Regarding the use of k^2-trees in the field of DJQs, in the recent work [11], K Nearest Neighbors Query and KCPQ are proposed for k^2-trees representing points of interest. It is the first algorithm in the literature for KCPQ using k^2-trees (ALBA-KCPQ). One of the main drawbacks of their paper is that the authors used very small synthetic and real datasets in the experimentation, because the maximum number of points is only 1 million for each synthetic dataset and for real data the combination was 76451 × 20480 and 4499454 × 196902. Moreover, the total response time of the KCPQ experiments is questionable because their implementation of (Classic) plane-sweep-based KCPQ algorithm needed hours to solve the query, when it should be answered in an order of *ms*. These surprising performance results also make KCPQ implementations on k^2-trees and their results questionable. Note also that their implementation, basically of the same algorithm, is in Java and it is not publicly available, and the high-level pseudo-codes provided in the journal paper omit low-level details that are key to performance. This makes it difficult to accurately reproduce their results from the available information. Finally, to the best of our knowledge, εDJQ has never been tackled using k^2-trees. Here we present efficient implementations of KCPQ and εDJQ to show the interest of the strategy of using k^2-trees to represent spatial data. Our algorithms were coded in C++ and are available to the community, and they were tested with large real-world datasets, comparing them with *Classic* and *Reverse-Run* plane-sweep DJQs on main memory.

3 Our Approach to DJQs Using k^2-trees

We describe in this section the new algorithms for KCPQ (Algorithm 1) and εDJQ (Algorithm 2) using k^2-trees. For all the distance calculations, we have used the Euclidean distance.

The input for Algorithm 1 consists on two matrices A and B, stored as k^2-trees (they would correspond to the $\mathcal{P}$ and $\mathcal{Q}$ datasets in the definitions in Sect. 2.1), and the number of pairs of closest points (although we use the name *NumPairs* instead of K in the pseudocode to avoid confusion with the k parameter of k^2-trees), We denote $A[i]$ the ith bit value of the bitmap for A. $A.lastLevel$ is the last level of the tree, corresponding to its leaves. Levels range from 0 to $\lceil lg(n) \rceil - 1$, being n the width of the original matrix.

The following data structures are used by Algorithm 1:

- A priority Queue *PQueue* that stores pairs of nodes to be processed. Each entry in *PQueue* contains:
    - A (sub)matrix of A, including its top-left coordinate and the offset of the associated node in the $T|L$ bitmap of the k^2-tree.
    - A (sub)matrix of B with the same information.
    - The level both matrices belong to in the k^2-tree conceptual trees.
    - The minimum possible distance between the points of A and B. It is computed as shown in Algorithm 3.

PQueue is a min priority queue over the distance (that is, pairs with lower minimum distance come first). It uses the standard methods: *isEmpty*(), *enqueue*() and *dequeue*().

- An ordered list *OutList* with capacity for $NumPairs$ elements (we actually use a max binary heap to manage these elements), each one storing a pair of

points (one coming from each input matrix), and the distance between them. The elements (pairs of points) are sorted according to their distance. It uses the methods: *length*(), *maxDist*() and *insert*().
- *MinDist(pA,pB,size)*, shown in Algorithm 3, obtains the minimum possible (Euclidean) distance between points of the matrices A and B that have their origins in pA and pB and are squares of $size \times size$.

Note that the pairs of (A,B) matrices that are generated in Algorithm 1 have a special property: their origin coordinates are always multiples of $size$. This allows us to compute the minimum distance more efficiently, but it would not work to get the minimum distance between two matrices in general.

The idea behind Algorithm 1 is to recursively partition the matrices A and B into k^2 submatrices each and compare each possible pair of submatrices (down to when they are not actually submatrices but really individual cells or points). One of the strong points of this algorithm is that, at some point, we can stop without processing all the remaining pairs of submatrices. This happens when the required number of closest pairs has already been obtained, and the largest distance between them is not larger than the minimum possible distance between the pairs of submatrices not yet processed.

The algorithm follows a Best-First (BF) traversal. It starts by enqueuing the whole matrices (which correspond to the level 0 of the k^2-tree and have 0 as the minimum possible distance between them). The output list *OutList* is also initialized, with room for at most $NumPairs$ elements.

Then, the priority queue is processed until it is empty, or the stop criteria are reached. Lines $6-8$ check if the output list already has the target $NumPairs$ elements. If so, and the minimum distance of the current pair of matrices is at least the maximum distance in *OutList*, we can be sure that the current and remaining matrices can be safely discarded, and the algorithm returns the current output list.

In other case, the current matrices are partitioned (lines $11-20$), but only if they have children (which is tested by directly using the k^2-tree bitmaps in lines 12 for matrix A and 15 for matrix B). For each pair of child submatrices, if they are in the last level of the k^2-tree then they are actually points. So, if there is room in *OutList* or its maximum distance is greater than the distance between the current pair of points, then the pair and its distance are inserted in order in *OutList* (lines $21-25$). Recall that the *insert* operation may need to remove the element with the largest distance if the output list already contains *NumPairs* elements.

If the submatrices are not in the last level of the k^2-tree, and if they meet the conditions to contain candidate pairs of points (*OutList* is not full or the minimum distance between the matrices is less than the maximum distance in *OutList*) they are enqueued in the priority queue (lines $28 \div 30$).

Algorithm 2 (εDJQ) uses the same scheme as the previous one, but with some key differences. The input consists now of the two k^2-trees A and B, plus the distance threshold ε. Since the algorithm does not limit the number of output pairs, *OutList* is now an unlimited-size, unordered list. For the same reason,

Algorithm 1. GetKCPQ: Get the *NumPairs* closest points.

```
1: function GetKCPQ(A, B, NumPairs)
2:     PQueue.enqueue({ {(0,0), 0}, {(0,0), 0}, 0, 0})
3:     OutList = new OrderedList(NumPairs)
4:     while not PQueue.isEmpty() do
5:         Node = PQueue.dequeue()
6:         if OutList.length() == NumPairs
7:                 and Node.minDist ≥ OutList.maxDist() then
8:             return OutList
9:         chLevel = Node.Level + 1
10:        chSize = n/k^chLevel
11:        for i = 0 to k^2 − 1 do                    ▷ Directly access the k^2-tree bitmap
12:            if A[Node.A.ptr + i] == 1 then
13:                chPtrA = getChild(A, Node.A.ptr + i)
14:                for j = 0 to k^2 − 1 do
15:                    if B[Node.B.ptr + j] == 1 then
16:                        chPtrB = getChild(B, Node.B.ptr + j)
17:                        childA = {(Node.A.x + chSize · (i mod k),
18:                                Node.A.y + chSize · ⌊i/k⌋), chPtrA}
19:                        childB = {(Node.B.x + chSize · (j mod k),
20:                                Node.B.y + chSize · ⌊j/k⌋), chPtrB}
21:                        if chLevel == A.lastLevel then                    ▷ Leaf nodes
22:                            distance = Dist((childA.x, childA.y), (childB.x, childB.y))
23:                            if OutList.length() < NumPairs
24:                                or OutList.maxDist() > distance then
25:                                OutList.insert((childA.x, childA.y), (childB.x, childB.y), distance)
26:                        else
27:                            minDist = MinDist((childA.x, childA.y), (childB.x, childB.y), chSize)
28:                            if OutList.length() < NumPairs
29:                                or OutList.maxDist() > minDist then
30:                                PQueue.enqueue({childA, childB, chLevel, minDist})
31:    return OutList
```

the algorithm does not have an "early exit", and it exits only after the priority queue is empty. Additionally, each element in the priority queue stores not only the minimum possible distance between the matrices, but also the maximum possible distance, computed by the function *MaxDist* (shown in comments in the pseudocode of Algorithm 3). The initial *MaxDist* for the whole matrices is $\sqrt{2}n$, where n is the width of each matrix.

The partitioning is done the same way, but for every pair of child submatrices the process is different:

- At leaf level of the k^2-trees (lines 18 – 21) the pair of nodes is inserted in *OutList* if the distance between them is at most ε.
- If the maximum distance (*MaxDist*) between the two matrices is at most ε, then all the combinations of points between the two matrices meet the criteria. We use the *rangeQuery* operation of the k^2-trees to get the points and insert all possible pairs into *OutList* (lines 23 – 31).
- Otherwise, if the minimum distance is at most ε, we enqueue the submatrices with the minimum and maximum distances between them (lines 32 – 33).

Algorithm 2. εDJQ: Get all pairs with a distance threshold of ε

```
function εDJQ(A, B, ε)
    PQueue.enqueue({ {(0,0), 0}, {(0,0), 0}, 0, 0, √2n})
    OutList = new List()
    while not PQueue.isEmpty() do
        Node = PQueue.dequeue()
        chLevel = Node.Level + 1
        chSize = n/k^chLevel
        for i = 0 to k^2 − 1 do                          ▷ Directly access the k^2-tree bitmap
            if A[Node.A.ptr + i] == 1 then
                chPtrA = getChild(A, Node.A.ptr + i)
                for j = 0 to k^2 − 1 do
                    if B[Node.B.ptr + j] == 1 then
                        chPtrB = getChild(B, Node.B.ptr + j)
                        childA = {(Node.A.x + chSize · (i mod k),
                                  Node.A.y + chSize · ⌊i/k⌋), chPtrA}
                        childB = {(Node.B.x + chSize · (j mod k),
                                  Node.B.y + chSize · ⌊j/k⌋), chPtrB}
                        if chLevel == A.lastLevel then                    ▷ Leaf nodes
                            distance = Dist((childA.x, childA.y), (childB.x, childB.y))
                            if distance ≤ ε then
                                OutList.insert((childA.x, childA.y), (childB.x, childB.y), distance)
                        else
                            minDist = MinDist((childA.x, childA.y), (childB.x, childB.y), chSize)
                            maxDist = MaxDist((childA.x, childA.y), (childB.x, childB.y), chSize)
                            if maxDist ≤ ε then
                                        ▷ All pairs in the range satisfy the distance condition
                                rangeA = A.rangeQuery(A.x, A.x+chSize-1, A.y, A.y+chsize-1)
                                rangeB = B.rangeQuery(B.x, B.x+chSize-1, B.y, B.y+chsize-1)
                                for pA ∈ rangeA do
                                    for pB ∈ rangeB do
                                        OutList.insert(pA, pB, Dist(pA, pB))
                            else if minDist ≤ ε then
                                PQueue.enqueue({childA, childB, chLevel, minDist, maxDist})
    return OutList
```

Algorithm 3. MinDist/MaxDist: min/max possible distance between 2 matrices

```
function MinDist(pA,pB,size)
  ▷ Also MaxDist(pA,pB,size)
  if pA.x = pB.x then
      hdist = 0
  else
      hdist = |pA.x − pB.x| − (size − 1)
      ▷ For MaxDist: hdist = |pA.x − pB.x| + (size − 1)
  if pA.y = pB.y then
      vdist = 0
  else
      vdist = |pA.y − pB.y| − (size − 1)
      ▷ For MaxDist: vdist = |pA.y − pB.y| + (size − 1)
  return √(hdist² + vdist²)
```

4 Experimental Results

We have tested our *DJQ* algorithms using the following real-world 2D point datasets, obtained from OpenStreetMap[1]: *LAKES* (L), that contains boundaries of water areas (polygons); *PARKS* (P), that contains boundaries of parks or green areas (polygons); *ROADS* (R), which contains roads and streets around the world (line-strings); and *BUILDINGS* (B), which contains boundaries of all buildings (polygons). For each source dataset, we take all the points extracted from the geometries of each line-string to build a large point dataset. Additionally, we round coordinates to 6 decimal positions, in order to be able to transform these values to k^2-tree coordinates in a consistent manner. Table 1 summarizes the characteristics of the original datasets and the generated point sets obtained from them. Note that all the datasets represent worldwide data, and points are stored as (*longitude*, *latitude*) pairs.

Table 1. Source datasets and generated point sets

Name	Source dataset		Generated dataset	
	#Records (M)	Size (GiB)	#Points (M)	Size (GiB)
LAKES (L)	8.4	8.6	345	8.6
PARKS (P)	10	9.3	305	7.5
ROADS (R)	72	24	682	17
BUILDINGS (B)	115	26	615	14

The main performance measures that we have used in our experiments are the space required by the data structure vs. the plain representation, and the total execution time to run a given DJQ. We measure elapsed time, and only consider the time necessary to run the query algorithm. This means that we ignore time necessary to load the files, as well as time required to sort the points for the plane-sweep algorithms.

All experiments were executed on an HP ProLiant DL380p Gen8 server with two 6-core Intel® Xeon® CPU E5-2643 v2 @ 3.50 GHz processors with 256 GiB RAM (Registered @1600 MHz), running Oracle Linux Server 7.9 with kernel Linux 4.14.35 (64bits). Our algorithms were coded in C++ and are publicly available[2]. For the k^2-tree algorithms, the SDSL-Lite[3] library was used.

First, we build the k^2-tree for each dataset. We use the simplest variant of k^2-tree with no optimizations. In order to insert the points in the k^2-tree, they are converted to non-negative integer values. Since we are considering worldwide coordinates in degrees, with 6 decimal places, each coordinate (x, y) is converted to matrix coordinates (r, c) using $(r, c) = ((x + 180) \cdot 10^6, (y + 90) \cdot 10^6)$. In this

[1] Available at http://spatialhadoop.cs.umn.edu/datasets.html.
[2] Available at https://gitlab.lbd.org.es/public-sources/djq/k2tree-djq.
[3] Available at https://github.com/simongog/sdsl-lite.

way, the points fit into a binary matrix with 360 million rows and 160 million columns, that is finally stored as a k^2-tree.

Table 2. Space required by k^2-tree representations

Dataset	Plain (GiB)	Binary (GiB)	k^2-tree (GiB)	Compression ratio
LAKES (L)	8.6	2.7	1.8	0.67
PARKS (P)	7.5	2.4	1.6	0.67
ROADS (R)	17	5.3	2.3	0.43
BUILDINGS (B)	14	4.8	2.3	0.48

Table 2 shows the space required by the k^2-tree representation of each dataset. We display as a reference the plain size of the dataset, as well as a "binary size" estimated considering that each coordinate can be represented using two 32-bit words. Note that each coordinate component can be stored using 28–29 bits for our datasets, but this would make data access slower, so we consider 32 bits to be the minimum cost for a reasonable plane-sweep algorithm that works with uncompressed data. We also display the compression ratio of the k^2-tree relative to the binary input size. Results show that the k^2-tree representation is able to efficiently represent the collection, and the compression obtained improves with the size of the dataset. Notice also that the k^2-tree version we use in our experiments does not include any of the existing optimizations for the k^2-tree to improve compression.

We compared the performance of our KCPQ algorithm with 4 different implementations based on plane-sweep: two implementations of *Classic* plane-sweep, with Sliding Strip (PS-CS) and with Sliding Semi-Circle (PS-CC) respectively, and the equivalent implementations of *Reverse-Run*, with Sliding Strip (PS-RRS) and Sliding Semi-Circle (PS-RRC). We performed our experiments checking all the pairwise combinations of our datasets. Due to space constraints, we display only the results for some combinations, denoted LxP, LxB, PxR, RxB, and PxB. The remaining combinations yielded similar comparison results. For each combination of datasets, we run the KCPQ algorithm for varying $K \in \{1, 10, 10^2, 10^3, 10^4, 10^5\}$.

Figure 2 displays the query times obtained by our algorithm and the four variants of plane-sweep studied. The first five plots display the results for all variants for 5 different dataset combinations. Results clearly show that the *Classic* variants (PS-CS and PS-CC) are much slower than the other alternatives in all cases (as in [10]). Therefore, we will focus on the comparison between our proposal and the *Reverse-Run* variants that are competitive with it.

The point datasets used have a significantly different amount of points, and correspond to different features, which leads to very different query times among the plots in Fig. 2. However, results show that our algorithm always achieves the best query times for large values of K. Particularly, for $K = 10^5$, our algorithm

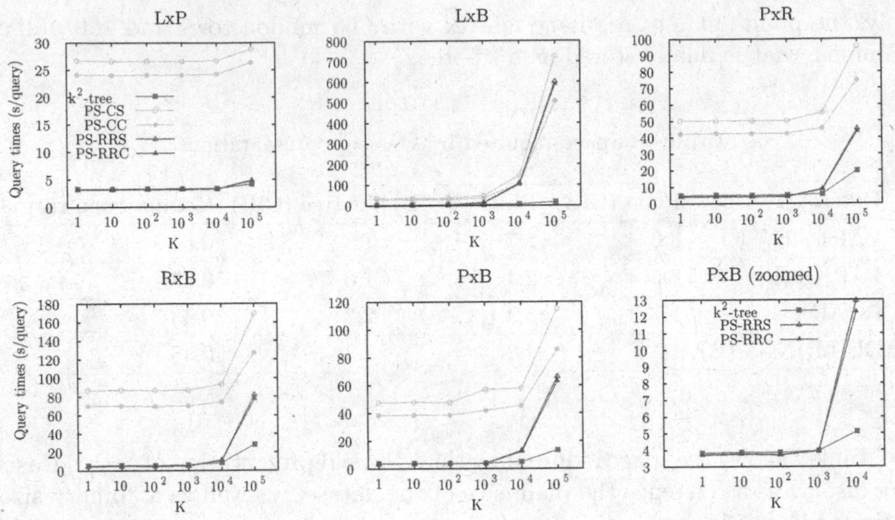

Fig. 2. Query times for KCPQ in k^2-trees and plane-sweep variants, changing K.

is between 1.15 and 33 times faster than the best alternative, PS-RRC, depending on the joined datasets. Additionally, we are always the fastest option for $K \geq 10^4$, and in some datasets from $K = 10^3$. For smaller K, our proposal is competitive but slightly slower than the *Reverse-Run* plane-sweep algorithms. The lower right chart of Fig. 2 shows a subset of the results for the PxB join, to better display the differences in performance for these smaller values of K. Results are similar in the remaining experiments: for smaller K, the k^2-tree algorithm is 3–15% slower than PS-RRC, depending on the dataset. This evolution with K is due to the characteristics of our algorithm: independently of K, we need to traverse a relatively large number of regions in both k^2-trees, even if many of these regions are eventually discarded, so the base complexity of our algorithm is comparable to that of *Classic* plane-sweep. On the other hand, this means that many candidate pairs have already been expanded and enqueued, so they can be immediately processed if more results are needed, making our algorithm more efficient for larger values of K.

Next, we compare our algorithm for εDJQ with two plane-sweep variants, *Classic* plane-sweep with Sliding Strip (εDJQ-CS) and *Reverse-Run* with Sliding Semi-Circle (εDJQ-RRC). We select a representative subset of joined datasets, namely LxP, PxR, RxB and PxB. In order to measure the scalability of the algorithms, we perform tests for varying $\varepsilon \in (7.5, 10, 25, 50, 75, 100) \times 10^{-5}$ (these values of ε are associated with the original coordinates in degrees, but recall that in the k^2-tree coordinates are scaled to integer values, so values of ε are also scaled accordingly).

Figure 3 displays the results obtained for each join query. Our algorithm based on k^2-trees is slower for LxP, but much faster in most cases for PxR, RxB and PxB (notice the logarithmic scale for query times). We attribute this difference

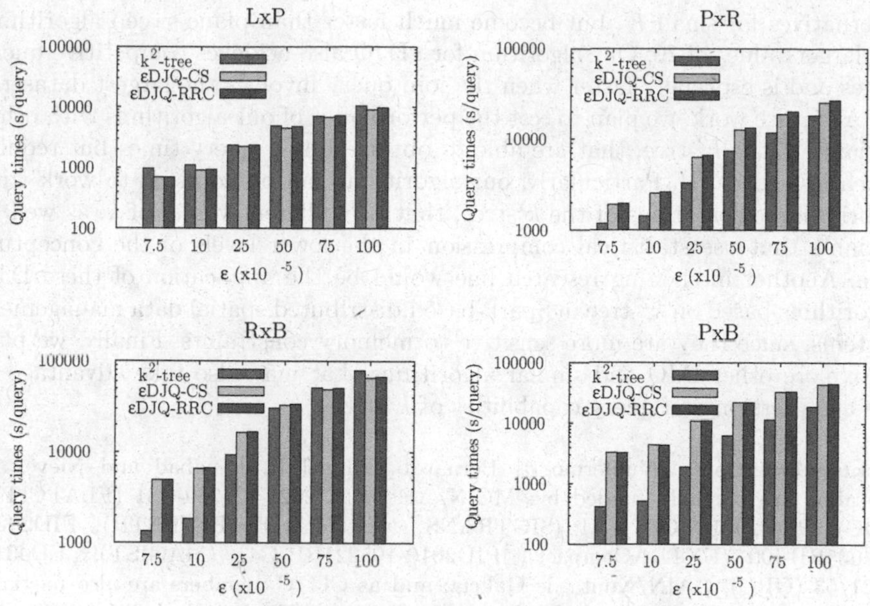

Fig. 3. Query times for εDJQ in k^2-trees and plane-sweep variants, changing ε.

mainly to the size of the datasets: LxP joins the two smallest datasets, whereas the remaining configurations involve one or two of the larger datasets. For these 3 larger joins, our algorithm is always much faster for the smaller values of ε. In this case, our algorithm does not improve for larger ε, as for KCPQ, because no early stop condition exists: we must traverse all candidate pairs as long as their minimum distance is below ε, and for very large ε the added cost to traverse the k^2-trees to expand many individual pairs makes our proposal slower, even if we are able to efficiently filter out many candidate regions. These queries with smaller values of ε, in which we are much faster than plane-sweep algorithms, are precisely the ones that would most benefit from our approach based on compact data structures, since the number of query results increases with ε: for $\varepsilon = 100 \times 10^{-5}$ we obtain over 10^9 results, and these results would become the main component of memory usage. Notice that, in practice, in our experiments we measure the time to retrieve and count the query results, but do not store them in RAM to avoid memory issues in some query configurations.

5 Conclusions and Future Work

We have introduced two algorithms to solve DJQs on top of the k^2-tree representation of point datasets. Our proposal takes advantage of the compression and indexing capabilities of the k^2-tree to efficiently answer KCPQ and εDJQ queries in competitive time and with significantly lower memory requirements. Our results show that our algorithms for KCPQ queries are competitive with the alternatives for small K, but become much faster than plane-sweep algorithms for larger values of K. Our algorithm for εDJQ also achieves competitive query times and is especially faster when the join query involves the largest datasets.

As future work, we plan to test the performance of our algorithms with other variants of the k^2-tree, that are able to obtain similar query times but require much less space [1]. Particularly, our algorithms can be adjusted to work with hybrid implementations of the k^2-tree, that use different values of k, as well as variants that use statistical compression in the lower levels of the conceptual tree. Another interesting research line would be the application of these DJQ algorithms based on k^2-tree in Spark-based distributed spatial data management systems, since they are more sensitive to memory constraints. Finally, we plan to explore other DJQ and similar algorithms that may also take advantage of the compression and query capabilities of k^2-trees.

Acknowledgments. Guillermo de Bernardo, Miguel R. Penabad and Nieves R. Brisaboa are partially funded by: MCIN/AEI [PDC2021-121239-C31 (FLATCITY-POC), PDC2021-120917-C21 (SIGTRANS, NextGenerationEU/PRTR), PID2020-114635RB-I00 (EXTRACompact), PID2019-105221RB-C41 (MAGIST)]; ED431C 2021/53 (GRC), GAIN/Xunta de Galicia; and as CITIC members are also partially funded by ED431G 2019/01 (CSI), Xunta de Galicia, FEDER Galicia 2014–2020. The work by Antonio Corral was partially funded by the EU ERDF and the Andalusian Government (Spain) under the project UrbanITA (ref. PY20_00809) and the Spanish Ministry of Science and Innovation under the R&D project HERMES (ref. PID2021-124124OB-I00).

References

1. Brisaboa, N.R., Ladra, S., Navarro, G.: Compact representation of web graphs with extended functionality. Inf. Syst. **39**(1), 152–174 (2014)
2. Brisaboa, N.R., Cerdeira-Pena, A., de Bernardo, G., Navarro, G.: Óscar Pedreira: extending general compact querieable representations to GIS applications. Inf. Sci. **506**, 196–216 (2020)
3. Corral, A., Manolopoulos, Y., Theodoridis, Y., Vassilakopoulos, M.: Algorithms for processing k-closest-pair queries in spatial databases. Data Knowl. Eng. **49**(1), 67–104 (2004)
4. Álvarez García, S., Brisaboa, N., Fernández, J.D., Martínez-Prieto, M.A., Navarro, G.: Compressed vertical partitioning for efficient RDF management. Knowl. Inf. Syst. **44**(2), 439–474 (2015)
5. García-García, F., Corral, A., Iribarne, L., Vassilakopoulos, M., Manolopoulos, Y.: Efficient distance join query processing in distributed spatial data management systems. Inf. Sci. **512**, 985–1008 (2020)
6. Gutiérrez, G., Sáez, P.: The k closest pairs in spatial databases - when only one set is indexed. GeoInformatica **17**(4), 543–565 (2013)
7. Kim, Y.J., Patel, J.M.: Performance comparison of the R*-tree and the quadtree for kNN and distance join queries. IEEE Trans. Knowl. Data Eng. **22**(7), 1014–1027 (2010)
8. Mamoulis, N.: Spatial Data Management. Synthesis Lectures on Data Management. Morgan & Claypool Publishers (2012)

9. Navarro, G.: Compact Data Structures: A Practical Approach. Cambridge University Press, USA (2016)
10. Roumelis, G., Vassilakopoulos, M., Corral, A., Manolopoulos, Y.: A new plane-sweep algorithm for the k-closest-pairs query. In: SOFSEM, pp. 478–490 (2014)
11. Santolaya, F., Caniupán, M., Gajardo, L., Romero, M., Torres-Avilés, R.: Efficient computation of spatial queries over points stored in k^2-tree compact data structures. Theoret. Comput. Sci. **892**, 108–131 (2021)

Towards a Complete Direct Mapping from Relational Databases to Property Graphs

Abdelkrim Boudaoud(✉), Houari Mahfoud, and Azeddine Chikh

Abou-Bekr Belkaid University & LRIT Laboratory, Tlemcen, Algeria
{abdelkrim.boudaoud,houari.mahfoud,azeddine.chikh}@univ-tlemcen.dz

Abstract. It is increasingly common to find complex data represented through the graph model. Contrary to relational models, graphs offer a high capacity for executing analytical tasks on complex data. Since a huge amount of data is still presented in terms of relational tables, it is necessary to understand how to translate this data into graphs. This paper proposes a *complete mapping* process that allows transforming any relational database (schema and instance) into a property graph database (schema and instance). Contrary to existing mappings, our solution preserves the three fundamental mapping properties, namely: *information preservation, semantic preservation* and *query preservation*. Moreover, we study mapping any *SQL* query into an equivalent *Cypher* query, which makes our solution practical. Existing solutions are either incomplete or based on non-practical query language. Thus, this work is the first complete and practical solution for mapping relations to graphs.

Keywords: Direct mapping · Complete mapping · Relational database · Graph database · SQL · Cypher

1 Introduction

Relational databases (RDs) have been widely used and studied by researchers and practitioners for decades due to their simplicity, low data redundancy, high data consistency, and uniform query language (SQL). Hence, the size of web data has grown exponentially during the last two decades. The interconnections between web data entities (e.g. interconnection between YouTube videos or people on Facebook) are measured by billions or even trillions [6] which pushes the relational model to quickly reach its limits as querying high interconnected web data requires complex SQL queries which are time-consuming. To overcome this limit, the graph database model is increasingly used on the Web due to its flexibility to present data in a normal form, its efficiency to query a huge amount of data and its analytic powerful. This suggests studying a mapping from RDs to graph databases (GDs) to benefit from the aforementioned advantages. This kind of mapping has not received more attention from researchers since only a few works [4,5,13,14] have considered it. A real-life example of this mapping has been discussed in [13]: "*investigative journalists have recently found, through graph analytics, surprising social*

P. Fournier-Viger et al. (Eds.): MEDI 2022, LNAI 13761, pp. 222–235, 2023.
https://doi.org/10.1007/978-3-031-21595-7_16

relationships between executives of companies within the Offshore Leaks financial social network data set, linking company officers and their companies registered in the Bahamas. The Offshore Leaks PG was constructed as a mapping from relational database (RDB) sources". In a nutshell, the proposed mappings suffer from at least one of the following limits: a) they do not study fundamental properties of mapping; b) they do not consider a practical query language to make the approach more useful; c) they generate an obfuscated schema.

This paper aims to provide a *complete mapping* (*CM*) from RDs to GDs by investigating the fundamental properties of mapping [14], namely: *information preservation* (IP), *query preservation* (QP), and *semantic preservation* (SP). In addition to data mapping, we study the mapping of SQL queries to Cypher queries which makes our results more practical since SQL and Cypher are the most used query languages for relational and graph-data respectively.

Contributions and Road-Map. Our main contributions are as follows: *1)* we formalize a *CM* process that maps RDs to GDs in the presence of schema; *2)* we propose definitions of *schema graph* and *graph consistency* that are necessary for this mapping; *3)* we show that our *CM* preserves the three fundamental mapping properties (*IP*, *SM*, and *QP*); *4)* in order to prove *QP*, we propose an algorithm to map SQL queries into equivalent Cypher queries. To our knowledge, this work is the first complete effort that investigates mapping relations to graphs.

Related Work. We classify previous works as follows:

Mapping RDs to RDF Data. Squeda et al. [12] studied the mapping of RDs to RDF graph data and relational schema to OWL ontology's. Moreover, SQL queries are translated into SPARQL queries. They were the first to define a set of mapping properties: information preservation, query preservation, semantic preservation and monotonicity preservation. They proved first that their mapping is information preserving and query preserving, while when it comes to the two remaining properties, preserving semantics makes the mapping not monotonicity preserving.

Mapping RDs to Graph Data. De Virgilio et al. [4,5] studied mapping a) RDs to property graph data (PG) by considering schema both in source and target; and b) any SQL query into a set of graph traversal operations that realize the same semantic over the resulted graph data. We remark that the proposed mapping obfuscates the relational schema since the resulted graph schema is difficult to understand. Moreover, the mapping does not consider typed data. From the practical point of view, the graph querying language considered is not really used in practice and the proposed query mapping depends on the syntax and semantics of this language which makes hard the application of their proposal for another query language. In addition, they apply an aggregation process that maps different relational tuples to the same graph vertex in order to optimize graph traversal operations. However, this makes the mapping not information preserving and can skew the result of some analytical tasks that one would like to apply over the resulted data graph.

Table 1. Comparative table of related works.

Type	Work	Mapping		Preserved properties			Mapping rules
		Schema	Instance	IP	QP	SP	
RDs → RDF	Sequeda et al. [12]	✓	✓	✓	✓	✓	✓
	De vergillio et al. [4,5]	✓	✓				✓
	Stoica et al. [13,14]	✓	✓	✓	✓	✓	✓
RDF → PG	Angeles et al. [2]	✓	✓	✓		✓	✓
RDs → PG	O.Orel et al. [11]		✓				✓
	S.Li et al. [9]		✓				
	Our work	✓	✓	✓	✓	✓	✓

Stoica et al. [13,14] studied the mapping of RDs to GDs and any relational query (formalized as an extension of relational algebra) into a G-core query. Firstly, the choice of source and destination languages hinders the practicability of the approach. Moreover, it is hard to see if the mapping is semantic preserving since no definition of graph data consistency is given. Attributes, primary and foreign keys are verbosely represented by the data graph, which makes this later hard to understand and to query.

Orel et al. [11] discussed mapping relational data only into property graphs without giving attention neither to schema nor mapping properties.

The Neo4j system provides a tool called Neo4j-ETL [1], which allows users to import their relational data into Neo4j to be presented as property graphs. Notice that the relational structure (both instance and schema) is not preserved during this mapping since some tuples of the relational data (resp. relations of the relational schema) are represented as edges for storage concerns (as done in [5]). However, as remarked in [13], this may skew the results of some analytical tasks (e.g. density of the generated graphs). Moreover, Neo4j-ETL does not allow the mapping of queries. S. Li et al. [9] study an extension of Neo4j-ETL by proposing mapping of SQL queries to Cypher queries. However, their mapping inherits the limits of Neo4j-ETL. In addition, no detailed algorithm is given for the query mapper which makes impossible the comparison of their proposal with other ones. This is also the limit of [10].

Finally, Angles et al. [2] studied mappings RDF databases to property graphs by considering both data and schema. They proved that their mapping ensures both information and semantic preservation properties.

Table 1 summarizes most important features of related works.

2 Preliminaries

This section defines the several notions that will be used throughout this paper.

Let $\mathcal{R}$ be an infinite set of relation names, $\mathcal{A}$ is an infinite set of attribute names with a special attribute *tid*, $\mathcal{T}$ is a finite set of attribute types (*String*, *Date*, *Integer*, *Float*, *Boolean*, *Object*), $\mathcal{D}$ is a countably infinite domain of data values with a special value NULL.

2.1 Relational Databases

A *relational schema* is a tuple $S = (R, A, T, \Sigma)$ where:

1. $R \subseteq \mathcal{R}$ is a finite set of relation names;
2. A is a function assigning a finite set of attributes for each relation $r \in R$ such that $A(r) \subseteq \mathcal{A} \setminus \{tid\}$;
3. T is a function assigning a type for each attribute of a relation, i.e. for each $r \in R$ and each $a \in A(r) \setminus \{tid\}$, $T(a) \subseteq \mathcal{T}$;
4. Σ is a finite set of *primary* (PKs) and *foreign* keys (FKs) defined over R and A. A *primary key* over a relation $r \in R$ is an expression of the form $r[a_1, \cdots, a_n]$ where $a_{1 \leq i \leq n} \in A(r)$. A *foreign key* over two relations r and s is an expression of the form $r[a_1, \cdots, a_n] \rightarrow s[b_1, \cdots, b_n]$ where $a_{1 \leq i \leq n} \in A(r)$ and $s[b_1, \cdots, b_n] \in \Sigma$.

An *instance* I of S is an assignment to each $r \in R$ of a finite set $I(r) = \{t_1, \cdots, t_n\}$ of *tuples*. Each *tuple* $t_i : A(r) \cup \{tid\} \rightarrow \mathcal{D}$ is identified by $tid \neq$ NULL and assigns a value to each attribute $a \in A(r)$. We use $t_i(a)$ (resp. $t_i(tid)$) to refer to the value of attribute a (resp. *tid*) in tuple t_i. Moreover, for any tuples $t_i, t_j \in I(r)$, $t_i(tid) \neq t_j(tid)$ if $i \neq j$.

For any instance I of a relational schema $S = (R, A, T, \Sigma)$, we say that I satisfies a primary key $r[a_1, \cdots, a_n]$ in Σ if: 1) for each tuple $t \in I(r)$, $t(a_{1 \leq i \leq n}) \neq$ NULL; and 2) for any $t' \in I(r)$, if $t(a_{1 \leq i \leq n}) = t'(a_{1 \leq i \leq n})$ then $t = t'$ must hold. Moreover, I satisfies a foreign key $r[a_1, \cdots, a_n] \rightarrow s[b_1, \cdots, b_n]$ in Σ if: 1) I satisfies $s[b_1, \cdots, b_n]$; and 2) for each tuple $t \in I(r)$, either $t(a_{1 \leq i \leq n}) =$ NULL or there exists a tuple $t' \in I(s)$ where $t(a_{1 \leq i \leq n}) = t'(b_{1 \leq i \leq n})$. The instance I satisfies all integrity constraints in Σ, denoted by $I \models \Sigma$, if it satisfy all primary keys and foreign keys in Σ.

Finally, a *relational database* is defined with $D_R = (S_R, I_R)$ where S_R is a relational schema and I_R is an instance of S_R.

2.2 Property Graphs

A *property graph (PG)* is a multi-graph structure composed of labeled and attributed vertices and edges defined with $G = (V, E, L, A)$ where: 1) V is a finite set of vertices; 2) $E \subseteq V \times V$ is a finite set of directed edges where $(v, v') \in E$ is an edge starting at vertex v and ending at vertex v'; 3) L is a function that assigns a label to each vertex in V (resp. edge in E); and 4) A is a function assigning a nonempty set of key-value pairs to each vertex (resp. edge). For any edge $e \in E$, we denote by $e.s$ (resp. $e.d$) the starting (resp. ending) vertex of e.

2.3 SQL Queries and Cypher Queries

We study in this paper the mapping of relational data into PG data. In addition, we show that any relational query over the source data can be translated into an equivalent graph query over the generated data graph. To this end, we model relational queries with the SQL language [8] and the graph queries with the Cypher language [7] since each of these languages is the most used in its category. To establish a compromise between the expressive power of our mapping and its processing time, we consider a simple but very practical class of SQL queries and we define its corresponding class of Cypher queries. It is necessary to understand the relations between basic SQL queries and basic Cypher queries before studying all the expressive power of these languages.

The well-known syntax of SQL queries is "Select I from R where C" where: a) I is a set of items; b) R is a set of relations names; and c) C is a set of conditions. Intuitively, an SQL query selects first some tuples of relations in R that satisfy conditions in C. Then, the values of some records (specified by I) of these tuples are returned to the user.

On the other side, the Cypher queries considered in this paper have the syntax: "*Match* patterns *Where* conditions *Return* items". Notice that a Cypher query aims to find, based on edge-isomorphism, all subgraphs in some data graph that match the pattern specified by the query and also satisfy conditions defined by this latter. Once found, only some parts (i.e. vertices, edges, and/or attributes) of these subgraphs are returned, based on items specified by *Return* clause. Therefore, the *Match* clause specifies the structure of subgraphs we are looking for on the data graph; the *Where* clause specifies conditions (based on vertices, edges and/or attributes) these subgraphs must satisfy; and finally, the *Return* clause returns some parts of these subgraphs to the user as a table.

Example 1. Figure 1 depicts a data graph where vertices represent entities (i.e. *Doctor*, *Patient*, *Diagnostic* and *Admission*); inner information (called attributes) represent properties of this vertex (e.g. *Speciality*); and edges represent relationships between these entities. For instance, an *Admission* vertex may be connected to some *Patient*, *Doctor* and *Diagnostic* vertices to specify for some patient: a) his doctor; b) information about his admission at the hospital; and c) diagnostics made for this patient.

The following Cypher query returns the name of each patient who is admitted at some date:

$MATCH\ (a : Admissions) < -[: Admissions - Patients] - (p : Patients)$
$WHERE\ a.Admi\_date\ =\ "30/11/2021"$
$RETURN\ p.Name$

□

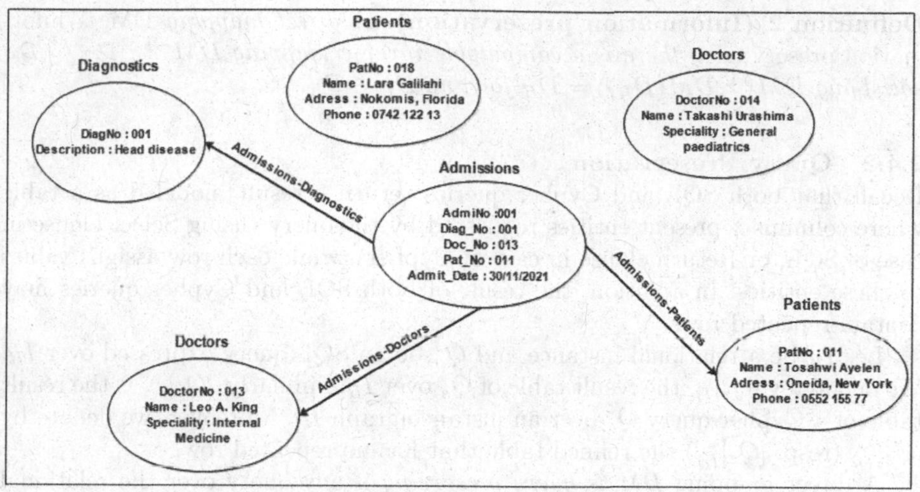

Fig. 1. Example of data graph.

2.4 Direct Mapping (*DM*)

Inspired from [12–14], We define in this section the *direct mapping* from a relational database into a graph database and we discuss its properties. Given a relational database D_R composed of S_R and I_R, a direct mapping consists of translating each entity in D_R into a graph database without any user interaction. That is, any $D_R = (S_R, I_R)$ (with possibly empty S_R), is translated automatically into a pair of property graphs (S_G, I_G) (with possibly empty S_G), that we call a graph database. Let $\mathcal{D}_R$ be an infinite set of relational databases, and $\mathcal{D}_G$ be an infinite set of graph databases. Based on these notions, we give the next definition of direct mapping and its properties.

Definition 1. *A* direct mapping *is a total function* $DM : \mathcal{D}_R \rightarrow \mathcal{D}_G$. □

Intuitively, for each $D_R \in \mathcal{D}_R$, $DM(D_R)$ produces a graph database $D_G \in \mathcal{D}_G$ that aims to represent the source relational database (i.e. instance and optionally schema) in terms of a graph.

We define next fundamental properties that a direct mapping must preserve [12], namely: *information preservation*, *query preservation* and *semantic preservation.* The two first properties ensure that the direct mapping does not lose neither information nor semantic of the relational database being translated. The last property ensures that the mapping does not hinder the querying capabilities as any relational query can be translated into a graph query.

2.4.1 Information Preservation

A direct mapping *DM* is *information preserving* if no information is lost during the mapping of any relational database.

Definition 2 (Information preservation). *A direct mapping* DM *is* information preserving *if there is a computable inverse mapping $DM^{-1} : \mathcal{D}_G \rightarrow \mathcal{D}_R$ satisfying $DM^{-1}(DM(D_R)) = D_R$ for any $D_R \in \mathcal{D}_R$.* □

2.4.2 Query Preservation

Recall that both SQL and Cypher queries return a result modeled as a table where columns represent entities requested by the query (using Select clause in case of SQL, or Return clause in case of Cypher), while each row assigns values to these entities. In addition, the result of both SQL and Cypher queries may contain repeated rows.

Let I_R be a relational instance and Q_s be an SQL query expressed over I_R. We denote by $[Q_s]_{I_R}$ the result table of Q_s over I_R. Similarly, $[Q_c]_{I_G}$ is the result table of a Cypher query Q_c over an instance graph I_G. Moreover, we denote by $[Q_s]^*_{I_R}$ (resp. $[Q_c]^*_{I_G}$) the refined table that has no repeated row.

A direct mapping *DM* is *query preserving* if any query over the relational database D_R can be translated into an equivalent query over the graph database D_G that results from the mapping of D_R. That is, query preservation ensures that every relational query can be evaluated using the mapped instance graph.

Since SQL and Cypher languages return results in different forms, proving query preservation consists to define a mapping from the SQL result to the Cypher result. This principle was proposed first in [14] between relational and RDF queries. Therefore, we revise the definition of query preservation as follows:

Definition 3 (Query preservation). *A direct mapping* DM *is* query preserving *if, for any relational database D_R=(S_R,I_R) and any SQL query Q_s, there exists a Cypher query Q_c such that: each row in $[Q_s]^*_{I_R}$ can be mapped into a row in $[Q_c]^*_{I_G \in DM(D_R)}$ and vice versa. By mapping a row r into a row r', we assume that r and r' contain the same data with possibly different forms.* □

2.4.3 Semantics Preservation

A direct mapping *DM* is *semantics preserving* if any consistent (resp. inconsistent) relational database is translated into a consistent (resp. inconsistent) graph database.

Definition 4 (Semantic preservation). *A direct mapping* DM *is semantic preserving if, for any relational database $D_R = (S_R, I_R)$ with a set of integrity constraints Σ, $I_R \models \Sigma$ iff: $DM(D_R)$ produces a consistent graph database.* □

Notice that no previous work have considered semantic preservation over data graphs. That is, no definition of graph database consistency have been given. We shall give later our own definition.

3 Complete Mapping (*CM*)

In this section, we propose a complete mapping *CM* that transforms a complete relational database (schema and instance) into a complete graph database (schema and instance). We call our mapping *Complete* since some proposed mappings (e.g. [11]) deal only with data and not schema.

Definition 5 (*Complete Mapping*). *A complete mapping is a function $CM : \mathcal{D}_R \rightarrow \mathcal{D}_G$ from the set of all RDs to the set of all GDs such that: for each complete relational database $D_R = (S_R, I_R)$, $CM(DR)$ generates a complete graph database $D_G = (S_G, I_G)$.* □

In order to produce a complete graph database, our *CM* process is based on two steps, *schema mapping* and *instance mapping*, which we detail hereafter.

3.1 Schema Graph and Instance Graph

Contrary to relational data, graph data still have no schema definition standard. Hence, we extend the property graph definition in order to introduce our schema graph definition.

Definition 6 (*Schema Graph*). *A* schema graph *is an extended property graph defined with $S_G = (V_S, E_S, L_S, A_S, Pk, Fk)$ where: 1) V_S is a finite set of vertices; 2) $E_S \subseteq V_S \times V_S$ is a finite set of directed edges where $(v, v') \in E_S$ is an edge starting at vertex v and ending at vertex v'; 3) L_S is a function that assigns a label to each vertex in V_S (resp. edge in E_S); 4) A_S is a function assigning a nonempty set of pairs $(a_i : t_i)$ to each vertex (resp. edge) where $a_i \in \mathcal{A}$ and $t_i \in \mathcal{T}$; 5) Pk is a partial function that assigns a subset of $A_s(v)$ to a vertex v; and finally 6) for each edge $e \in E_S$, $Fk(e, s)$ (resp. $Fk(e, d)$) is a subset of $A_S(e.s)$ (resp. $A_S(e.d)$).* □

The functions Pk and Fk will be used later to incorporate integrity constraints over graph databases.

Definition 7 (*Instance Graph*). *Given a schema graph $S_G = (V_S, E_S, L_S, A_S, Pk, Fk)$, an instance I_G of S_G, called an* instance graph, *is given by a property graph $I_G = (V_I, E_I, L_I, A_I)$ where:*

1. *V_I and E_I are the set of vertices and the set of edges as defined for schema graph;*
2. *for each vertex $v_i \in V_I$, there exists a vertex $v_s \in V_S$ such that: a) $L_I(v_i) = L_S(v_s)$; and b) for each pair $(a : c) \in A_I(v_i)$ there exists a pair $(a : t) \in A_S(v_s)$ with* type*(c)=t. We say that v_i corresponds to v_s, denoted by $v_i \sim v_s$.*
3. *for each edge $e_i = (v_i, w_i)$ in E_I, there exists an edge $e_s = (v_s, w_s)$ in E_S such that: a) $L_I(e_i) = L_S(e_s)$; b) for each pair $(a : c) \in A_I(e_i)$ there exists a pair $(a : t) \in A_S(e_s)$ with* type*(c)=t; and c) $v_i \sim v_s$ and $w_i \sim w_s$. We say that e_i corresponds to e_s, denoted by $e_i \sim e_s$.* □

As for relational schema, a schema graph determines the structure, meta-information and typing that instance graphs must satisfy. It is clear that an instance I_G of S_G assigns a set of vertices (resp. edges) to each vertex v_s (resp. edge e_s) in S_G that have the same label as v_s (resp. e_s). Moreover, a vertex v_i

(resp. edge e_i) in I_G corresponds to a vertex v_s (resp. edge e_s) in S_G if the value c, attached to any attribute a of v_i (resp. edge e_i), respects the type t given for a within v_s (resp. e_s).

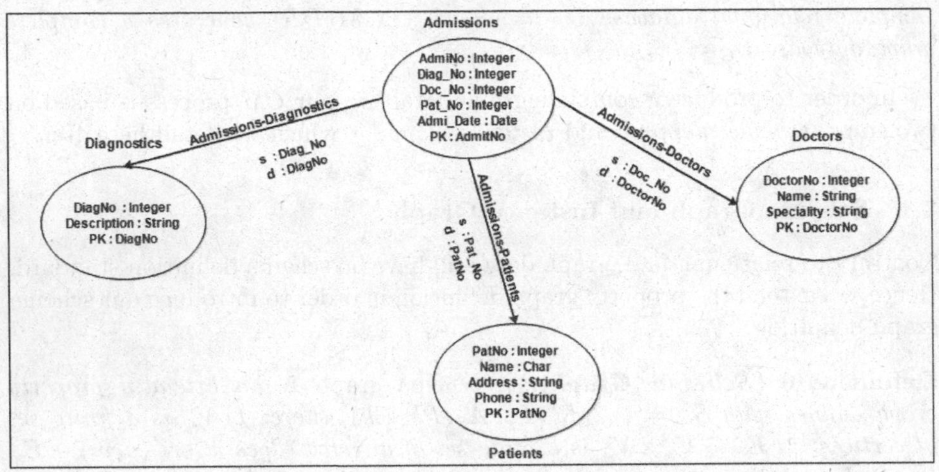

Fig. 2. Example of schema graph.

Example 2. Figure 2 depicts an example of a schema graph where each vertex (resp. edge) is represented naturally with its label (e.g. vertex *Admissions*, edge *Admissions-Doctors*) and a list of typed attributes (e.g. *AdmiNo:Integer*). As a special case, the value of the attribute Pk on some vertex refers to the value of the function Pk on this vertex (e.g. $Pk{:}AdmiNo$ on vertex *Admissions*). Moreover, the values of attributes s and d on some edge e refer to the values of the function $Fk(e, s)$ (resp. $Fk(e, d)$) at this edge (e.g. s : *Doc_No* and d : *DoctorNo* on edge *Admissions – Doctors*). The use of these special attributes (Pk, s and d) will be detailed later. One can see that the data graph of Fig. 1 is an instance graph of the schema graph of Fig. 2 since each vertex (resp. edge) of the former corresponds to some vertex (resp. edge) of the latter. □

Finally, a *graph database* is defined with $D_G = (S_G, I_G)$ where S_G is a schema graph and I_G is an instance of S_G.

3.2 Schema Mapping (*SM*)

Given a relational schema $S_R = (R, A, T, \Sigma)$, we propose a schema mapping (*SM*) process that produces a schema graph $S_G = (V_S, E_S, L_S, A_S, Pk, Fk)$ as follows:

1. For each relation name $r \in R$ in S_R, there exists a vertex $v_r \in V_S$ such that $L_S(v_r) = r$;

2. For each $a \in A(r)$ with $T(a) = t$, we have a pair $(a : t) \in A_S(v_r)$;
3. For each *primary key* $r[a_1, \cdots, a_n] \in \Sigma$, we have $Pk(v_r) =$ "$a_1, \ldots, a_n$";
4. For each *foreign key* $r[a_1, \cdots, a_n] \rightarrow s[b_1, \cdots, b_n] \in \Sigma$, we have an edge $e = (v_r, v_s) \in E_s$ such that : a) $L_s(e) = (s - r)$; b) $Fk(e, s)$= "$a_1, \ldots, a_n$"; and c) $Fk(e, d)$ = "$b_1, \ldots, b_n$".
5. A special pair (*vid:Integer*) is attached to each vertex of S_G for storage concerns.

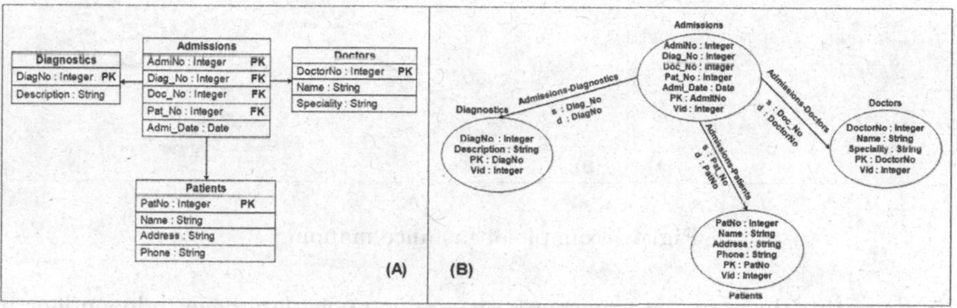

Fig. 3. Example of schema mapping.

Example 3. Figure 3 depicts (A) a relational schema and (B) its corresponding schema graph. One can see that our schema mapping rules are respected: 1) each relation is mapped to a vertex that contains the label of this relation, its primary key and a list of its typed attributes; 2) each foreign key between two relations (e.g. relations *Admissions* and *Patients* in part A) is represented by an edge between the vertices of these two relations (e.g. edge *Admissions-Patients* in part B). □

3.3 Instance Mapping (*IM*)

Given a relational database $D_R = (S_R, I_R)$, we propose an instance mapping (*IM*) process that maps the relational instance I_R into an instance graph $I_G = (V_I, E_I, L_I, A_I)$ as follows:

1. For each tuple $t \in I(r)$, there exists a vertex $v_t \in V_I$ with $L_I(v_t) = r$. We denote by v_t the vertex that corresponds to the tuple t;
2. For each tuple $t \in I(r)$ and each attribute a with $t(a) = c$, we have: $(a : c) \in A_I(v_t)$ if $a \neq tid$; and $(vid : c) \in A_I(v_t)$ otherwise.
3. for each *foreign key* $r[a_1, \cdots, a_n] \rightarrow s[b_1, \cdots, b_n]$ defined with S_G and any tuples $t \in I(r)$ and $t' \in I(s)$, if $t(a_i) = t'(b_i)$ for each $i \in [1, n]$, then: there is an edge $e = (v_t, v_{t'}) \in E_I$ with $L_I(e) = r - s$.

Example 4. An example of our *IM* process is given in Fig. 4 where part (A) is the relational instance and part (B) is its corresponding instance graph. □

It is clear that the special attribute *vid* is used to preserve the tuples identification (i.e. the value of attribute *tid*) during the mapping process.

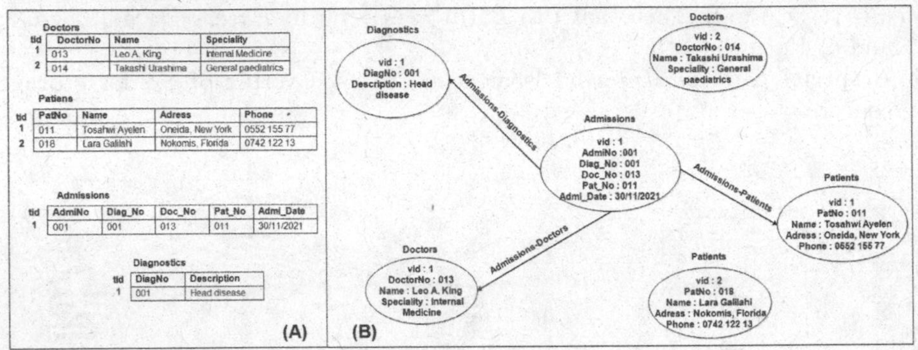

Fig. 4. Example of instance mapping.

We notice that our data mapping process is query language independent in the sense that any query language (e.g. Cypher, Gremlin, PGQL) can be applied over the resulting data graph.

4 Properties of *CM*

We show that our *CM* satisfies the three fundamental mapping properties [14]: *information preservation*, *query preservation* and *semantics preservation*.

4.1 Information Preservation

First, we explain that *CM* does not lose any part of the information in the relational instance being translated:

Theorem 1. *The direct mapping* CM *is information preserving.*

Proof. Theorem 1 can be proved easily by showing that there exists a computable mapping $CM^{-1} : D_G \rightarrow D_R$ that reconstructs the initial relational database from the generated graph database. Since our mapping CM is based on two steps (schema and instance mappings), then CM^{-1} requires the definition of SM^{-1} and IM^{-1} processes. Due to the space limit, the definition of CM^{-1} is given in the extended version of this paper [3].

4.2 Query Preservation

Second, we show that the way *CM* maps relational instances into instance graphs allows one to answer the SQL query over a relational instance by translating it into an equivalent Cypher query over the generated graph instance.

Algorithm 1. S2C Algorithm

Input: A simple SQL query Q_s, A relational schema S_R
Output: Its equivalent Cypher query Q_c.

1: Rename relations (via *AS*-clause) in Q_s if not applied;
2: Extract the Select-clause (SC), the From-clause (FC) and the Where-clause (WC) from Q_s;
3: Generate a Match-clause (mc) from FC:
   a) by translating each relation name r in SC into a vertex with label r; and
   b) by representing each join in Q_s with an edge in mc basing on S_R;
4: Generate a Where-clause (wc) from WC by translating each condition over a relation (attribute) in WC into a condition over the corresponding vertex (attribute);
5: Generate a Return-clause (rc) from SC by translating each relation (attribute) in SC into a corresponding vertex (attribute);
6: Generate a Cypher query Q_c by combining mc, wc and rc;
7: Return Q_c;

Theorem 2. *The direct mapping* CM *is query Preserving.*

Proof. Proving Theorem 2 can be done by providing an algorithm $S2C$ that, for any SQL query Q_s, produces an equivalent Cypher query Q_c such that: for any relation database $D_R = (S_R, I_R)$ and any SQL query Q_s over I_R, each row in $[Q_s]^*_{I_R}$ can be mapped to a row in $[Q_c]^*_{I_G}$ where $I_G \in CM(D_R)$ and $Q_c = S2C(Q_s)$. Our algorithm $S2C$ is summarized in Algorithm 1. Given an SQL query Q_s in input, $S2C$ proceeds as follows: a) analyze and extract clauses from Q_s; b) compute for any SQL clause their equivalent Cypher clause; c) combine the resulted Cypher clauses in Q_c; and d) return the final Cypher query Q_c. Due to space limit, we give a reduced version of our query mapping algorithm that deals only with simple queries. However, one can easily extend our algorithm to deal with composed versions (e.g. queries with IN clause). A running example of query mapping is given in the extended version of this paper [3].

4.3 Semantic Preservation

Finally, we show that CM is semantic preserving by checking consistency (resp. inconsistency) of relational database and graph database. Recall that a direct mapping is semantic preserving if any consistent (resp. inconsistent) relational database is translated into a consistent (resp. inconsistent) graph database. While consistency of relational database is well-known, no definition is given for graph database since there exists no standard for (schema) graph definition. To overcome this limit, we added the functions Pk and Fk to our schema graph definition in order to make possible the consistency checking for graph databases.

Definition 8 (*Graph consistency*). *For any graph database $D_G = (S_G, I_G)$, the instance I_G is said to be consistent w.r.t S_G if:*

- *For each vertex $v_s \in V_s$ with $Pk(v_s) =$ "$a_1, ..., a_n$" and each vertex $v_i \in V_I$ that corresponds to v_s: there exists no pair $(a_i : NULL) \in A_I(v_i)$ with $i \in [1, n]$. Moreover, for each $v_i' \in V_I \backslash \{v_i\}$ that corresponds to v_s, the following condition must not hold: for each $i \in [1, n]$, $(a_i : c) \in A_I(v_i) \cap A_i(v_i')$.*

– *For each edge* $e_s \in E_s$ *with* $Fk(e_s, s) =$ "$a_1, ..., a_n$" *and* $Fk(e_s, d) =$ "$b_1, ..., b_n$", *if* $e_i = (v_1, v_2) \in E_I$ *is an edge that corresponds to* e_s *then we have:* $(a_i : c) \in A_I(v_1)$ *and* $(b_i : c) \in A_I(v_2)$ *for each* $i \in [1, n]$. □

Intuitively, the consistency of graph databases is inspired from that of relational databases.

Theorem 3. *The direct mapping* CM *is semantic Preserving.*

Proof. The proof of Theorem 3 is straightforward and can be done by contradiction based on the mapping rules of *IM* (Sect. 3.3). Given a relational database $D_R = (S_R, I_R)$ and let $D_G = (S_G, I_G)$ be its equivalent graph database generated by *CM*. We suppose that *CM* is not semantic preserving. This means that either (A) I_R is consistent and I_G is inconsistent; or (B) I_R is inconsistent while I_G is consistent. We give only proof of case (A) since that of case (B) can be done in a similar way.

We suppose that I_R is consistent w.r.t S_R while I_G is inconsistent w.r.t S_G. Based on Definition 8 I_G is inconsistent if one of the following conditions holds:

1) There exists a vertex $v_i \in V_I$ that corresponds to a vertex $v_s \in V_s$ where: a) $Pk(v_s) =$ "$a_1, ..., a_n$"; and b) $(a_i : NULL) \in A_I(v_i)$ for some attribute $a_{i\in[1,n]}$. Based on mapping rules of *IM* process, v_i corresponds to some tuple t in I_R and attributes $a_1, ..., a_n$ correspond to a primary key defined over I_R by S_G. Then (b) implies that the tuple t assigns a $NULL$ value to the attribute a_i which makes I_R inconsistent. However, we supposed that I_R is consistent.
2) There are two vertices $v_1, v_2 \in V_I$ that correspond to a vertex $v_s \in V_s$ where: a) $Pk(v_s) =$ "$a_1, ..., a_n$"; and b) $(a_i : c) \in A_I(v_1) \cap A_I(v_2)$ for each $i \in [1, n]$. Based on mapping rules of *IM* process, v_1 (resp. v_2) corresponds to some tuple t_1 (resp. t_2) in I_R and attributes $a_1, ..., a_n$ correspond to a primary key defined over I_R by S_G. Then (b) implies that the tuples t_1 and t_2 assign the same value to each attribute a_i which makes I_R inconsistent. However, we supposed that I_R is consistent.
3) There exists an edge $e_i = (v_1, v_2)$ in E_I that corresponds to an edge $e = (v_s, v_d)$ in E_s where: a) $Fk(e, s) =$ "$a_1, ..., a_n$"; b) $Fk(e, d) =$ "$b_1, ..., b_n$"; and c) there exists some attribute $a_{i\in[1,n]}$ with $(a_i : c_1) \in A_I(v_1)$, $(a_i : c_2) \in A_I(v_2)$, and $c_1 \neq c_2$. Based on mapping rules of *IM* process, v_1 (resp. v_2) corresponds to some tuple t_1 (resp. t_2) in I_R, v_s (resp. v_d) corresponds to some relation s (resp. d) in S_R, the function Fk over edge e refers to a foreign-key $s[a_1, ..., a_n] \to d[b_1, ..., b_n]$ defined over I_R by S_G. Then (b) implies that the tuple t_1 assigns a value to some attribute $a_{i\in[1,n]}$ that is different to that assigned by tuple t_2 to attribute $b_{i\in[1,n]}$. This means that there is a violation of foreign-key by tuple t_1 which makes I_R inconsistent. However, we supposed that I_R is consistent.

Therefore, each case of I_G inconsistency leads to a contradiction, which means that if I_R is consistent then its corresponding I_G cannot be inconsistent.

By doing proof of part (B) in a similar way, we conclude that if I_R is consistent (resp. inconsistent) then its corresponding instance graph I_G must be consistent (resp. inconsistent). This completes the proof of Theorem 3. □

5 Conclusion and Future Works

In this paper, we proposed a complete mapping process that translates any relational database into an equivalent graph database by considering both schema and data mapping. Our mapping preserves the information and semantics of the relational database and maps any SQL query, over the relational database, into an equivalent Cypher query to be evaluated over the produced graph database. We plan to extend our model to preserve another mapping property, called monotonicity [14], which ensures that any update to the relational database will not require generating the corresponding graph database from scratch. Also, we will enrich the definition of the relational schema with more integrity constraints. We are conducting an experimental study based on real-life databases to check our approach's efficiency and effectiveness.

References

1. Neo4j ETL. https://neo4j.com/developer/neo4j-etl/
2. Angles, R., Thakkar, H., Tomaszuk, D.: Mapping RDF databases to property graph databases. IEEE Access **8**, 86091–86110 (2020)
3. Boudaoud, A., Mahfoud, H., Chikh, A.: Towards a complete direct mapping from relational databases to property graphs. CoRR abs/2210.00457 (2022). https://doi.org/10.48550/arXiv.2210.00457
4. De Virgilio, R., Maccioni, A., Torlone, R.: Converting relational to graph databases, p. 1 (2013)
5. De Virgilio, R., Maccioni, A., Torlone, R.: R2G: a tool for migrating relations to graphs, pp. 640–643 (2014)
6. Fan, W., Wang, X., Yinghui, W.: Answering pattern queries using views. IEEE Trans. Knowl. Data Eng. **28**, 326–341 (2016)
7. Francis, N., et al.: Cypher: an evolving query language for property graphs, pp. 1433–1445 (2018)
8. Guagliardo, P., Libkin, L.: A formal semantics of SQL queries, its validation, and applications. Proc. VLDB Endow. **11**, 27–39 (2017)
9. Li, S., Yang, Z., Zhang, X., Zhang, W., Lin, X.: SQL2Cypher: automated data and query migration from RDBMS to GDBMS. In: Zhang, W., Zou, L., Maamar, Z., Chen, L. (eds.) WISE 2021. LNCS, vol. 13081, pp. 510–517. Springer, Cham (2021). https://doi.org/10.1007/978-3-030-91560-5_39
10. Matsumoto, S., Yamanaka, R., Chiba, H.: Mapping RDF graphs to property graphs, pp. 106–109 (2018)
11. Orel, O., Zakošek, S., Baranović, M.: Property oriented relational-to-graph database conversion. Automatika **57**, 836–845 (2017)
12. Sequeda, J.F., Arenas, M., Miranker, D.P.: On directly mapping relational databases to RDF and OWL, pp. 649–658 (2012)
13. Stoica, R., Fletcher, G., Sequeda, J.F.: On directly mapping relational databases to property graphs, pp. 1–4 (2019)
14. Stoica, R.-A.: R2PG-DM: a direct mapping from relational databases to property graphs. Master's thesis, Eindhoven University of Technology (2019)

A Matching Approach to Confer Semantics over Tabular Data Based on Knowledge Graphs

Wiem Baazouzi[1(✉)], Marouen Kachroudi[2], and Sami Faiz[3]

[1] Ecole Nationale des Sciences de l'Informatique, Laboratoire de Recherche en génIe logiciel, Application Distribuées, Systèmes décisionnels et Imagerie intelligente, Université de la Manouba, LR99ES26 Manouba, 2010 Tunis, Tunisie
wiem.baazouzi@ensi-uma.tn

[2] Faculté des Sciences de Tunis, Informatique Programmation Algorithmique et Heuristique, Université de Tunis El Manar, LR11ES14, 2092 Tunis, Tunisie
marouen.kachroudi@fst.rnu.tn

[3] Université de Tunis El Manar, Ecole Nationale d'Ingénieurs de Tunis, Laboratoire de Télédétection et Systèmes d'Information à Référence Spatiale, 99/UR/11-11, 2092 Tunis, Tunisie
sami.faiz@insat.rnu.tn

Abstract. In this article, we present KEPLER-ASI, a matching approach to overcome possible semantic gaps in tabular data by referring to a Knowledge Graph. The task proves difficult for the machines, which requires extra effort to deploy the cognitive ability in the matching methods. Indeed, the ultimate goal of our new method is to implement a fast and efficient approach to annotate tabular data with features from a Knowledge Graph. The approach combines search and filter services combined with text pre-processing techniques. The experimental evaluation was conducted in the context of the SemTab 2021 challenge and yielded encouraging and promising results referring to its performance and ranks held.

Keywords: Tabular data · Knowledge graph · Fair principles

1 Introduction

Consolidating and implementing the FAIR[1] principles[2] for data conveyed on the Web is a real need to facilitate their management and use. Indeed, the added value of such a process is the generation of new knowledge through the tasks of data integration, data cleaning, data mining and machine learning. Thus, the successful implementation of FAIR principles drastically improves the value of data by making it: findable, accessible while overcoming semantic ambiguities.

[1] FAIR stands for Findability, Accessibility, Interoperability, and Reuse.
[2] https://www.go-fair.org/fair-principles/.

P. Fournier-Viger et al. (Eds.): MEDI 2022, LNAI 13761, pp. 236–249, 2023.
https://doi.org/10.1007/978-3-031-21595-7_17

In this register, we highlight that good data management, is not an objective in itself, but rather the global approach leading to knowledge discovery and acquisition, as well as to the integration and the subsequent reuse of the data by the stakeholder community after the data publication process. Consequently, semantic annotation is considered a specific knowledge acquisition task. Thus, the semantic annotation process can use formal metadata resources described in a semantic framework (*i.e.*, one or more ontologies) based on semantic repositories use. The latest contributions on this subject demonstrate that tabular data are carefully conveyed to the Web in various formats.
The preponderant one is tabular form (*e.g.*, CSV (Comma-Separated Values)). Tables on the Web are a very valuable data source, thus, injecting semantic information into these last ones has the potential to boost a wide range of applications, such as Web searching, answering queries, and building Knowledge Bases ($\mathcal{KB}$). Research reports that there are various issues with tabular data available on the Web, such as learning with limited labelled data, defining or updating ontologies, exploiting prior knowledge, or scaling up existing solutions. Therefore, this task is often difficult in practice due to missing, incomplete or ambiguous metadata (*e.g.*, table and column names). Over the last few years, we have identified several works that can be mainly classified as supervised (in the form of annotated tables to carry out the learning task) [5,10] or unsupervised (tables whose data is not dedicated to learning) [5,14]. To solve these problems, we propose a global approach named KEPLER-ASI, which addresses the challenge of matching tabular data to Knowledge Graphs ($\mathcal{KG}$).
Data annotation is a fundamental process in tabular data analysis [3], it allows to infer the meaning of other information. Then deduce the tabular data meaning relating to a Knowledge Graph. The data we used was based both on Wikidata and DBPedia. In a broader context, the data used and manipulated obey the triples format representation: subject ($\mathcal{S}$), a predicate ($\mathcal{P}$), and an object ($\mathcal{O}$). This notation ensures semantic navigability across data and makes all data manipulation more fluid, explicit, and reliable. Recent years have seen an increasing number of works on Semantic Table Interpretation. In this context, SemTab 2021[3] has emerged as an initiative that aims at benchmarking systems dealing with tabular annotation based on $\mathcal{KG}$ entities, referred to as table annotation. SemTab is organised into three tasks, each one with several evaluation rounds. For the 2021 edition, for instance, it involves: (*i*) assigning a semantic type (*e.g.*, a $\mathcal{KG}$ class) to a column (`CTA`); (*ii*) matching a cell to a $\mathcal{KG}$ entity (`CEA`); (*iii*) assigning a $\mathcal{KG}$ property to the relationship between two columns (`CPA`).

We aim for automatic on-the-fly annotation of tabular data. Thus, our annotation approach is fully automated, as it does not collect upstream information regarding entities or metadata standards. Our method is quick and easy to deploy since it leverages existing resources like Wikidata and Dbpedia to access entities. The paper is thus organized as follows: Sect. 2 presents some key concepts, Sect. 3 is dedicated to the state of the art, Sect. 4 presents our contribution, Sect. 5 draws up a experimental report before concluding with the Sect. 6.

[3] http://www.cs.ox.ac.uk/isg/challenges/sem-tab/2021/index.html.

2 Key Notions

In what follows, some key notions relating to our studied context supported by some examples and illustrations.

- ***Tabular Data*** : S is a two-dimensional tabular structure made up of an ordered set of N rows and M columns, as depicted by Fig. 1. n_i is a row of the table (i = 1 ... N), m_j is a column of the table (j = 1 ... M). The intersection between a row n_i and a column m_j is $c_{i,j}$, which is a value of the cell $S_{i,j}$. The table contents can have different types (string, date, float, number, etc.). Thereby, we will have these structures : Target Table (S): M × N, Subject Cell: $S_{(i,0)}$ (i = 1, 2 ... N) and Object Cell: $S_{(i,j)}$ (i = 1, 2 ... M),(j = 1, 2 ... N).

$$
\begin{array}{c} \\ Row_1 \\ \\ Row_j \\ \\ Row_M \end{array}
\begin{array}{c}
\begin{array}{ccccc} Col_0 & & Col_i & & Col_N \end{array} \\
\begin{pmatrix}
S_{1,0} & \cdots & \cdots & \cdots & S_{1,N} \\
\vdots & \ddots & \ddots & \ddots & \vdots \\
S_{j,0} & \cdots & S_{j,i} & \cdots & S_{j,N} \\
\vdots & \ddots & \ddots & \ddots & \vdots \\
S_{M,0} & \cdots & \cdots & \cdots & S_{M,N}
\end{pmatrix}
\end{array}
$$

Fig. 1. Tabular data at a glance.

- ***Knowledge Graph***: Knowledge Graphs have been the focus of research since 2012, resulting in a wide variety of published descriptions and definitions. The lack of a common core is a fact that was also reported by Paulheim [7] in 2015. Paulheim listed in his survey of Knowledge Graph refinement the minimum set of characteristics that must be present to distinguish Knowledge Graphs from other knowledge collections, which restricts the term to any graph-based knowledge representation. In the online review [7], authors agreed that a more precise definition was hard to find at that point. This statement points out the need for a closer investigation and a reflection in this area. Farber and *al.* [8] defined a Knowledge Graph as a Resource Description Framework (RDF) graph. Also, the authors stated that the $\mathcal{KG}$ term was coined by Google to describe any graph-based Knowledge Base ($\mathcal{KB}$).

3 Literature Review

Various research works have tackled the issue of semantic tables annotation. They vary according to the deployed techniques as well as the adopted approach. The CSV2KG system [13] consists of six phases. First, raw annotations are assigned to each cell of the table considered as input. Candidates undergo disambiguation by similarity measures applied to each candidate's label. Then the column types and the properties between the columns are inferred using the seed annotations. In the next step, the inferred column types and properties are used to create more refined header cell annotations (the cells in the first column of a table). Further processing uses the newly generated header cells to correct other table cells, using property annotations. Finally, new column types are inferred using all available corrected cells. The source code for this system is in Python. DAGOBAH [2] is a system implemented as a set of sequential complementary tools. The three main functionalities are: (*i*) the identification of the semantic relationships between tabular data and Knowledge Graphs, (*ii*) Knowledge Graphs enrichment by transforming the informational volume contained in the array into triples, (*iii*) Metadata production that can be used for reference, research and recommendation processes. DAGOBAH determines the list of candidate annotations using a SPARQL query. As for the LinkingPark [4] system, it takes a table as input, which is passed to an editor to extract entity links as well as property links. From the entity links, said method generates candidate entities via a cascading pipeline which becomes the input for both the disambiguation module and the property link detection module. The authors also integrated the property characteristics to determine the relationship between the different rows of the starting table. Furthermore, the JenTab [1] system operates according to 9 modules, each of which has a well-defined objective. The first module constitutes the system core and is responsible for most matching operations. Based on the search for annotation tags, this module generates candidates for the three tasks (`CTA`, `CEA`, and `CPA`) and removes unlikely candidates. The second module attempts to retrieve missing `CEA` matches based on the context of a row and a column. Indeed, this processing only applies to the cells which have not received any matching during the first module. Subsequently, in the third module, and the absence of candidates in the `CTA` task, the processing (`CEA` by Row) relaxes the requirements. The fourth module focuses on the selection of solutions. After a new stage of filtering on the `CEA` candidates using the context of each row, the authors opted to select the solutions with very high confidence values. Modules 5 and 7 attempt to fill in the gaps relating to the failure to identify potential candidates. In case new candidates are identified, modules 6 and 8 screen them again. Module 9 represents the last resort to generating solutions for features without candidate annotations. At this level, the authors assume that certain parts of the context are false, to re-examine each assertion. Authors reconsider all the candidates discarded in the previous stages to find the best solution among them. The LexMa system [12] starts with a preprocessing phase cropping the text in the cell and converting the resulting strings to uppercase. After that, the system retrieves the top 5 entities for each cell value from the Wikidata search service. Subsequently, the lexical match is evaluated based on cosine similarity. This similarity measure is applied to vectors coded and formed from labels derived from cell values. Cell labels and values are tokenized, then stop words are removed before creating input vectors. At this point, the authors trigger an identical search on DBpedia via its dedicated search service operating with SPARQL queries.

To sum up, all of the above approaches rely on a learning strategy. Moreover, for the real-time context, the applications become greedy, which imposes obtaining the result as quickly as possible. This scenario requires the deployment of more logistical and technical efforts. Moreover, the applicability of these solutions will strengthen semantic interoperability across all domains. In the following, we present a detailed description of our contribution, namely KEPLER-ASI.

4 The KEPLER-ASI Approach

In this section, we describe the different stages of our system while presenting some basic notions to highlight the technical issues identified. To address the SemTab challenge tasks, KEPLER-ASI operates according to the workflow depicted by Fig. 2. There are five major complementary modules, namely: Preprocessing, Query Engine (or eventually External Resource Consultation), $\mathcal{KG}$_Candidates Filtering and Annotation and File Generation. These steps are the same for each round, but the changes remain minimal depending on the variations observed in each case.

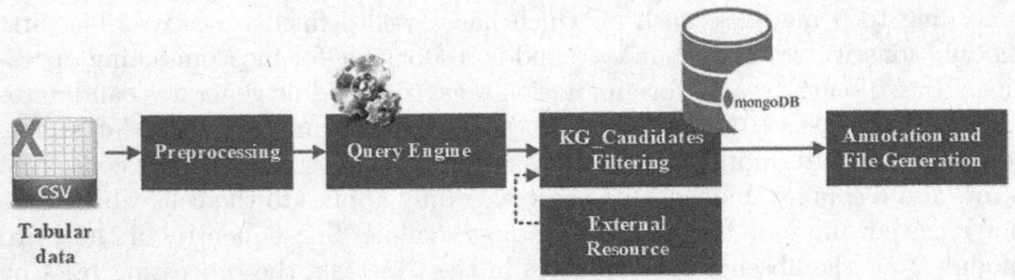

Fig. 2. Overview of our approach.

4.1 Preprocessing Module

It should be noted that the content of each table can be expressed according to different types and formats, namely: numeric, character strings, binary data, date/time, boolean, addresses, etc. Indeed, with the great diversity of data types, the preprocessing step is crucial. Therefore, the goal of preprocessing is to ensure that the processing of each table is triggered without errors. The effort is especially accentuated when the data contain spelling errors. In other words, these issues must be resolved before we apply our approach. In order to well carry out this step, we used several techniques and libraries such as (Textblob[4], Pyspellchecker[5], etc.) to rectify and correct all the noisy textual data in the

[4] https://textblob.readthedocs.io/en/dev/.
[5] https://pypi.org/project/pyspellchecker/.

considered tables. As an example, we detect punctuation, parentheses, hyphen and apostrophe, and also stop words by using the `Pandas`[6] library to remove them. Like a classic treatment in this register, we ended this phase by transforming all the upper case letters into lower case. The priming is carried out by an analysis of the processing columns, which aims to understand and delimit the set of regular expressions which contains a set of units: the area, the currency, the density, the electric current, the energy, flow rate, force, frequency, energy efficiency, unit of information, length, density, mass, numbers, population density, power, pressure, speed, temperature, time, torque, voltage and volume. This step allows identifying multiple Regextypes using regular expressions (*e.g.* numbers, geographic coordinates, address, code, colour, URL). Since all values of type text are selected, preprocessing for natural languages is performed using the `langrid`[7] library to detect 26 languages in our data. By the way, it's a novelty for this year's SemTab campaign, *i.e.*, which makes the task more difficult with the introduction of natural language barriers. The `langrid` library is a stand-alone language detection tool. It is performed in a large number of languages (97 currently). Doing so, correction, data type and language detection are performed. This treatment considerably reduces execution effort and cost by avoiding the massive repetition of these treatments for all the table cells, and this in each subtask.

4.2 Query Engine Module

This module is the core of our contribution. We manage through a SPARQL query to extract the candidate annotations from the Knowledge Graphs, namely Wikidata and DBpedia. This phase allows the annotation process candidates extraction.

Example 1. Starting from an English entity description, in what follows an example of a SPARQL query to retrieve the label, the class name as well as the properties from Wikidata (or eventually Dbpedia) :

```
   endpoint_url = "https://query.wikidata.org/sparql"
   query = """
   SELECT ?itemLabel ?class  ?property
   WHERE {
?item   ?itemDescription "%s"@en .
?item wdt:P31 ?class
   }"""
```

4.3 External Resource Consultation Module

External resources consultation is a complementary process to that of querying Knowledge Graphs. Indeed, in the case where the result of the query does not

[6] https://pandas.pydata.org.
[7] https://github.com/openlangrid.

provide candidates to perform the annotation, an external resource is used, in search of a possible answer. Note that this consultation is generically modelled in our system. In other words, we can couple our system with any resource in RDF format. This step gives our contribution a great adaptability to the studied context.

4.4 $\mathcal{KG}$_Candidates Filtering Module

Candidate annotations filtering is based on an efficient and fast Information Retrieval technique. Indeed, any identified annotations are indexed and saved in a NoSQL database, namely, MongoDB. Subsequently, we identify the final annotation considered as the result of the matching process after querying this database through its integrated search engine. Thus, the candidate annotation retained is the one with the first rank and having the highest score, in accordance with lines 10, 10 and 8 in respectively Algorithms 1, 2 and 3. We opted for the MongoDB database for the considerable gain in execution times thanks to its processing capabilities, *i.e.*, scaling and search efficiency.

4.5 Annotation and File Generation Module

Assigning a Semantic Type to a Column (CTA). As depicted by Fig. 3 and following Algorithm 1, the task is to annotate each entity column with elements from Wikidata (or possibly DBPedia) as its type identified during the preprocessing phase.

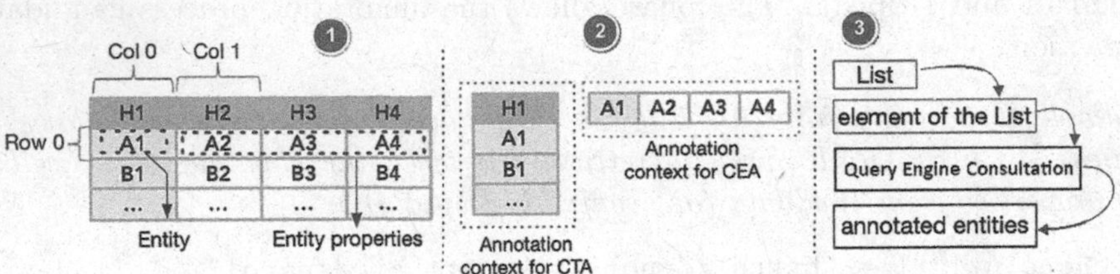

Fig. 3. CTA task.

Each item is marked with the tag in Wikidata or DBPedia. This treatment allows semantics identification. The `CTA` task is performed based on Wikidata or DBPedia APIs to look for an item according to its description. The collected information about a given entity used in our approach is an instance list (expressed by the `instanceOf` primitive and accessible by the P31 code), the subclass of (expressed by the `subclassOf` primitive and accessible by code P279) and overlaps (expressed by the `partOf` primitive and accessible by code P361). At this point, we can process the `CTA` task using a SPARQL query. The SPARQL query is our interrogation mean fed from the entity information that governs the choice of each data type since they are a list of instances (P31), of subclasses (P279) or a part of a class (P361). The result of the SPARQL query may return a single type, but in some cases, the result is more than one type, so in this case, no annotation is produced for the CTA task.

Algorithm 1: CTA task

```
Data: Table T
Result: Annotated Table T'
i ← 0
while col_i ∈ T do
    class_annot ← ∅
    while cell ∈ col do
        Label ← cell.expressionValue
        CorrectedLabel ← SpellCheckEngine(Label)
        KG_candidates ← QueryEngine(CorrectedLabel)
        class_annot ← KG_candidates
    end
    Annotate(T'.col_i, getBestRankedClass(class_annot))
end
```

Matching a Cell to a $\mathcal{KG}$ Entity (CEA). The CEA task aims to annotate the cells of a given table to a specific entity listed on Wikidata or DBPedia. Figure 4 and Algorithm 2 gather the CEA task that is performed based on the same principle of the CTA task.

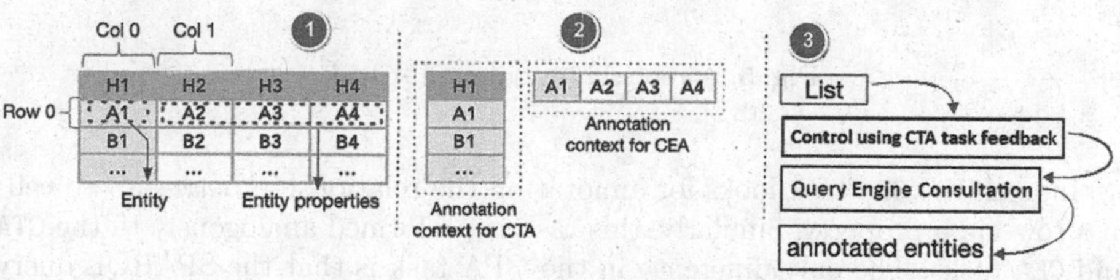

Fig. 4. Descriptive model of CEA task.

Our approach reuses the results of the CTA task process by introducing the necessary modifications to the SPARQL query. If the operation returns more than one annotation, we run a treatment based on examining the context of the considered column, relative to what was obtained with the CTA task, to overcome the ambiguity problem.

Algorithm 2: CEA task

```
Data: Table T
Result: Annotated Table T'
i ← 0
while row_i ∈ T do
    entity_annot ← ∅
    while cell ∈ row do
        Label ← cell.expressionValue
        CorrectedLabel ← SpellCheckEngine(Label)
        KG_candidates ← QueryEngine(CorrectedLabel)
        entity_annot ← KG_candidates;
    end
    Annotate(T'.row_i, getBestRankedEntity(entity_annot))
end
```

Matching a Property to a $\mathcal{KG}$ Entity (CPA). After having annotated the cell values as well as the different types of each of the considered entities, we will identify the relationships between two cells appearing on the same row via a property using a SPARQL query, as detailed by Fig. 5 and Algorithm 3.

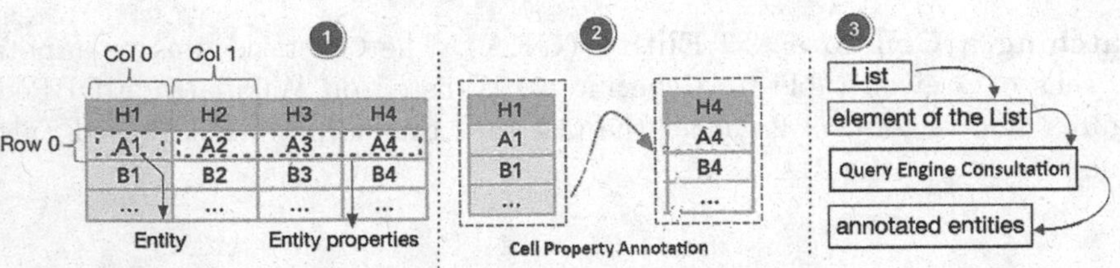

Fig. 5. A representation of CPA task.

Indeed, the CPA task looks for annotating the relationship between two cells in a row via a property. Similarly, this task is performed analogously to the CTA and CEA tasks. The only difference in the CPA task is that the SPARQL query must select both the entity and the corresponding attributes. The properties are easy to match since we have already determined them during CEA and CTA task processing.

Algorithm 3: CPA task

Data: Table $\mathcal{T}$
Result: Annotated Table $\mathcal{T}'$
$i \leftarrow 0$ $j \leftarrow 0$
while $(col_i, col_j) \in \mathcal{T} and i \neq j$ **do**
  $property\_annot \leftarrow \emptyset$
  $\mathcal{KG}\_class\_Label_1 \leftarrow$
    $Annotate(\mathcal{T}'.col_i, getMostFrequentClass(class\_annot))$
  $\mathcal{KG}\_class\_Label_2 \leftarrow$
    $Annotate(\mathcal{T}'.col_j, getMostFrequentClass(class\_annot))$
  $\mathcal{KG}\_candidates \leftarrow QueryEngine(\mathcal{KG}\_class\_Label_1, \mathcal{KG}\_class\_Label_2)$
  $property\_annot \leftarrow \mathcal{KG}\_candidates;$
  Annotate($\mathcal{T}'$.col$_i$,T'.col_j, $getBestRankedProperty(property\_annot)$)
end

5 KEPLER-ASI Performance and Results

In this section, we will present the results of KEPLER-ASI for the different matching tasks in the three rounds of SemTab 2021. We report that results are presented according to two scenarios, *i.e.*, before the deadline and after the deadline (since the organizers allow participants 1 month before freezing the values). Values are improved after the deadlines as we finish the investigating work about the data specifics, thus adjusting our filters for the candidates' identification. These results highlight the strengths of KEPLER-ASI with its encouraging performance despite the multiplicity of issues[8]. The mention *"after deadline"* means that there have been optimization efforts in terms of preprocessing and the choice of the used resource, even after the end of the round in question.

5.1 Round 1

In this first round of SemTab 2021, we have four tasks, namely: `CTA-WD`, `CEA-WD`, `CTA-DBP` and `CEA-DBP`. The column type annotation (`CTA -WD`) assigns a Wikidata semantic type (a Wikidata entity) to a column. Cell Entity Annotation (`CEA-WD`) maps a cell to a $\mathcal{KG}$ entity. The processing carried out to search for correspondence on Wikidata is similarly carried out on Dbpedia. Data for the `CTA-WD` and `CEA-WD` tasks focus on Wikidata. As we explained in Sect. 1, Wikidata is structured according to the RDF formalism, *i.e.*, subject ($\mathcal{S}$), predicate ($\mathcal{P}$) and Object ($\mathcal{O}$). Each element considered is marked with a label in Wikidata, thus guaranteeing maximum advantage of its semantics. The `CTA-WD` and

[8] All the official experimental values obtained and presented within the framework of this study (and challenge) are available and searchable via this link: https://www.aicrowd.com/challenges/semtab-2021. Please refer to the first author profile for a clear and detailed overview of all metrics. Note that there are 3 Rounds.

CEA-WD task data contain 180 tables. In Table 1, we provide an input table example. The first column contains an entity label, while the other columns contain the associated attributes.

Table 1. An example of data fragment for a table to match with Wikidata

Col0	Col1	Col2	Col3	Col4
Libertarian Party	Libertarianism	1971	4,489,221 (3.28%)	1 (0.01%)
Green Party	Green Politics	2001	1,457,216 (1.07%)	0

The column type annotation (CTA -DBP) assigns a DBPedia semantic type (a DBPedia entity) to a column. Cell Entity Annotation (CEA-DBP) matches a cell to a Knowledge Graph entity. The CTA-DBP and CEA-DBP task data also contain 180 tables. Results are summarized in Table 2.

Table 2. Results for Round 1

	F1 Score	Precision	Rank
CTA- WD	0.464	0.481	4
CTA-WD (after deadline)	0.746	0.758	3
CEA-WD	0.194	0.760	5
CEA-WD (after deadline)	0.663	0.818	2
CTA- DBP	0.027	0.133	5
CTA-DBP (after deadline)	0.503	0.521	1
CEA-DBP	0.110	0.644	5
CEA-DBP (after deadline)	0.602	0.604	4

In Round 1, we focused on the preprocessing phase to choose and validate the spellchecker according to textual information, which can significantly improve the relative results of the CEA and CTA tasks. In summary, our review led to the use of two correctors, namely, Textblob and Pyspellchecker. Both tools are intuitive, easy to use, and perform well in terms of Natural Language Processing (NLP). During Round 1, the data size factor was impacting. We recognize that this round highlights the limits of machines in the face of such information volumes. Therefore, we can conclude that faced with this situation, the computing power and the speed of access to the external resources representing the Knowledge Graphs (*i.e.*, Wikidata and DBPedia) are decisive. In addition, we consider that the introduction of the cross-lingual aspect of this campaign has accentuated the challenge and allowed us to approach real scenarios that open and unlock the eventualities of the different proposed approaches applicability. Indeed, to support the cross-lingual aspect, we acted at the level of the SPARQL query, as indicated in Example 1, to automatically change the language label, and collect the candidates in any language. Thus, we have ensured the genericity of our SPARQL query.

5.2 Round 2

In Round 2, despite the distinction of the data and their grouping into two different families, they have a biology tint. Due to advances in biological research techniques, new data are generated in the biomedical field and published in unstructured or tabular form. These data are delicate to be integrated semantically due to their size and the complexity of biological relationships maintained between the entities. Summary of metrics for this round is in Table 3.

Table 3. Results for Round 2

	F1 Score	Precision	Rank
BioTable-CTA- WD	0.811	0.811	6
BioTable-CEA-WD	0.347	0.811	6
BioTable-CEA-WD (after deadline)	0.677	0.798	6
BioTable-CPA-WD	0.853	0.880	4
HardTable-CTA-WD	0.894	0.931	5
HardTable-CEA-WD	0.707	0.919	6
HardTable-CPA-WD	0.915	0.989	5

Specifically, for tabular data annotation, the data representation can have a significant impact on performance since each entity can be represented by alphanumeric codes (*e.g.* chemical formulas or gene names) or even have multiple synonyms. Therefore, the studied field would greatly benefit from automated methods to map entities, entity types, and properties to existing datasets to speed up the new data integrating process through the domain. In this round, the focus was on Wikidata through two test cases: BioTable and HardTable. The different tasks: `BioTable-CTA-WD`, `BioTable-CEA-WD` and `BioTable-CPA-WD` on the one hand, to which we add `Hard-CTA-WD`, `Hard-CEA-WD` and `Hard-CPA-WD`, are all carried out on 110 tables. During Round 2, we focused on the disambiguation problem. We have to decide when to obtain several candidates after querying the $\mathcal{KG}$s. Indeed, our approach during Round 1 was useful and allowed us to reuse certain achievements. At this stage, we affirm that automatic elements disambiguation processing remains a tedious task, given what it generates as an effort of semantic analysis and interpretation. Indeed, we have opted for the use of an external resource, namely Uniprot[9] [11]. UniProt integrates, interprets and standardizes data from multiple selected resources to add biological knowledge and associated metadata to protein records and acts as a central

[9] https://www.uniprot.org.

hub. UniProt was recognized as an ELIXIR core data resource in 2017 [6] and received CoreTrustSeal certification in 2020. The data resource fully supports Findable, Accessible, Interoperable and Reusable, thus concretizing the FAIR data principles [9].

5.3 Round 3

Round 3 has 3 main test families: `BioDiv`: represented by 50 tables, `GitTables`: represented by 1100 tables and `HardTables`: represented by 7207 tables. Note that the stakes are the same for this round. Moreover, the evaluation is blind, *i.e.*, the participants do not have access to the evaluation platform and its options. In other words, there is no test opportunity to adjust the approach parameters according to the characteristics of the input. In this round, we opted for Uniprot to carry out treatments similar to those described in Round 2. Out of the 7 proposed tasks, KEPLER-ASI managed to process 3. In the `CTA-BioDiv` task, we ranked first. For the `GIT-DBP` base test, we ranked second and for `CTA-HARD` we ranked sixth. For the other cases, our method produces outputs containing duplications, whereas these correspondences do not allow us to obtain evaluation metrics.

6 Conclusion and Outlooks

To summarize and conclude, we have presented in this paper our KEPLER-ASI approach. Our system is approaching maturity and achieving very encouraging performance. We have succeeded in combining several strategies and treatment techniques, which is also the strength of our system. We boosted the preprocessing and spellchecking steps that got the system up and running. In addition, despite the data size, which is quite large, we managed to get around this problem by using a kind of local dictionary, which allows us to reuse already existing matches. Thus, we realized a considerable saving of time, which allowed us to adjust and rectify after each execution. We also participated in all the tasks without exception, which allowed us to test our system on all facets, *i.e.*, to identify its strengths and weaknesses. We tackled several proposed tasks. Our solution is based on a generic SPARQL query using the cell content, as a given item description. In each round, despite the time allocated by the organizers running out, we continued the work and the improvements, having the conviction that each effort counts and brings us closer to better control of the studied field. KEPLER-ASI is a promising approach, but which will be further improved: First, we will apply several methods yet to correct spelling mistakes and other typos in the source data (since it is considered as a limitation, in addition to the natural language barrier). Finally, we will develop our system by integrating new data processing techniques (some Big Data-oriented paradigms). Indeed, the parallel implementation will allow us to circumvent the data size problem, which is the gap for our current machines. Eventually, the idea of moving to a data representation using indexes would be an interesting track to investigate to master the search space, formed by the considered tabular data.

References

1. Abdelmageed, N., Schindler, S.: JenTab: matching tabular data to knowledge graphs. In: Proceedings of the Semantic Web Challenge on Tabular Data to Knowledge Graph Matching (SemTab 2020) co-located with the 19th International Semantic Web Conference (ISWC 2020), Virtual conference (originally planned to be in Athens, Greece), 5 November 2020. CEUR Workshop Proceedings, vol. 2775, pp. 40–49 (2020)
2. Chabot, Y., Labbé, T., Liu, J., Troncy, R.: DAGOBAH: an end-to-end context-free tabular data semantic annotation system. In: Proceedings of the Semantic Web Challenge on Tabular Data to Knowledge Graph Matching Co-located with the 18th International Semantic Web Conference, SemTab@ISWC 2019, Auckland, New Zealand, 30 October 2019. CEUR Workshop Proceedings, vol. 2553, pp. 41–48 (2019)
3. Chen, J., Jiménez-Ruiz, E., Horrocks, I., Sutton, C.: Learning semantic annotations for tabular data. arXiv preprint arXiv:1906.00781 (2019)
4. Chen, S., et al.: Linkingpark: an integrated approach for semantic table interpretation. In: Proceedings of the Semantic Web Challenge on Tabular Data to Knowledge Graph Matching (SemTab 2020) co-located with the 19th International Semantic Web Conference (ISWC 2020), Virtual conference (originally planned to be in Athens, Greece), 5 November 2020. CEUR Workshop Proceedings, vol. 2775, pp. 65–74 (2020)
5. Cremaschi, M., De Paoli, F., Rula, A., Spahiu, B.: A fully automated approach to a complete semantic table interpretation. Futur. Gener. Comput. Syst. **112**, 478–500 (2020)
6. Drysdale, R., et al.: The ELIXIR core data resources: fundamental infrastructure for the life sciences. Bioinformatic **38**, 2636–2642 (2020)
7. Ehrlinger, L., Wöß, W.: Towards a definition of knowledge graphs. SEMANTiCS (Posters, Demos, SuCCESS) **48**, 1–4 (2016)
8. Färber, M., Bartscherer, F., Menne, C., Rettinger, A.: Linked data quality of dbpedia, freebase, opencyc, wikidata, and yago. Semantic Web **9**(1), 77–129 (2018)
9. Garcia, L., Bolleman, J., Gehant, S., Redaschi, N., Martin, M.: Fair adoption, assessment and challenges at UniProt. Sci. Data **6**(1), 1–4 (2019)
10. Pham, M., Alse, S., Knoblock, C.A., Szekely, P.: Semantic labeling: a domain-independent approach. In: Groth, P., et al. (eds.) ISWC 2016. LNCS, vol. 9981, pp. 446–462. Springer, Cham (2016). https://doi.org/10.1007/978-3-319-46523-4_27
11. Ruch, P., et al.: Uniprot. Tech. rep. (2021)
12. Tyagi, S., Jiménez-Ruiz, E.: LexMa: tabular data to knowledge graph matching using lexical techniques. In: Proceedings of the Semantic Web Challenge on Tabular Data to Knowledge Graph Matching (SemTab 2020) co-located with the 19th International Semantic Web Conference (ISWC 2020), Virtual conference (originally planned to be in Athens, Greece), 5 November 2020. CEUR Workshop Proceedings, vol. 2775, pp. 59–64 (2020)
13. Vandewiele, G., Steenwinckel, B., Turck, F.D., Ongenae, F.: CVS2KG: transforming tabular data into semantic knowledge. In: Proceedings of the Semantic Web Challenge on Tabular Data to Knowledge Graph Matching Co-located with the 18th International Semantic Web Conference, SemTab@ISWC 2019, Auckland, New Zealand, 30 October 2019. CEUR Workshop Proceedings, vol. 2553, pp. 33–40 (2019)
14. Zhang, Z.: Effective and efficient semantic table interpretation using tableminer+. Seman. Web **8**(6), 921–957 (2017)

τJUpdate: A Temporal Update Language for JSON Data

Zouhaier Brahmia[1(✉)], Fabio Grandi[2], Safa Brahmia[1], and Rafik Bouaziz[1]

[1] University of Sfax, Sfax, Tunisia
zouhaier.brahmia@fsegs.rnu.tn, rafik.bouaziz@usf.tn
[2] University of Bologna, Bologna, Italy
fabio.grandi@unibo.it

Abstract. Time-varying JSON data are being used and exchanged in various today's application frameworks like IoT platforms, Web services, cloud computing, online social networks, and mobile systems. However, in the state-of-the-art of JSON data management, there is neither a consensual nor a standard language for updating (i.e., inserting, modifying, and deleting) temporal JSON data, like the TSQL2 or SQL:2016 language for temporal relational data. Moreover, existing JSON-based NoSQL DBMSs (e.g., MongoDB, Couchbase, CouchDB, OrientDB, and Riak) and both commercial DBMSs (e.g., IBM DB2 12, Oracle 19c, and MS SQL Server 2019) and open-source ones (e.g., PostgreSQL 15, and MySQL 8.0) do not provide any support for maintaining temporal JSON data. Also in our previously proposed temporal JSON framework, called τJSchema, there was no feature for temporal JSON instance update. For these reasons, we propose in this paper a temporal update language, named τJUpdate (Temporal JUpdate), for JSON data in the τJSchema environment. We define it as a temporal extension of our previously introduced non-temporal JSON update language, named JUpdate (JSON Update). Both the syntax and the semantics of the data modification operations of JUpdate have been extended to support temporal aspects. τJUpdate allows (i) to specify temporal JSON updates in a user-friendly manner, and (ii) to efficiently execute them.

Keywords: JSON · Temporal JSON · JUpdate · Temporal JSON data manipulation · JSON update operation · τJSchema · Conventional JSON instance · Temporal JSON instance

1 Introduction

The lightweight format JavaScript Object Notation (JSON) [15], which is endorsed by the Internet Engineering Task Force (IETF), is currently being used by various networked applications to store and exchange data. Moreover, many of these applications running in IoT, cloud-based and mobile environments, like Web services, online social networks, e-health, smart-city and smart-grid applications, require bookkeeping of the full history of JSON data updates so that

P. Fournier-Viger et al. (Eds.): MEDI 2022, LNAI 13761, pp. 250–263, 2023.
https://doi.org/10.1007/978-3-031-21595-7_18

they can handle temporal JSON data, audit and recover past JSON document versions, track JSON document changes over time, and answer temporal queries.

However, in the state-of-the-art of JSON data management [1,6,17,20–22,27], there is neither a consensual nor a standard language for updating (i.e., inserting, modifying, and deleting) temporal JSON data, like the TSQL2 (Temporal SQL2) [28] or SQL:2016 [24] language for temporal relational data. It is worth mentioning here that the extension of the SQL language, named SQL/JSON [18,23,29] and standardized by ANSI to empower SQL to manage queries and updates on JSON data, has no built-in support for updating time-varying JSON data. In fact, even for non-temporal data, SQL/JSON is limited since it does not support the update of a portion of a JSON document through the SQL UPDATE statement [26].

Moreover, existing JSON-based NoSQL database management systems (DBMSs) (e.g., MongoDB, Couchbase, CouchDB, DocumentDB, MarkLogic, OrientDB, RethinkDB, and Riak) and both commercial DBMSs (e.g., IBM DB2 12, Oracle 19c, and Microsoft SQL Server 2019) and open-source ones (e.g., PostgreSQL 15, and MySQL 8.0) do not provide any support for maintaining temporal JSON data [3,11,13].

In this context, with the aim of having an infrastructure that allows efficiently creating and validating temporal JSON instance documents and inspired by the τXSchema design principles [9], we have proposed in [2] a comprehensive framework, named τJSchema (Temporal JSON Schema). In this environment, temporal JSON data are produced from conventional (i.e., non temporal) JSON data, by applying a set of temporal logical and physical characteristics that have been already specified by the designer on the conventional JSON schema, that is a JSON Schema [14] file that defines the structure of the conventional JSON data:

- the temporal logical characteristics [2] allow designers to specify which components (e.g., objects, object members, arrays, ...) of the conventional JSON schema can vary over valid and/or transaction time;
- the temporal physical characteristics [2] allow designers to specify where timestamps should be placed and how the temporal aspects should be represented.

A temporal JSON schema is generated from a conventional JSON schema and the set of temporal logical and physical characteristics that have been specified for this non-temporal JSON schema. Thus, by using temporal JSON schemas and temporal characteristics and by making a separation between conventional JSON data and temporal JSON data, from one hand, and between conventional JSON schema and temporal JSON schema, from the other hand, τJSchema offers the following advantages: (i) it extends the traditional JSON world to temporal aspects in a systematic way; (ii) it guarantees logical and physical data independence [8] for temporal JSON data (i.e., a temporal JSON document, having some physical representation, could be automatically transformed into a different temporal document with a different physical representation while conserving the semantics of the temporal JSON data, that is keeping the same temporal logical characteristics); (iii) it does not require changes to existing JSON instance/schema files nor revisions of the JSON technologies (e.g., the IETF specification of the JSON format [15], the IETF specification of the JSON Schema language [14], JSON-based NoSQL DBMSs, JSON editors/validators, JSON Schema editors/generators/validators, etc.). However, there is no feature for temporal JSON instance update in τJSchema.

With the aim of overcoming the lack of an IETF standard or recommendation for updating JSON data, we have recently proposed a powerful SQL-like language, named JUpdate (JSON Update) [6], to allow users to perform updates on (non-temporal) JSON data. It provides fourteen user-friendly high-level operations (HLOs) to fulfill the different JSON update requirements of users and applications; not only simple/atomic values but also full portions (or chunks) of JSON documents can be manipulated (i.e., inserted, modified, deleted, copied or moved) The semantics of JUpdate is based on a minimal and complete set of six primitives (i.e., low-level operations, which can be easily implemented) for updating JSON documents. The data model behind JUpdate is the IETF standard JSON data model [15]. Thus, from one hand, it is independent from any underlying DBMSs, which simplifies its use and implementation, and, from the other hand, it can be used to maintain generic JSON documents.

Taking into account the requirements mentioned above, we considered very interesting to fill the evidenced gap and to propose a temporal JSON update language that would help users in the non-trivial task of updating temporal JSON data. Moreover, based on our previous work, we think that (i) the JUpdate language [6] can be a good starting point for deriving such a temporal JSON update language, and (ii) the τJSchema framework can be used as a suitable environment for defining the syntax and semantics of a user-friendly temporal update language, mainly due to its support of logical and physical data independence.

For all these reasons, we propose in this paper a temporal update language for JSON data named τJUpdate (Temporal JUpdate) and define it as a temporal extension of our JUpdate language, to allow users to update (i.e., insert, modify, and delete) JSON data in the τJSchema environment. To this purpose, both the syntax and the semantics of the JUpdate statements have been extended to support temporal aspects. The τJUpdate design allows users to specify in a friendly manner and efficiently execute temporal JSON updates. In order to motivate τJUpdate and illustrate its use, we will provide a running example.

The rest of the paper is structured as follows. The next section presents the environment of our work and motivates our proposal. Section 3 proposes τJUpdate, the temporal JSON instance update language for the τJSchema framework. Section 4 illustrates the use of some operations of τJUpdate, by means of a short example. Section 5 provides a summary of the paper and some remarks about our future work.

2 Background and Motivation

In this section, first we briefly describe the τJSchema framework (more details can be found in [2]), and then we present a motivating example that (i) recalls how temporal JSON data are represented under τJSchema, (ii) presents problems and difficulties of dealing with temporal data management using a JUpdate-like language, and (iii) introduces our contributions.

2.1 The τJSchema Framework

τJSchema allows a NoSQL database administrator (NSDBA) to create a temporal JSON schema for temporal JSON instances, from a conventional JSON schema, some temporal logical characteristics, and some temporal physical characteristics. It uses the following two principles: (i) separation between the conventional JSON schema and the temporal JSON schema, and also between the conventional JSON instances and the temporal JSON instances; (ii) use of temporal logical and physical characteristics to specify temporal logical and physical aspects, respectively, at schema level.

Since there are many techniques to make a (non-temporal) JSON document temporal, the logical and physical independence supported by the τJSchema framework represents a real breakthrough in temporal JSON data management, as it separates temporal JSON data design (specified via temporal logical characteristics) from implementation details (specified via temporal physical characteristics). Notice that this aspect is emphasized when dealing with updating, through a JUpdate-like language, temporal JSON documents (in the next subsection). In fact, in JSON documents, some JSON structuring conforming to the conventional JSON schema is devoted to modeling the non-temporal structure of data, whereas some additional JSON structuring is needed to encode the temporal aspects of the data modeling, actually based on some timestamped multi-version representation. Hence, by adopting a τJSchema-based approach, we want to also separate temporal data update specification from implementation details. We want to enable users to manipulate (i.e., insert, modify, and delete) temporal JSON data by reasoning at the level of their conventional JSON schema, abstracting from the knowledge of additional JSON structuring needed to encode low-level data versioning and timestamping details. In practice, we want the users to express their JUpdate updates exactly as if their JSON data were not temporal. The only thing they have to add to their update high-level statements, when dealing with valid-time data, is a VALID clause to specify the "applicability period" of the update in case they want to explicitly manage it. It should be mentioned that our approach in this paper is similar to that proposed in our previous work [7], where an "XQuery Update Facility"-like language is used to support update operations on temporal data that are recorded in XML format, abstracting from their implementation details.

2.2 Motivating Example

We assume that a company uses a JSON repository for the storage of the information about the devices that it manufactures and sells, where each device is described by its name and cost price. For simplicity, let us consider a temporal granularity of one day for representing the data change events (and, therefore, for temporal data timestamping). We assume that the initial state of the device repository, valid from February 1, 2022, can be represented as shown in Fig. 1: it contains, in a JSON file named device1.json, data about one device called CameraABC costing €35.

```
{ "devices":[
   { "device":{
       "name":"CameraABC",
       "costPrice":35 } } ] }
```

Fig. 1. The initial state of the device repository (file device1.json, on February 01, 2022).

Then, we assume that, effective from April 15, 2022, the company starts producing a new device named CameraXYZ with a cost price of €42 and CameraABC's cost price is raised by 8%. The new state of the device repository can be represented in a JSON file named device2.json as shown in Fig. 2. Changed parts are presented in red color.

```
{ "devices":[
   { "device":{
       "name":"CameraABC",
       "costPrice":37.8} },
   { "device":{
       "name":"CameraXYZ",
       "costPrice":42} }] }
```

Fig. 2. A new state of the device repository (file device2.json, on April 15, 2022). (Color figure online)

Consequently, we consider that the device repository is implemented in the τJSchema framework and that the conventional JSON schema for our JSON device data has been annotated so that "device" is a time-varying object for representing the history of devices along valid time. As a result, the entire history of the device repository can be represented in the temporal JSON document shown in Fig. 3, composed of two slices corresponding to the repository states of Fig. 1 and Fig. 2.

```
{ "temporalJSONDocument":{
   "conventionalJSONDocument":{
    "sliceSequence":[
     {"slice":{
        "location":"device1.json",
        "begin":"2022-02-01" } },
     {"slice":{
        "location":"device2.json",
        "begin":"2022-04-15" } } ] } } }
```

Fig. 3. The temporal JSON document representing the entire history of the device repository (file deviceTJD.json, on April 15, 2022).

The temporal JSON document can also be "squashed" to obtain a self-contained temporal JSON document, conformant to the temporal JSON schema that can be derived from both the conventional JSON schema and the temporal logical and physical characteristics, representing the whole devices' history, as shown in Fig. 4. The valid-time timestamps are presented in blue color.

```
{ "devices":[
   { "device":[
     { "name":"CameraABC",
       "costPrice":35,
       "VTbegin":"2022-02-01",
       "VTend":"2022-04-14" },
     { "name":"CameraABC",
       "costPrice":37.8,
       "VTbegin":"2022-04-15",
       "VTend":"Forever" } ],
   { "device":[
     { "name":"CameraXYZ",
       "costPrice":42,
       "VTbegin":"2022-04-15",
       "VTend":"Forever" } ] } ] }
```

Fig. 4. The squashed JSON document corresponding to the entire history of the device repository (file deviceSJD.json, on April 15, 2022).

Notice that the squashed JSON document deviceSJD.json in Fig. 4 also corresponds to one of the manifold possible representations of our temporal JSON [3] data without the τJSchema approach.

After that, let us consider that we have to record in the device repository that the company has stopped manufacturing the device CameraABC effective from May 25, 2022. At the state-of-the-art of JSON technology, we could use JUpdate HLOs to directly perform the required updates on the deviceSJD.json file in Fig. 4. A skilled developer, expert in both temporal databases and JUpdate, and aware of the precise structure of the squashed document, will satisfy such requirements via the following JUpdate statement:

```
UPDATE deviceSJD.json
PATH $.devices[@.device[@.name="CameraABC"
          && @.VTend="Forever"].VTend]
VALUE "2022-05-24"                                            (S1)
```

In practice, deleting the device CameraABC effective from 2022-05-25 means ending with 2022-05-24 the valid timestamp of the last version of such a device, assuming for simplicity (and without checking) that the version valid at 2022-05-25 is the last CameraABC's version and, therefore, there are no future versions to delete. Anyway, we think that this is a complex solution for what it is a simple problem (for example, in temporal relational databases).

Thus, our **first contribution** is a temporal extension of the JUpdate language. JUpdate statements will be equipped with a new VALID clause to specify the so-called "applicability period" of the update, that is the time period in which the update has to be in effect (e.g., from 2022-05-25 on, in our example). This solution will allow the developer to formulate the required update as a JUpdate deletion valid from 2022-05-25 of CameraABC's data, relying on the temporal semantics of the language for its correct execution, including version and timestamp management. Nevertheless, working on the temporal JSON document in Fig. 4, this will mean to specify the following DeleteValue operation:

```
DELETE FROM deviceSJD.json
PATH $.devices[@.device[@.name="CameraABC"
          && @.VTend="Forever"]]
VALID from "2022-05-24"                                    (S2)
```

Although this solution is simpler than solution (S1), it requires from the developer a detailed knowledge of the specific temporal structuring of the JSON file including version organization and timestamping. Another consequence is that such a solution template would not be portable to another setting in which a different temporal structuring of JSON data is adopted.

Moreover, our **second contribution** is to integrate the temporal JUpdate extension into the τJSchema framework, in order to enjoy the logical and physical independence property. In this framework, the required update will be specified via the following τJUpdate DeleteValue statement:

```
DELETE FROM deviceSJD.json
PATH $.devices[@.device[@.name="CameraABC"]]
VALID from "2022-05-24"                                    (S3)
```

The update could be applied either to the temporal JSON document (i.e., deviceTJD.json) or to its squashed version (i.e., deviceSJD.json); the system using the temporal logical and physical characteristics can manage both ways correctly. Notice that, ignoring the VALID clause, the solution (S3) represents exactly the same way we would specify the deletion of the device CameraABC's data in a non-temporal environment (e.g., executing it on the device2.json file in Fig. 2). In practice, we want to allow the developer to focus on the structuring of data simply as defined in the conventional JSON schema and not on the temporal JSON schema, leaving the implementation details and their transparent management to the system (e.g., the mapping to a squashed JSON document, being aware of the temporal characteristics). This means, for example, that in order to specify a cost price update, we want τJUpdate users be able to deal with updates to the "device.costPrice" value instead of dealing with updates to the "device.costPrice" array of objects, where each object represents a version of a cost price and has three properties: "VTbegin" (the beginning of the valid-time timestamp of the version), "VTend" (the end of the valid-time timestamp of the version), and "value" (the value of the version).

Notice that such a way in which temporal updates of JSON data will be specified with our τJUpdate language, corresponds exactly to the way updates of temporal relational data can be specified using a temporal query language like TSQL2 [28] or SQL:2016 [24], that is using the same update operations that are used in a non-temporal context augmented with a VALID clause to specify the applicability period of each update operation.

In sum, the motivation of our approach is twofold: from one hand, (i) leveraging the logical/physical independence supported by the τJSchema framework to the JUpdate language and, from the other hand, (ii) equipping τJSchema with a user-friendly update language, which is consistent with its design philosophy.

3 The τJUpdate Language

In this section, we propose the τJUpdate language, by showing how the JUpdate specification [6] has to be extended. More precisely, in Sect. 3.1, we start by presenting the syntax of τJUpdate high-level operations (HLOs) before defining their semantics while considering temporal JSON documents in unsquashed form.

3.1 Syntax and Semantics of τJUpdate Update HLOs

The management of transaction time does not require any syntactic extension to the JUpdate language: owing to the transaction time semantics, only current data can be updated and the "applicability period" of the update is always [Now, UntilChanged], which is implied and cannot be overridden by users. On the contrary, the management of valid time is under the user's responsibility. Hence, syntactic extensions of the JUpdate language are required to allow users to specify a valid time period representing the "applicability period" of the update. To this purpose, the JUpdate update HLOs [6] are augmented with a VALID clause as shown in Fig. 5.

```
τJUpdateHLO   ::=  JUpdateHLO  "VALID"  validTimePeriod
JUpdateHLO    ::=  ValueChangeHLO |  MemberChangeHLO
                |  ObjectChangeHLO
ValueChangeHLO  ::=  InsertValue |  DeleteValue |  UpdateValue
                  |  CopyValue   |  MoveValue
MemberChangeHLO  ::=  InsertMember  |  DeleteMember
                   |  RenameMember  |  ReplaceMember
                   |  CopyMember   |  MoveMember
ObjectChangeHLO  ::  UpdateObject
validTimePeriod  ::= "in [" validTimeBegin "," validTimeEnd "]"
                   | "from"  validTimeBegin
                   | "to"  validTimeEnd
validTimeBegin  ::=  "Beginning" | "Now" | temporalValue
validTimeEnd  ::=  "Forever" | "Now" | temporalValue
```

Fig. 5. The syntax of τJUpdate HLOs.

Due to space limitations, we do not consider here other JUpdate HLOs (e.g., InsertMember, ReplaceMember, UpdateObjects) as they are used for specifying complex updates; they will be investigated in a future work. Temporal expressions "from T" and "to T", while T is a temporal value, are used as syntactic sugar for the temporal expressions "in [T, Forever]" and "in [Beginning, T]", respectively.

As far as the semantics of τJUpdate is concerned, we can define it, for the sake of simplicity, by considering JSON update operations on the temporal JSON document in its unsquashed form. Based on the well-known theory developed in the temporal database field [12,19], the operational semantics of a τJUpdateHLO, equal to a JUpdateHLO augmented with the VALID clause, can be defined as follows:

- validTimePeriod is evaluated. The result must be a valid period specification; otherwise a type error is raised. Let [vts, vte] be the period resulting from the evaluation.
- Let jdoc be the temporal JSON document involved in the update; find in jdoc all the temporal slices jdoc_vers having a timestamp VTimestamp which overlaps [vts, vte].
- For each such slice jdoc_vers:
  - let jdoc_vers′ the result of the evaluation of JUpdateHLO on jdoc_vers;
  - if VTimestamp $\subset$ [vts, vte] then remove the whole slice jdoc_vers from the temporal JSON document jdoc (and delete the corresponding JSON file) else restrict to VTimestamp $\setminus$ [vts, vte] the timestamp of jdoc_vers in the temporal JSON document jdoc;
  - add jdoc_vers′ to the temporal JSON document jdoc as a new slice with timestamp VTimestamp $\cap$ [vts, vte].
- Coalesce the resulting slices in the temporal JSON document.

The last step aims at limiting the unnecessary proliferation of slices, giving rise to redundant JSON files in the unsquashed setting. Two slices, jdoc_vers1 and jdoc_vers2 with timestamps VTimestamp1 and VTimestamp2, respectively, can be coalesced when jdoc_vers1 and jdoc_vers2 are equal and VTimestamp1 meets VTimestamp2 [28]. In this case, coalescing produces one slice jdoc_vers1 with timestamp VTimestamp1 $\cup$ VTimestamp2.

This definition of the τJUpdate HLO semantics, which can be easily extended to the transaction-time or bitemporal case, is in line with the τJSchema principles, considering a temporal JSON document as representing a sequence of conventional JSON documents, and reuses the standard (non-temporal) JUpdate HLOs.

Even if the temporal JSON document jdoc is physically stored in squashed form, the above semantics can still be used to evaluate a τJUpdate HLO after the document has been explicitly unsquashed. The results of the evaluation can then be squashed back to finally produce an updated temporal JSON document. Although correct from a theoretical point of view, such a procedure could be inefficient in practice, in particular when the temporal JSON document is composed of several slices. To resolve this problem, a different method can be applied for updating temporal JSON documents that are stored in squashed form. To this end, the semantics of τJUpdate HLOs can be defined in an alternative way, as shown in the next subsection (the solution is inspired from our previous work on updates to temporal XML data [7]).

4 Running Example Reprise

In this section, we resume the motivating example introduced in Sect. 2.2 to illustrate some of the functionalities of τJUpdate.

First of all, starting from the initial state of the device repository containing only the slice in Fig. 1, the second slice in Fig. 2 can be added via the execution of the following sequence of τJUpdate HLOs:

```
INSERT INTO deviceTJD.json
PATH $.devices[last]
VALUE { "device":{ "name":"CameraXYZ", "costPrice":42 } }
VALID from "2022-04-15";
UPDATE deviceTJD.json
PATH $.devices[device.name="CameraABC"].costPrice
VALUE $.devices[device.name="CameraABC"].costPrice * 1.08
VALID from "2022-04-15"
```

The first one is an example of InsertValue HLO that inserts CameraXYZ's data, while the second one is an example of UpdateValue HLO that increases CameraABC's cost price. The result of this HLO sequence corresponds to the temporal JSON document in Fig. 3 completed by the slices in Fig. 1 and Fig. 2, and which has been shown in squashed form in Fig. 4.

As an example of DeleteValue HLO, we can consider the τJUpdate HLO (S3) in Sect. 2.2, deleting CameraABC's data effective from 2022-05-25. As an example of RenameMember HLO, we can consider changing the name of the "devices" object to "products", also valid from 2022-05-25. Notice that such an operation could be more properly considered as a conventional JSON schema change, as it acts on metadata rather than on data and, thus, could be better effected using the high-level JSON schema change operation RenameProperty, acting on the conventional JSON schema, we previously defined in [5], which is automatically propagated to extant conventional JSON data. However, as part of τJUpdate, we can also consider it a JSON data update that propagates indeed to the JSON schema by means of the implicit JSON schema change mechanism that we have proposed in [4]. The global effects in the τJSchema framework, anyway, are exactly the same. Such updates can be performed via the following τJUpdate HLOs:

```
DELETE FROM deviceTJD.json
PATH $.devices[device.name="CameraABC"]
VALID from "2022-05-25";
ALTER DOCUMENT deviceTJD.json
OBJECT $.devices
RENAME MEMEBER devices TO products
VALID from "2022-05-25"
```

The result of this HLO sequence is the new temporal JSON document shown in Fig. 6 with the new slice shown in Fig. 7.

```
{ "temporalJSONDocument":{
    "temporalJSONSchema":{
      "conventionalJSONSchema":{
        "sliceSequence":[
          {"slice":{
            "location":"deviceCJS1.json",
            "begin":"2022-02-01" } },
          {"slice":{
            "location":"deviceCJS2.json",
            "begin":"2022-05-25" } }
        ] } },
    "conventionalJSONDocument":{
      "sliceSequence":[
        {"slice":{
          "location":"device1.json",
          "begin":"2022-02-01" } },
        {"slice":{
          "location":"device2.json",
          "begin":"2022-04-15" } },
        {"slice":{
          "location":"device3.json",
          "begin":"2022-05-25" } } ] }
} }
```


Fig. 6. The new temporal JSON document representing the whole history of the device repository (file deviceTJD.json).


```
{ "products":[
    { "device":{
        "name":"CameraXYZ",
        "costPrice":42 } } ] }
```


Fig. 7. The final state of the device repository (file device3.json).

As a side effect of the RenameMember HLO requiring an implicit JSON schema change, two conventional JSON schema versions are included in the new temporal JSON document (without entering into the whole details, deviceCJS1.json is the conventional JSON schema version having "devices" as its root object, whereas deviceCJS2.json is the conventional JSON schema version having "products" as its root object). As a consequence, squashing of the temporal JSON document in Fig. 6 produces two squashed JSON documents: deviceSJD1.json shown in Fig. 8, which is conformant to the first conventional JSON schema version deviceCJS1.json, and deviceSJD2.json shown in Fig. 9, which is conformant to the second conventional JSON schema version deviceCJS2.json. Changes are evidenced with red color.

```
{ "devices":[
    { "device":[
      { "name":"CameraABC",
        "costPrice":35,
        "VTbegin":"2022-02-01",
        "VTend":"2022-04-14" },
      { "name":"CameraABC",
        "costPrice":37.8,
        "VTbegin":"2022-04-15",
        "VTend":"2022-05-24" } ] },
    { "device":[
      { "name":"CameraXYZ",
        "costPrice":42,
        "VTbegin":"2022-04-15",
        "VTend":"2022-05-24" } ] } ] }
```

Fig. 8. The squashed JSON document (file deviceSJD1.json) corresponding to the first conventional JSON schema version deviceCJS1.json. (Color figure online)

```
{ "products":[
    { "device":[
      { "name":"CameraXYZ",
        "costPrice":42,
        "VTbegin":"2022-05-25",
        "VTend":"Forever" } ] } ] }
```

Fig. 9. The squashed JSON document (file deviceSJD2.json) corresponding to the second conventional JSON schema version deviceCJS2.json. (Color figure online)

5 Conclusion

In this paper, we have proposed τJUpdate, a temporal extension of the JUpdate language by equipping JUpdate update HLOs with a VALID clause to specify the applicability period of the update operations, in the τJSchema framework. Ignoring the VALID clause, any τJUpdate HLO is exactly the same as the corresponding JUpdate HLO to be executed in a non-temporal environment. Indeed, by taking advantage of the τJSchema logical and physical independence feature, our goal was to help the users by allowing them to focus only on the data structure as defined in the conventional JSON schema, and ignore how data are structured in the temporal JSON schema. Hence, implementation details and their transparent management are left to the system. Moreover, any τJUpdate HLO could be specified either on the temporal JSON document or on its squashed version; the system is able to correctly manage both ways, via the use of temporal (logical and physical) characteristics. We have also shown in Sect. 3.1 how the τJUpdate semantics can be defined to correctly deal with temporal JSON documents physically stored according to both forms (i.e., unsquashed and squashed forms).

Moreover, since JSON databases [16] are document-oriented NoSQL databases [10,25], which are in general schemaless, a JSON instance document could be, at the end of an update operation, not conformant to its initial JSON schema. To cover this aspect, we have also dealt with JSON data updates that require implicit JSON schema changes (exemplified with the RenameMember

HLO). Hence, in such a situation, τJUpdate executes implicit changes to conventional JSON schema, in a way transparent to the user, before performing temporal updates on conventional JSON data.

In the future, we envisage to extend τJUpdate to also support updating transaction-time and bitemporal JSON data, in τJSchema, as in the present work we have dealt only with valid-time JSON data. Finally, we plan to develop a tool that supports τJUpdate, in order to show the feasibility of our proposal and to use it in the experimental evaluation of our language (e.g., involving usability, user-friendliness and performance).

References

1. Bourhis, P., Reutter, J., Vrgoč, D.: JSON: data model and query languages. Inf. Syst. **89**, 101478 (2020)
2. Brahmia, S., Brahmia, Z., Grandi, F., Bouaziz, R.: τJSchema: a framework for managing temporal JSON-based NoSQL databases. In: Hartmann, S., Ma, H. (eds.) DEXA 2016. LNCS, vol. 9828, pp. 167–181. Springer, Cham (2016). https://doi.org/10.1007/978-3-319-44406-2_13
3. Brahmia, S., Brahmia, Z., Grandi, F., Bouaziz, R.: A disciplined approach to temporal evolution and versioning support in JSON data stores. In: Emerging Technologies and Applications in Data Processing and Management, pp. 114–133. IGI Global (2019)
4. Brahmia, Z., Brahmia, S., Grandi, F., Bouaziz, R.: Implicit JSON schema versioning driven by big data evolution in the τJSchema framework. In: Farhaoui, Y. (ed.) BDNT 2019. LNNS, vol. 81, pp. 23–35. Springer, Cham (2020). https://doi.org/10.1007/978-3-030-23672-4_3
5. Brahmia, Z., Brahmia, S., Grandi, F., Bouaziz, R.: Versioning schemas of JSON-based conventional and temporal big data through high-level operations in the τJSchema framework. Int. J. Cloud Comput. **10**(5–6), 442–479 (2021)
6. Brahmia, Z., Brahmia, S., Grandi, F., Bouaziz, R.: JUpdate: a JSON update language. Electronics **11**(4), 508 (2022)
7. Brahmia, Z., Grandi, F., Bouaziz, R.: τXUF: a temporal extension of the XQuery update facility language for the τXSchema framework. In: Proceedings of the 23rd International Symposium on Temporal Representation and Reasoning (TIME 2016), Technical University of Denmark, Copenhagen, Denmark, 17–19 October 2016, pp. 140–148 (2016)
8. Burns, T., et al.: Reference model for DBMS standardization, database architecture framework task group (DAFTG) of the ANSI/X3/SPARC database system study group. SIGMOD Rec. **15**(1), 19–58 (1986)
9. Currim, F., Currim, S., Dyreson, C., Snodgrass, R.T.: A tale of two schemas: creating a temporal XML schema from a snapshot schema with τXSchema. In: Bertino, E., et al. (eds.) EDBT 2004. LNCS, vol. 2992, pp. 348–365. Springer, Heidelberg (2004). https://doi.org/10.1007/978-3-540-24741-8_21
10. Davoudian, A., Chen, L., Liu, M.: A survey on NoSQL stores. ACM Comput. Surv. (CSUR) **51**(2), 1–43 (2018)
11. Goyal, A., Dyreson, C.: Temporal JSON. In: 2019 IEEE 5th International Conference on Collaboration and Internet Computing (CIC 2019), pp. 135–144 (2019)
12. Grandi, F.: Temporal databases. In: Encyclopedia of Information Science and Technology, Third Edition, pp. 1914–1922. IGI Global (2015)

13. Hu, Z., Yan, L.: Modeling temporal information with JSON. In: Emerging Technologies and Applications in Data Processing and Management, pp. 134–153. IGI Global (2019)
14. Internet Engineering Task Force: JSON Schema: A Media Type for Describing JSON Documents, Internet-Draft, 19 March 2018. https://json-schema.org/latest/json-schema-core.html
15. Internet Engineering Task Force: The JavaScript Object Notation (JSON) Data Interchange Format, Internet Standards Track document, December 2017. https://tools.ietf.org/html/rfc8259
16. Irshad, L., Ma, Z., Yan, L.: A survey on JSON data stores. In: Emerging Technologies and Applications in Data Processing and Management, pp. 45–69. IGI Global (2019)
17. Irshad, L., Yan, L., Ma, Z.: Schema-based JSON data stores in relational databases. J. Database Manag. (JDM) **30**(3), 38–70 (2019)
18. ISO/IEC, Information technology – Database languages – SQL Technical Reports – Part 6: SQL support for JavaScript Object Notation (JSON), 1st Edition, Technical report ISO/IEC TR 19075-6:2017(E), March 2017. http://standards.iso.org/ittf/PubliclyAvailableStandards/c067367_ISO_IEC_TR_19075-6_2017.zip
19. Jensen, C., Snodgrass, R.: Temporal database. In: Liu, L., Özsu, M.T. (eds.) Encyclopedia of Database Systems, 2nd edn., pp. 3945–3949. Springer, New York (2018). https://doi.org/10.1007/978-1-4614-8265-9_395
20. Liu, Z.: JSON data management in RDBMS. In: Emerging Technologies and Applications in Data Processing and Management, pp. 20–44. IGI Global (2019)
21. Liu, Z., Hammerschmidt, B., McMahon, D.: JSON data management: supporting schema-less development in RDBMS. In: Proceedings of the 2014 ACM SIGMOD International Conference on Management of Data (SIGMOD 2014), Snowbird, UT, USA, 22–27 June 2014, pp. 1247–1258 (2014)
22. Lv, T., Yan, P., Yuan, H., He, W.: Linked lists storage for JSON data. In: 2021 International Conference on Intelligent Computing, Automation and Applications (ICAA 2021), pp. 402–405 (2021)
23. Melton, J., et al.: SQL/JSON part 1, DM32.2-2014-00024R1, 6 March 2014. https://www.wiscorp.com/pub/DM32.2-2014-00024R1_JSON-SQL-Proposal-1.pdf
24. Michels, J., et al.: The new and improved SQL:2016 standard. ACM SIGMOD Rec. **47**(2), 51–60 (2018)
25. NoSQL Databases List by Hosting Data – Updated 2020. https://hostingdata.co.uk/nosql-database/
26. Petković, D.: SQL/JSON standard: properties and deficiencies. Datenbank-Spektrum **17**(3), 277–287 (2017)
27. Petković, D.: Implementation of JSON update framework in RDBMSs. Int. J. Comput. Appl. **177**, 35–39 (2020)
28. Snodgrass, R.T., et al. (eds.): The TSQL2 Temporal Query Language. Kluwer Academic Publishing, New York (1995)
29. Zemke, F., et al.: SQL/JSON part 2 – Querying JSON, ANSI INCITS DM32.2-2014-00025r1, 4 March 2014. https://www.wiscorp.com/pub/DM32.2-2014-00025r1-sql-json-part-2.pdf

Author Index

Ait Wakrime, Abderrahim 119
Al-Atabany, Walid 58
Alnaggar, Yara Ali 16
Amer, Karim 16
Awad, Ahmed 43, 147

Baazouzi, Wiem 236
Barakat, Rameez 162
Belguidoum, Meriem 176
Belhassena, Amina 193
Benaini, Redouane 119
Bernardo, Guillermo de 207
Bertout, Antoine 133
Bouaziz, Rafik 250
Bouba, Khaoula 119
Boudaoud, Abdelkrim 222
Brahmia, Safa 250
Brahmia, Zouhaier 250
Brisaboa, Nieves R. 207

Chikh, Azeddine 222
Corral, Antonio 207

El Seknedy, Mai 102
Elattar, Mustafa 26
Elattar, Mustafa A. 89
Elden, Rana Hossam 58
Elhelw, Mohamed 16
Elkafrawy, Passent 43
Eltabey, Ayaalla 89

Faiz, Sami 236
Fawzi, Sahar Ali 102
Ferrarotti, Flavio 72
Fetteha, Marwan A. 3

Gad, Eyad 26
Gamal, Aya 26
Ghoneim, Vidan Fathi 58
Gourari, Aya 176
Grandi, Fabio 250
Grolleau, Emmanuel 133

Hadhoud, Marwa M. A. 58
Hassanein, Ehab E. 147
Hattenberger, Gautier 133
Helal, Iman M. A. 147
Hongzhi, Wang 193

Jamal, Salma 89

Kachroudi, Marouen 236
Kamni, Soulimane 133
Kassem, Aly M. 89
Kaufmann, Daniela 72
Khaled, Salma 89

Mahfoud, Houari 222
Mahgoub, Mahmoud 43
Medhat, Walaa M. 162
Mohamed, Samah 89
Moharram, Hassan 43

Naeem, ElSayed 16

Osama, Alaa 89
Ouhammou, Yassine 119, 133

Penabad, Miguel R. 207

Radwan, Ahmed G. 3
Radwan, Moataz-Bellah A. 162

Said, Lobna A. 3
Sayed, Wafaa S. 3
Sebaq, Ahmad 16
Sehili, Ines 176
Selim, Sahar 26
Sochor, Hannes 72

Yousef, Ahmed H. 162

Zaki, Nesma M. 147

Printed in the United States
by Baker & Taylor Publisher Services